Vearl
BROWNING

ars mechanica [ars mekanika] Latin phrase of Greek origin : (a) ars (Lat. ars, artis), manner of doing something according to the rules, (b) mechanica (from the Greek *mêkhanê*, machine). (1) Knowledge and expertise in the field of applied mechanics, devices and machines. (2) By extension: the manual trades in general. (3) *Hist.* It was said of the people of Liège as early as the eighteenth century : *"There is no other people which has pushed invention as far as they in the field of mechanical works"*. (4) The Herstal Group has its origins in this fertile soil which it has continued to make productive since 1889, by bringing to it all the improvements and talents demanded by the continual evolution of techniques and needs.

Auguste Francotte - Claude Gaier
Robert Karlshausen

ARS MECHANICA

The Ultimate FN Book

FN HERSTAL, BROWNING, WINCHESTER

HERSTAL group
FN HERSTAL - BROWNING

ARS MECHANICA

ISBN 978-2-87415-877-3
Legal cataloging : D/2007/6840/193

Preface

The history of a company is always a human history.
That of the FN, later to become the Herstal Group, is exemplary in this respect.

First, there is the intuition and reason of the founders, who understood that they had to abandon their individual identity as arms manufacturers and combine their forces if they were to achieve a critical size in a market undergoing huge expansion.

Then there are those who contributed – and still contribute – to making the Herstal Group and its "FN Herstal", "Browning" and "Winchester" brands a benchmark on the world stage.

In the course of its 120 year history, the company has sometimes seemed to be a veritable social laboratory. Its workers have battled within the company for improvements in working conditions, like the symbolic women's "Equal work, equal pay" movement, but also beyond its walls to ensure the company's survival and development, at a time when its days seemed numbered.

This story of men and women is also an expression of great skill. From the very first, in its core trade as in the other activities it has taken on, the name FN has always been synonymous with excellence and its products renowned for their superior qualities.

The long-term survival of a company rests on a clever cocktail of expertise and moving with the times. And in this respect too, FN is exemplary.

The knowledge and expertise in gunmaking in Liège, its transmission down the generations for over 500 years and the determination not to abandon this sum of experience and skill have played a part in overcoming the difficult times.
Today, thanks in particular to the intervention of the public authorities, the Herstal Group remains an international group run from its country of origin, a leader in its field, and presenting great vitality and healthy margins for future progress.

After diversifying its activities with varied success, today the Herstal Group is focused on its prime business - the manufacture of arms, whether for hunting and sporting or particularly in the field of defense, and perfectly in keeping with democracy.

"A nation can have no peace without an army" wrote Tacitus almost two thousand years ago. As an extension to this, it could be said today that a well-equipped army is indispensable to a democratic state.

With the ethical rules it has set itself, with the quality of its products, the expertise of its workers, the performance of its management and its industrial, sales and research teams the Herstal Group is the epitome of the «Made in Wallonia» dynamism and skill. An example to be followed.

Jean-Claude Marcourt,
Minister of Economy, Employment and Foreign Trade

Foreword

The creation of this commemorative book was undertaken on the initiative of Philippe Tenneson, Chairman and Managing Director of the Herstal Group. Robert Sauvage, the Group's Public Relations and Communication Manager, was its architect and coordinator. Claude Gaier was in charge of history, editing and iconography, assisted by Pascal Pruvost, Managing Director of New Look Communications SA, a long-standing collaborator and consultant to the Herstal Group who also designed the mock-up and layout.

The texts which make up this book were written at different periods by three authors. The first part, which deals with the origins of the arms industry up to 1964 was written by Auguste Francotte. The second part, which covers the period from 1964 to 1989, was written by Claude Gaier, mentioned above. Finally, the story of the following period up to 2007 is the work of Robert Karlshausen.

Each author has written in his own personal style, bringing to the work his own individuality, and covering periods which each had personal experience of within the company.

Over time, many members of the Group's personnel and management team have been asked to contribute by evoking their memories and their vision of the company, as well as revealing the philosophy and the techniques of their trade. Consequently, their share in the creation of this work has been significant.

Finally, a number of people within the Group, both past and present, have made a direct contribution to this work; principally the late Philippe Questienne, Joseph Balhan and Gaston Van Belleghem, and also Richard Bauter, David Bertinelli, Pierre Bourgeois, Philippe Chamoret, Georges Chauveheid, Philippe Claessens, Pat Cone, Becky Costello, Anne Devroye, Thierry Dumortier, Daniel Durbut, Scott Engen, André Fonsny, Marie-Jeanne Gaier-Ernotte, Charles Guèvremont, Chip Hewlett, André Jacquemin, Glen Jensen, Damien Kaivers, Marie-Louise Meuleneers, Georges Micheels, Pierre Minet, Fernando Muscinelli, Rob Platt, Bernadette Romainville, Todd Schaffer, Jean-Marie Soubras, Greg Stoddart, Jean Stuto, Anthony Vanderlinden, Jim Vanderperren, Joseph Vreyen, Mats Waktare, Michel Wynants, Ralph Young.

The authors

A descendant of one of FN's founders, Auguste Francotte belongs to a family of arms manufacturers who have distinguished Liège industry for almost two centuries. A Doctor of Law, he spent a part of his career with FN as a legal advisor.

A Doctor of History at the University of Liège and Director of the Liège Arms Museum, Claude Gaier, whose work is considered authoritative at international level particularly in the fields of industry and arms manufacturing, was a member of FN's management team between 1964 and 1991. He has overseen the editing of the whole book, in addition to the illustration captions.

A Doctor of Law and former Company Secretary for the Herstal Group, Robert Karlshausen provided the main part of the editorial work relating to the company's development since 1990.

Before FN : The arms industry in Liège

Saloon-pistol, Flobert system (Liège, 1854).
The barrel is fluted, blued, engraved and gilded. The stock is carved in Louis XV style.
Liège Arms Museum - 4.865 (the Grand Curtius).

Grandeur and misfortunes

The poetic city of Liège

If you discover Liège the proper way, that is to say, from the summit of the Citadel hill on a clear day of warm sunlight, you will find that the city is essentially a poetic creation. Only the desire to round off, following some law of sentimental geometry, the somewhat wavering windings of a beautiful river and the ambition to set off with buildings the outlines of a beautifully undulating countryside could have inspired men to establish their abode in the modest and marshy plain which surrounded the confluences of the rivers Meuse, Ourthe and Légia.

From the Citadel, then, you can see Liège, bristling with steeples and crowned with domes like those imaginary cities with which Gothic artists would adorn their paintings.

On the top of a rocky spur, St-Martin's basilica harmoniously divides the picture, a representation of the gilded beauty so dear to the old masters.
Lower down, to the left and to the right, rising up from the sea of roofs, the cupola and campanile of St-John's, the bulbous dome of St-Andrew's and that of St-Vincent's, the steeples of St-Bartholomew's, rising up as solid and perennial twins, the ragged tower of St-Denis's, the slender pinnacle of the Seminary and even the disorderly and bizarre mass of skyscrapers all give the town its particular tempo, sentimental yet reasonable, and accentuated from between the undulations of the landscape by the fleeting arabesque of the river Meuse and its long arcade of bridges.

Thus Lucie Delarue-Mardrus gave her impression of Liège :
"I can remember that green bluff whence I saw rising in tiers that industrious City of Liège..."

Old Register of Trades in Liège (eighteenth century). Section pertaining to the Métiers des Febvres – the 'Blacksmiths' – whose coat of arms with the crowned hammer and tongs is set in a piece of "rocaille" in rococo style. Liège University Library.

View of Liège in 1567. Old facsimile of a now destroyed painting by Lucas van Valckenborgh in Berlin. In the background, the meandering Meuse. To the right, St-Martin's basilica. Reims, Museum of Saint-Denis.

The industrious city of Liège

Liège the industrious. This trait goes back a long way. At first glance, Liège is indisputably a work of art. But the historian knows that it is also one of the European cities where genius could express itself naturally in what was called the mechanical arts. Of these, the metallurgical arts which constituted the "Bon Métier des Febvres" (the Liège Smiths' Guild) have always occupied an important position because of the number and ability of the craftsmen they employed. Moreover the Bon Métier des Febvres was a very diversified guild which at that time comprised no fewer than five subdivisions or sections. Several of these sections included craftsmen who were specialised in the manufacturing of firearms. In fact, this industry had been established from the fourteenth century in the principality of which Liège was the capital.

For example, one chronicler tells us that in 1346 – the year of the Battle of Crécy – the united forces of Liège, Huy and Tongres stormed the walls of Hamal Castle with bombard cannons.

Liège a conquered city

Due to the incompleteness of our sources we cannot be sure as to when exactly the arms industry started up. The only thing we know for certain is that the principal centers of arms manufacturing were Liège, Dinant and the marquisate of Franchimont, that both ordnance pieces and small arms were produced, as well as powder and ammunition, and that following steady development during the first half of the fifteenth century, its rise was abruptly interrupted by political events.

In 1467, Charles the Bold, the victor of Brustem, imposed the Treaty of St-Laurence upon the conquered city of Liège. One of the provisions of the treaty prohibited the *"forging in the said City and country of heavy and light artillery, of suits of armor, as well as the melting down of metal, iron, brass or bronze for the manufacturing of military equipment"*.

The following year, the people of Liège, stirred up by the agents of Louis XI, were once again crushed in spite of the sacrifices made by 600 men from Franchimont. The town was ransacked and according to an account by Comines *"the Duke of Burgundy... set up his quarters at Polleur and had all the houses pillaged and the ironworks in the region, upon which the people mostly depended for their livelihood, destroyed"*.

Imaginary representation of the 1468 siege of the town of Liège by the Duke of Burgundy and the King of France in 1468. Wood engraving illustrating "Die alder excellenste cronycke van Brabant", Antwerp, 1498, plate 62.

In the land of Vulcan's furnaces

Neutrality and renaissance

Half a century later, under the prosperous reign of the magnificent Erard de la Marck (1505-1538), Liège arose from its ashes. And this was accompanied by a renaissance in the arms manufacturing industry. In fact, thanks to the wisdom of Erard and of his successors, the City did not become involved in the conflicts between the great neighboring nations, but became rich by supplying them with arms. While the dynastic and religious wars raged at the frontiers of the principality , Liège, under the shelter of its neutrality, became one of the main arsenals of Europe.

"It is the land of Vulcan's furnaces", wrote Guichardini. But the blacksmith-god was an artist as well as an ingenious craftsman. Some say that it is to one of the citizens of the principality, a Captain Sébastien de Corbion, that we owe the invention of the pistol (1520). The question remains controversial today, but there can be no doubt that the proliferation of firearms in that region, dating back to the sixteenth century, is evidence of the existence of an active local industry.

Moreover, the engravers of Liège played a decisive role in the history of the decoration of fine firearms. Having fled to Frankfurt to escape religious persecution, some of those artists – in particular the great Théodore de Bry and his sons – would publish Italian-influenced ornamental models from which the most famous gunsmiths of Augsburg and Nuremberg drew inspiration.

Distribution of the Liège arms industry according to the 1896 census.

Number of gunsmiths per commune

1	Liège	4 357
2	Herstal	2 085
3	Trembleur	639
4	Cheratte	497
5	Wandre	387
6	Forêt	281
7	Jupille	277
8	Nessonvaux	274
9	Saint-Remy	256
10	Mortier	248
11	Housse	199
12	Ans	183
13	Barchon	181
14	Saive	166
15	Argenteau	139
16	Dalhem	134
17	Vivegnis	122
18	Fraipont	112
19	Olne	89
20	Hermalle s/Argenteau	69
21	Glain	66
22	Vottem	63
23	Cerexhe-Heuseux	52
24	Haccourt	48
25	Oupeye	44
26	Bolland	39
27	Saint-André	34
28	Chaudfontaine	31
29	Hermée	29
30	Feneur	28
31	Mortroux	23
32	Bressoux	22
33	Milmort	21
34	Bellaire	19
35	Richelle	18
36	Julémont	18
37	Vaux-Sous-Chèvremont	16
38	Visé	14
39	Heure-Le-Romain	13
40	Tignée	9
41	Queue-Du-Bois	8
42	Saint-Nicolas	7
43	Grivegnée	5
44	Evegnée	5
45	Neufchâteau	5
46	Houtain-St-Siméon	4
47	Charneux	4
48	Lixhe	3
49	Bombaye	3
50	Melen	2
51	Retinne	2
52	Battice	2
53	Dison	2
54	Lanaye	2
55	Beyne-Heusay	1
56	Rocourt	1
57	Romsée	1
58	Soumagne	1
59	Thimister	1
60	Anthisnes	1

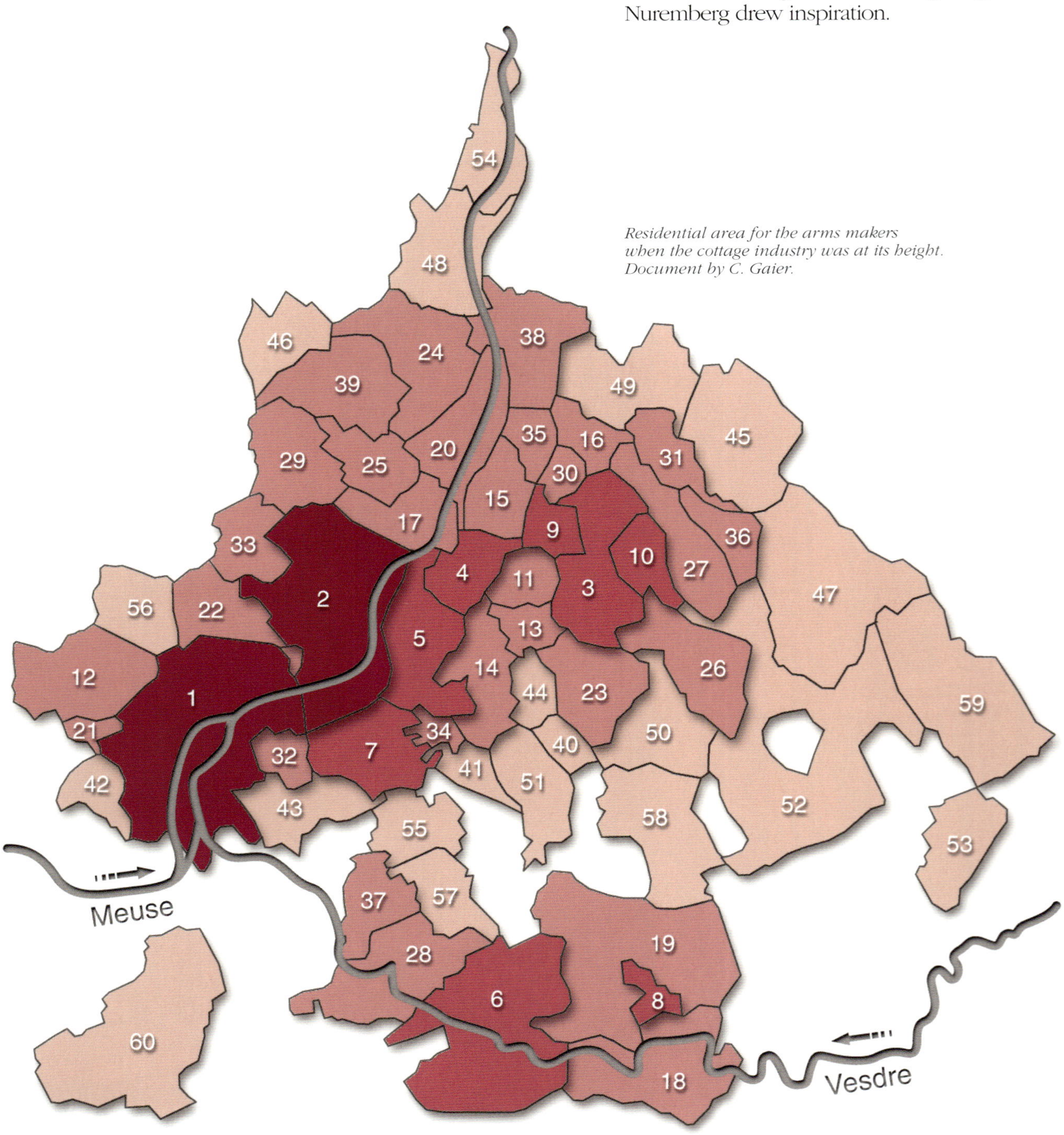

Residential area for the arms makers when the cottage industry was at its height. Document by C. Gaier.

The "bon métier des fèbvres"

However, the development of the arms manufacturing industry did not adjust very well to the corporative organization in force in the City. In fact, in contrast with the situation in certain other countries, the gunsmiths of Liège did not form independent guilds. As we have seen, they joined various sections of the "Bon Métier des Febvres" (literally "The good craft of the Iron workers"), whose regulations were poorly adapted to the needs of their profession.

In addition, corporative discipline paid too much attention to quality for it to be applicable to mass production. The necessity to produce large quantities of arms gave rise to the creation of a class of traders and businessmen, who engaged the services of craftsmen established in the country and therefore outside the jurisdiction of the guilds. Soon each region of the principality would specialize in certain types of equipment – the locks were mounted in Blégny, while the barrels were forged and bored in the valley of the Vesdre – so that the manufacturing of a weapon required a great deal of travelling between the villages around Liège.
The general characteristics of this type of organization would survive the "Ancien Régime" and even today some traces of it can still be found.

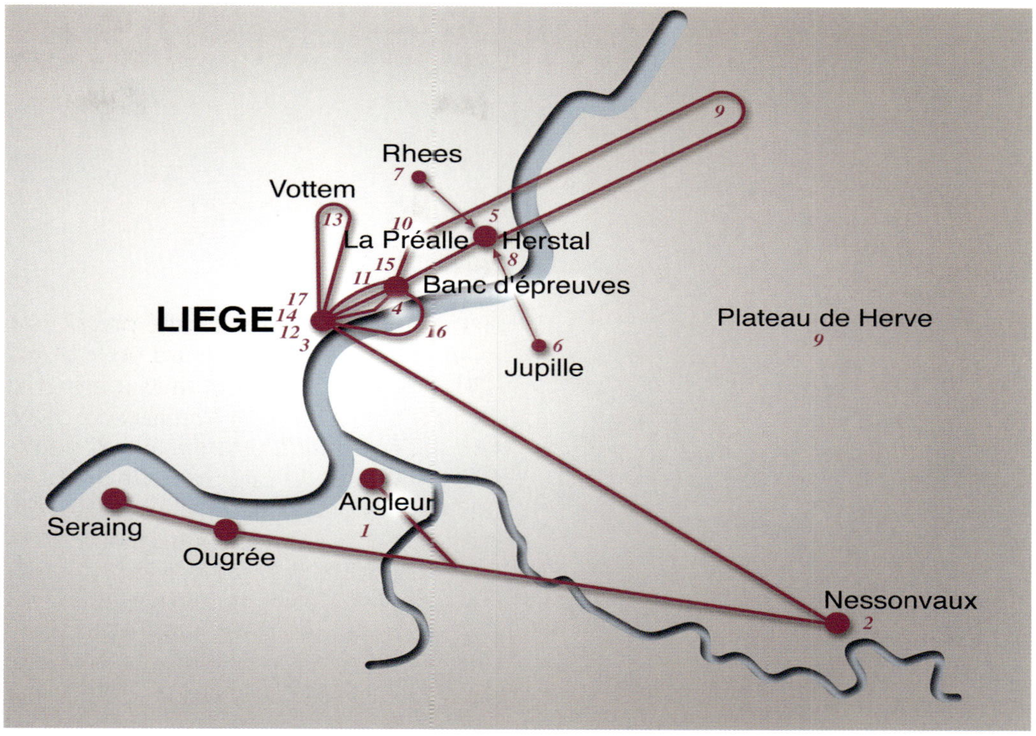

Production process of firearms in the region of Liège in the nineteenth century :

1. *Angleur, Ougrée, Seraing : steel production.*
2. *Valley of the Vesdre : manufacturing of barrels.*
3. *Liège : barrel finishing.*
4. *Liège : proof house, first and second tests.*
5. *Herstal : machining of the barrel under bolts.*
6. *Jupille : drop-forging of the action frame.*
7. *Rhees : preparing the trigger guard.*
8. *Herstal : assembling the barrel, the action-frame and the trigger guard.*
9. *Plateau of Herve or region of Oupeye : assembly, mounting of the firing mechanism, making-off.*
10. *La Préalle : shaping of the barrel chambers.*
11. *Liège : proof house, third test.*
12. *Liège : inletting of the stock and fore-end.*
13. *Vottem : stocking.*
14. *Liège : timing (lock finishing), fitting of the fore-end.*
15. *Liège : proof house, fourth test optional (smokeless powders).*
16. *Liège : precision shooting.*
17. *Liège : applying the finishing touches.*

Vulcan's Furnaces, print by Enea Vico (Parma 1523 - Ferrara 1567). As master of the furnaces, the god Vulcan has been considered to be the mythical personification of gunmakers since Greco-Roman Antiquity.

The dawn of the large enterprises

Portrait of Louis de Geer at the end of his life, from the painting by David Beck (1649). Born in Liège, he was the main arms supplier to Gustavus Adolphus, King of Sweden, and brought prosperity to the steel industry in that country, where he had emigrated together with many of his fellow countrymen.

Jean Curtius and Louis de Geer

The seventeenth century promised to be a glorious era for the principality of Liège and for its industrialists.

Among them the figure of Jean De Corte, alias Curtius, stands out. Since 1576 he had been the regular supplier to the Spanish armies who were either engaged in warfare or had a garrison in the north of Europe and more particularly in the Low Countries. Curtius, who at a certain moment had the monopoly for the production of saltpeter in the Spanish occupied territories between Burgundy and the North Sea, had set up a gunpowder factory in Vaux-sous-Chèvremont, in the valley of the Vesdre. Before he died, he had even dreamt of setting up metallurgical plants in Biscay in northern Spain.
But there is more. In addition to his enterprises and his calculations, the enormous fortune he had amassed and the influence he had held, the great munitions manufacturer displayed some sense of poetry which gave all those riches a human touch.

The sumptuous residence which Jean Curtius had built for himself on the bank of the river Meuse – a weighty galleon in stone and brick moored on firm ground – clearly evinces that the activities of a businessman can easily give rise to a particular "art de vivre".

The destiny of the financier Louis de Geer was no less remarkable. While still a child, he followed his father who had fled from religious persecution in Liège into Holland and had set up a banking business there. Louis de Geer succeeded his father in the business and continually extended it. In 1618, he took a lease of the fief of Fingspang near Norrköping in Sweden and, after having taken on several hundred workmen from Liège and Namur, bought a dozen forges, nail-smithies, foundries and arms manufacturing plants.

It was there that the famous Swedish pieces of ordnance were made which gave the great commander Gustavus Adolphus that superiority in military strength which ensured so many of his victories.

The former residence of Jean Curtius in Liège. Built around 1600 in the Meuse Renaissance style by the richest man of the former principality, this house is part of the town's museum complex, which includes in particular the Liège Arms Museum.

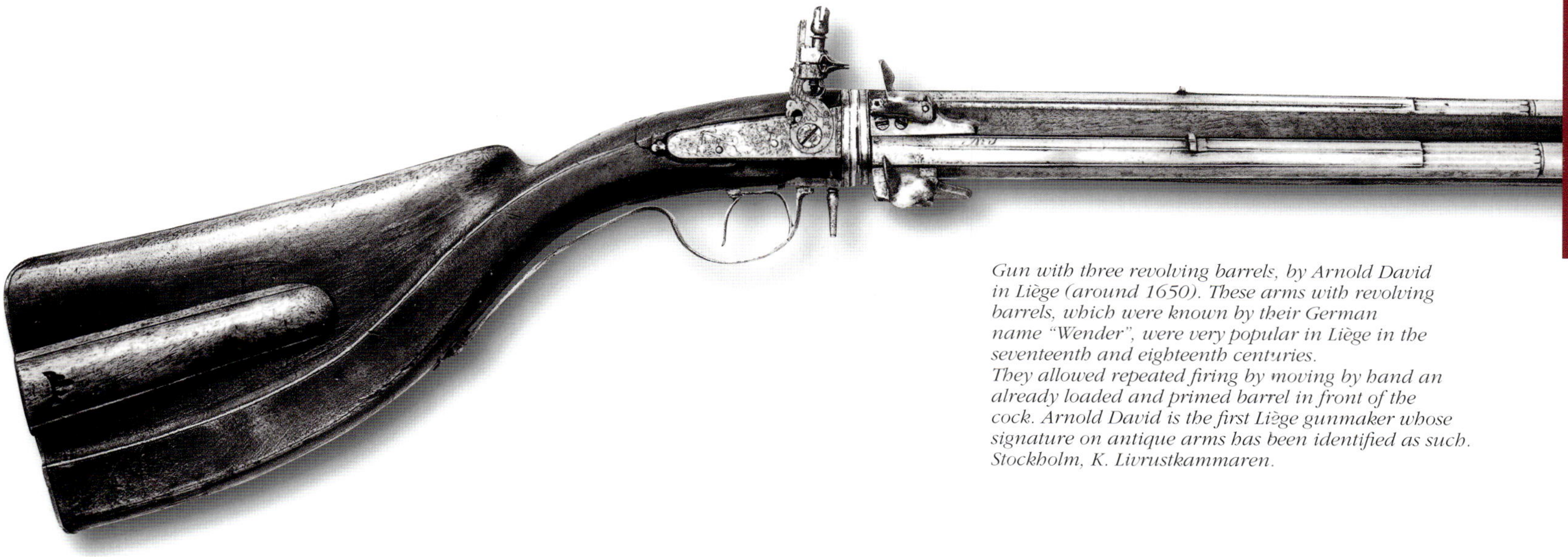

Gun with three revolving barrels, by Arnold David in Liège (around 1650). These arms with revolving barrels, which were known by their German name "Wender", were very popular in Liège in the seventeenth and eighteenth centuries. They allowed repeated firing by moving by hand an already loaded and primed barrel in front of the cock. Arnold David is the first Liège gunmaker whose signature on antique arms has been identified as such. Stockholm, K. Livrustkammaren.

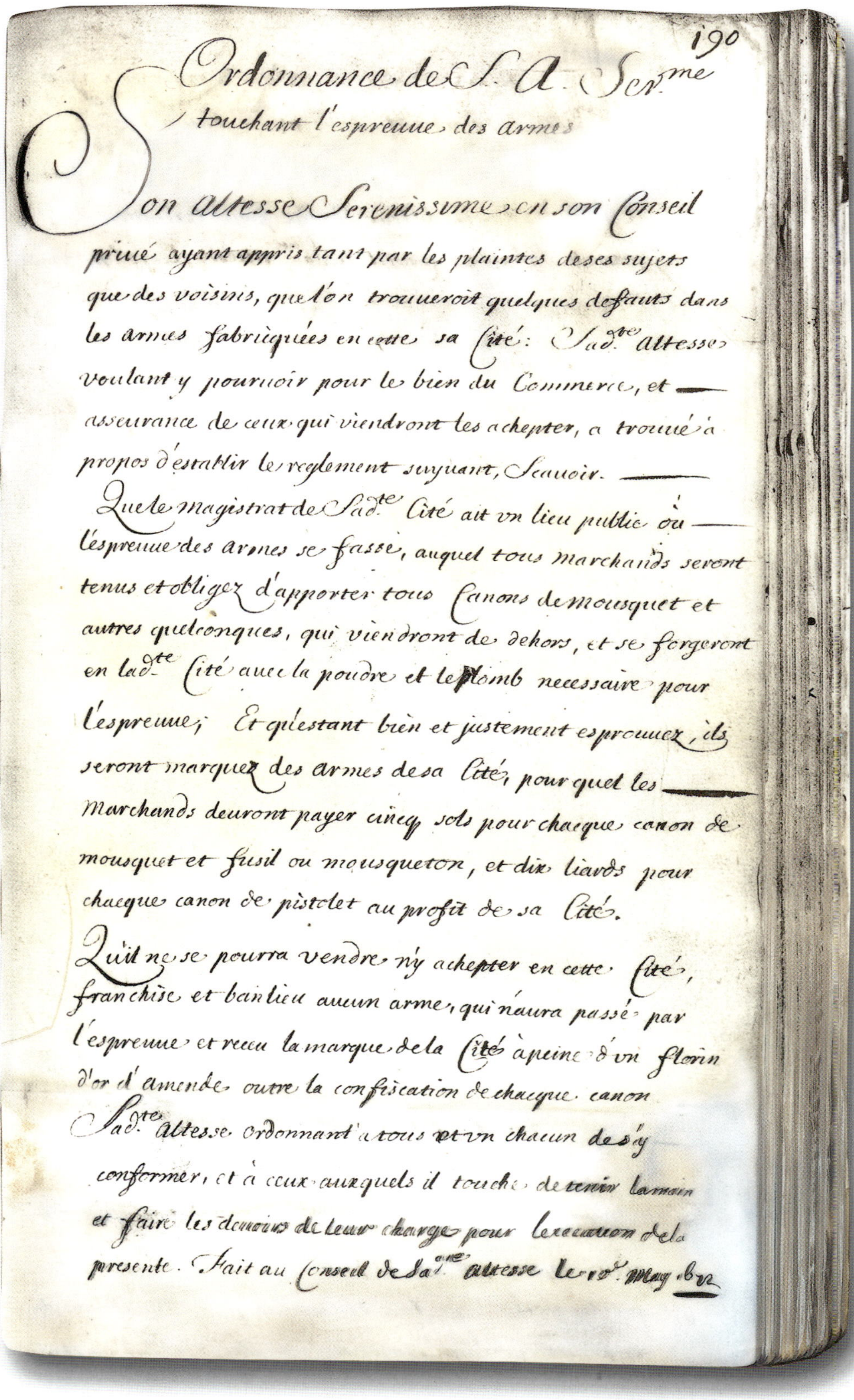

190

Ordonnance de S. A. Ser.me
touchant l'espreuue des armes

Son altesse Serenissime en son Conseil
priué ayant appris tant par les plaintes de ses sujets
que des voisins, que l'on trouueroit quelques defauts dans
les armes fabriquées en cette sa Cité: Sad.te altesse
voulant y pourvoir pour le bien du Commerce, et
assurance de ceux qui viendront les achepter, a trouué à
propos d'establir le reglement suyuant, Sçauoir.
Que le magistrat de Sad.te Cité ait vn lieu public où
l'espreuue des armes se fasse, auquel tous marchands seront
tenus et obligez d'apporter tous Canons de mousquet et
autres quelconques, qui viendront de dehors, et se forgeront
en lad.te Cité auec la poudre et le plomb necessaire pour
l'espreuue; Et qu'estant bien et justement esprouuez, ils
seront marquez des armes de sa Cité, pour quel les
marchands deuront payer cincq sols pour chaque canon de
mousquet et fusil ou mousqueton, et dix liards pour
chaque canon de pistolet au profit de sa Cité.
Qu'il ne se pourra vendre n'y achepter en cette Cité,
franchise et banlieu aucun arme, qui n'aura passé par
l'espreuue et receu la marque de la Cité à peine d'vn florin
d'or d'amende outre la confiscation de chaque canon
Sad.te altesse ordonnant à tous et vn chacun de s'y
conformer, et à ceux auxquels il touche de tenir la main
et faire les deuoirs de leur charge pour l'execution de la
presente. Fait au Conseil de Sad.te altesse le 10.e May 72

Chiselled pistol (one of a pair), ordered in Liège in 1651 by Carl Gustav Wrangel, Governor-General of Pomerania. Close-up of the trigger guard of one of the pistols. Skokloster, Sweden.

Ordinance by the Prince-Bishop Maximilian Henry of Bavaria, making compulsory the previously optional testing of firearms manufactured in Liège (10 May 1672). The obligation of testing, reinforced under Napoleon in 1810, was often dodged and evaded and could only be successfully imposed in 1845. National Archives in Liège, Privy Council, reg. 214, f° 190.

Hydraulic power and flint locks

Times were not only favorable to large commercial enterprises. The intellectual climate in the principality was also beneficial to technical progress. Hydraulic power was gradually being introduced into the steel industry, in wire-working as well as in foundrying and drop-forging. Workshops were set up in the valleys of the Ourthe, Vesdre and Amblève. However rudimentary they may still have been, the new machines allowed production to increase considerably and prices to come down. The manufacturers of muskets and pikes were the first to benefit from this situation at a time when the various wars in Europe opened up more new outlets for their trade.

It is well known that in the seventeenth century the art of gun making in Europe reached a summit. With the new flint locks which appeared in France around 1620, the firearms acquired new strength, reliability and precision which, both in warfare and hunting, ousted the former missile weapons and of course the old side arms.

Fine Arms

At the same time, arms manufacturing had become a real art. The manufacturing of beautifully decorated firearms, which had started in Germany towards the end of the sixteenth century, quickly spread to Italy, Spain and France where it had found a willing patron in Louis XIII. With the accession of Louis XIV, France had become the world center for the production of high-grade arms and attracted craftsmen from neighboring countries. One of the most famous, Adrien Reynier, known as The Dutchman, had in fact emigrated from Liège. The Dutchman, who was appointed Arquebusier Ordinary to the king, had a rather solid, rigorous and rich style which made the guns and pistols he turned out of his workshop masterpieces of classical arms manufacturing. His best productions were published around 1660, together with those of his Parisian colleague Thuraine, by the engraver Jacquinet. Their work influenced French production for three-quarters of a century.

Colbertism and mercantilism

In Liège, where court life was somewhat less brilliant than in Versailles, the manufacturing of arms had never known such splendor. The Prince-Bishops from Bavaria, open as they were to French ideas, seem nevertheless to have done their best to apply some of the principles of Colbertism to the principality, with a view to improving the quality of the arms being produced by craftsmen who did not belong to the guilds. In this respect, Maximilian-Henry of Bavaria decreed on 10 May 1672 an ordinance, which obliged the tradesmen to have their arms tested and hallmarked. Even though this ordinance was not always scrupulously observed, the extension of professional discipline to independent craftsmen appears to have had some beneficial effect. Also the reputation of the Liège gunsmiths extended across the borders in spite of the crisis which hit the principality from the second half of the seventeenth century. The Spanish Netherlands in fact followed France's example in applying a mercantilist policy which hampered foreign trade. The situation became worse from 1672 when Maximilian-Henry gave up his neutrality and sided with Louis XIV.

From that moment on, that is to say at the beginning of the War of the Netherlands, the principality was to become embroiled in all the conflicts which would tear Europe apart, right up to the 1713 treaties. Devastated several times in the course of the hostilities, its fields pillaged, its capital bombarded (1691), the country of Liège became engulfed in the whirlpool of conflicts. In actual fact, the gunsmiths had to a certain extent managed to escape all this turmoil. At that time, England, some German princes and even Peter the Great (1693) were arming their troops with guns made in Liège.

A gunsmith's workshop in the seventeenth century, from the collection of antique guns designs of Thuraine and Le Hollandois.

Map of the former principality of Liège by Sanson, Geographer to the King, Amsterdam (1696).
The yellow areas came under the Prince-Bishop of Liège. This political entity, with extraordinarily ragged contours, came under the Holy Germanic Empire, but in practice it enjoyed a considerable degree of independence. The principality of Liège was abolished in 1794 when it was annexed to the French Republic. In 1815 it passed under the sovereignty of the King of the Netherlands and has constituted part of the Kingdom of Belgium since 1830. Liège, Fonds U. Capitaine (C2, 2-45).

Peace and liberty in the principality

For the citizens of the principality, more than anywhere else, the eighteenth century was one of the most delightful and least irrational periods in history.
Their contemporaries had not failed to notice this, and in 1738 a French author, P.L. de Saumery, was able to compile a description of the "Delights of Liège".

With the Treaties of Rastadt (1713-1714) there came an era of peace and tranquillity for the principality. Its neutrality suffered no severe blows except for a brief occupation by France during the Seven Years' War (1756-1763).

The Statutes of 1684, amended by the Treaty of 1763, gave the City a constitution which combined tradition and progress in what was to become a model of political wisdom. Thus all the conditions for individual liberty were united, without which the spirit is incapable of creativity.

Plate with gun decorations, published in Paris in 1743 by the Liège artist Demarteau. From the sixteenth century up to the present day, the prosperity of the arms industry had brought about in Liège uninterrupted activity in arms decoration of international repute. These plates served as models and were passed on from one workshop to another. They contributed to spreading the talent of the Liège craftsmen throughout France, and French style throughout Europe.

Gun purported to have belonged to Louis XIV, built by Gilles Massin, arms manufacturer in Liège (about 1730). Detail of the lock and butt plate. Museum of Art and Industry in Saint-Etienne.

Arms are exported throughout the whole of Europe

Order and peace were of course beneficial to the arms industry. More particularly, on 24 July 1700, an ordinance decreed by the Prince-Bishop Joseph Clemence of Bavaria was to reinforce professional discipline as regards both the care taken in manufacturing as well as conditions of sale and particularly price-setting. The sequence of events abundantly shows the wisdom of this measure. In spite of the protectionism adhered to by the Austrian Low Countries, which surrounded the principality, and in spite of the measures taken by the French during their two years' occupation of Liège, the arms industry was to flourish in an almost uninterrupted fashion.

In the eighteenth century as in centuries past, the export of arms constituted one of the most secure foundations of this prosperity. In the course of time, we find practically all the armies of Europe being equipped with muskets, pistols and sabres, small cannon and cannonballs made in Liège. At the beginning of the eighteenth century, the Elector of Bavaria, the Emperor Charles VI, Frederick William I and the United Netherlands were clients of the manufacturers in Liège. A little later came Spain and Portugal (about 1740), then once again the United Netherlands (about 1775).

During the American War of Independence (1775-1783), the manufacturers in Liège supplied the rebels with apparently enormous quantities of arms. Although scant, the statistics available in this respect are nevertheless significant. Around the middle of the century, the annual production of muskets had already reached some 100,000, without counting pistols and other firearms. In 1788, at the dawn of the French Revolution, there were 70 to 80 manufacturers in Liège exporting more than 200,000 muskets each year, with a total value of three to four million Florins. And everything points to the fact that these figures are well below those reached during the War of Independence.

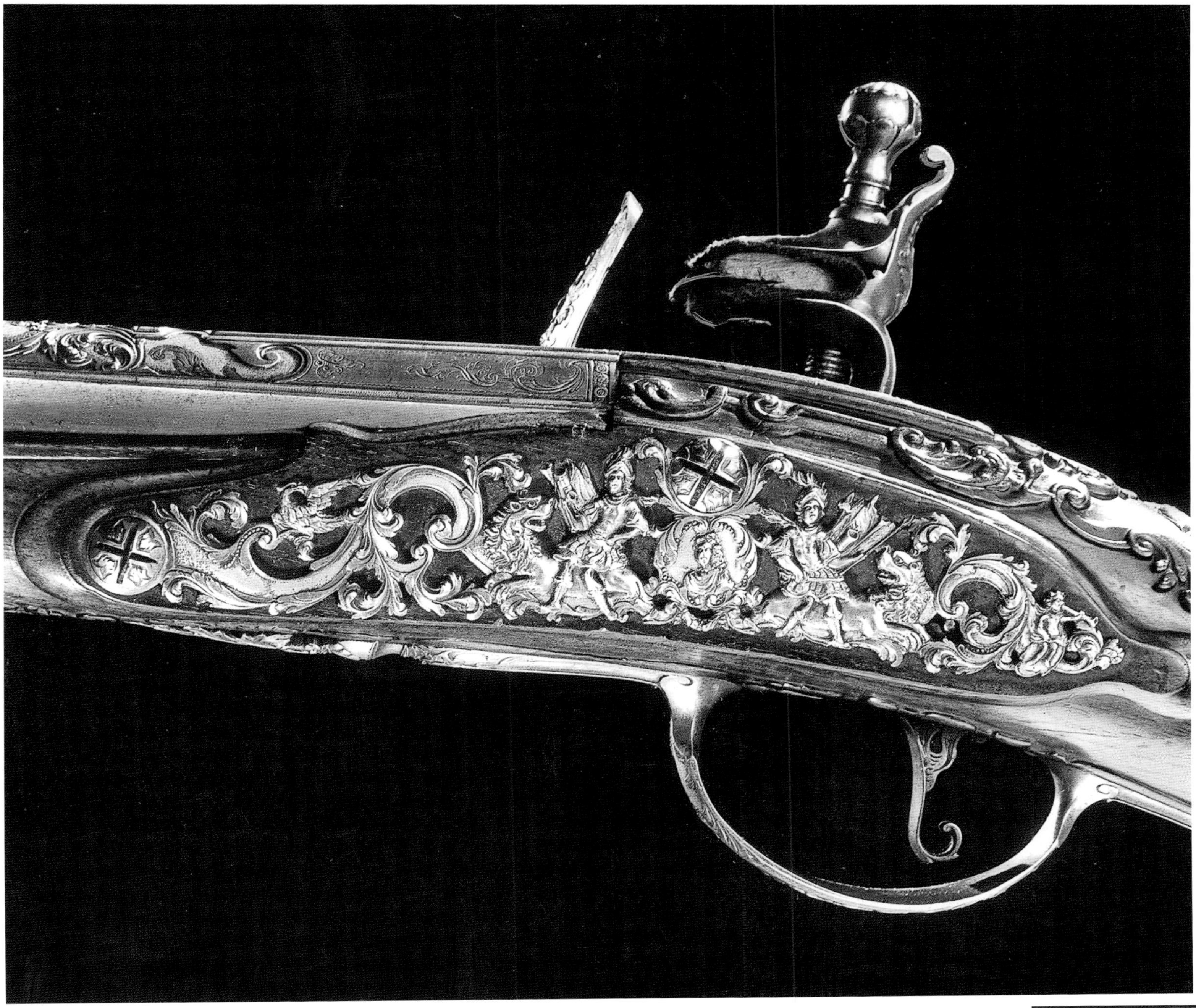

Craftsmen in high demand

The reputation of the craftsmen of Liège had led a number of nations to secure their co-operation more directly. In 1722, Frederick William I, having decided to establish an arsenal in Spandau, came to Liège to take on a large number of workmen of various specializations, and in 1788, according to Mirabeau, the craftsmen of Liège were still entrusted with the forging of cannon, both in Spandau and Potsdam. During the Seven Years' War, it was France who sought the cooperation of the principality's craftsmen. In fact some of them went to work in the factories of Maubeuge, Charleville, Saint-Etienne and Versailles.

With the climate of peace and economic progress, with an Episcopal court infatuated with luxury and the pleasures of hunting – successively John Theodore of Bavaria, Charles Nicholas d'Oultremont, the prince of Velbruck and Francis Anthony de Méan practiced this sport enthusiastically – with more refined and better trained craftsmen, conditions were beneficial to the manufacturing of small arms in Liège.

As a matter of fact, whereas in the course of the previous centuries the best craftsmen had emigrated, some, like G. and M. Berleur, G. and H. Devillers, J. Niquet and T. Thiermay, continued to apply their skills in the principality.

Thus Liège continued to exert an influence on its neighbors in the art of gun making. Two engravers from Liège, Joseph Demarteau, who was an apprentice of the designer Lacollombe, and his brother Gilles (b. Liège 1721 - d. Paris 1776), both established in Paris, used the most attractive rococo motifs in the decoration of arms. From the work of these artists, no less than three collections of gun designs have come down to us, of a very rich and ornate style which owes a lot to the French examples as well as to the models of Roman baroque.

So Liège was happier and freer than it had ever been. A now forgotten poet, Fabre d'Eglantine (1750-1794), has extolled the principality with a certain languid grace, but which still does not fail to stir :

"This ancient City where the silvered Meuse
Waters with joy its most beautiful vale
Where the happy citizen sees blossom abound
And the harvest of Ceres and Fruits of Pomone
And verdant vine of Erigone's lover
Where he finds in his homeland where
happiness reigns
That first of our Blessings, our Liberty sweet".

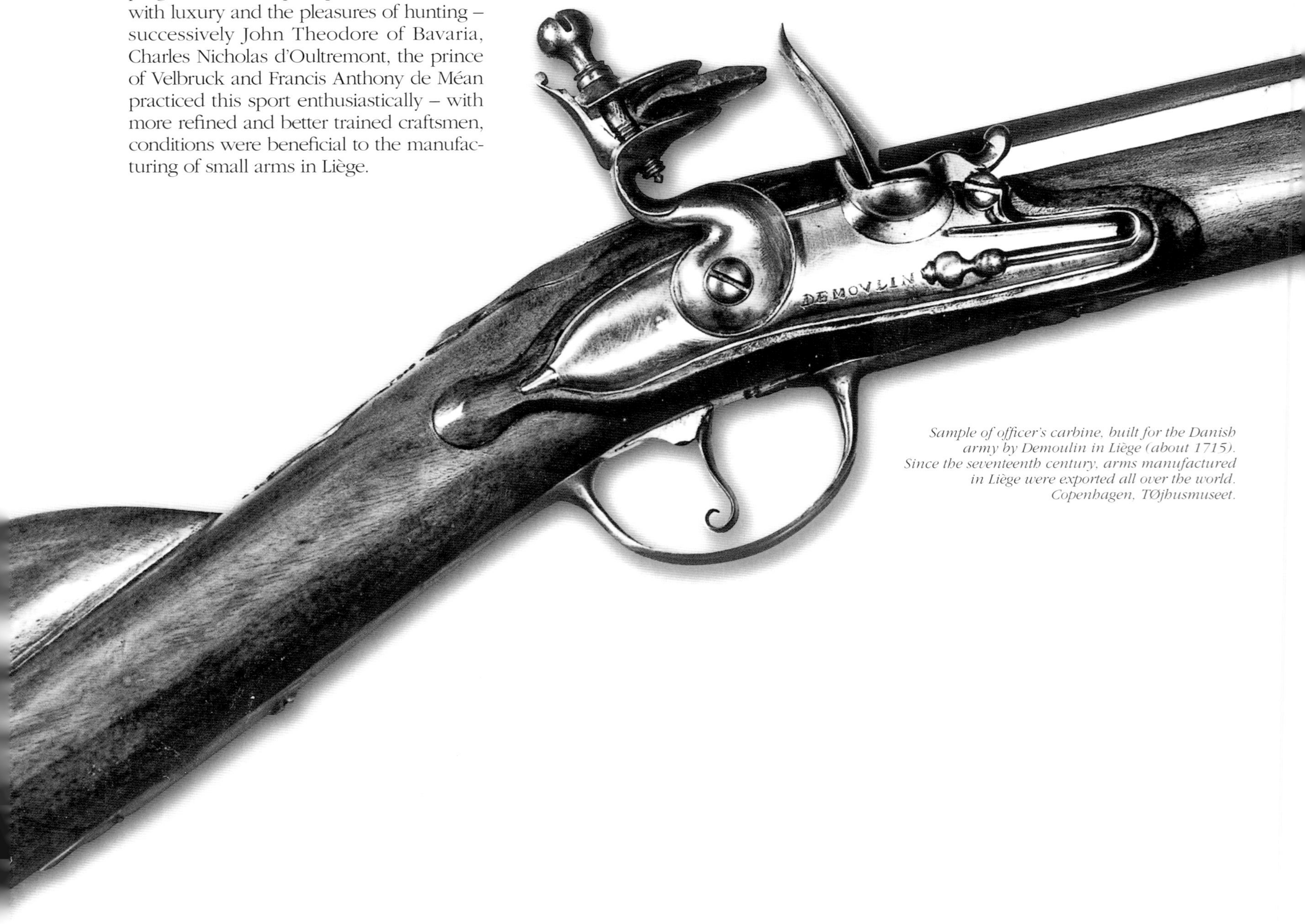

Sample of officer's carbine, built for the Danish army by Demoulin in Liège (about 1715). Since the seventeenth century, arms manufactured in Liège were exported all over the world. Copenhagen, TØjhusmuseet.

The city in political and economic turmoil

People grow tired of everything, even of being a citizen of Salente. The citizens of Liège demonstrated this on 18 August 1789 when, after a sudden revolution, they overthrew the Ancien Regime.

In February 1793, after a brief restoration, an assembly elected in the principality voted for annexation to the French Republic. This marked the end of a nation which had existed for a thousand years.

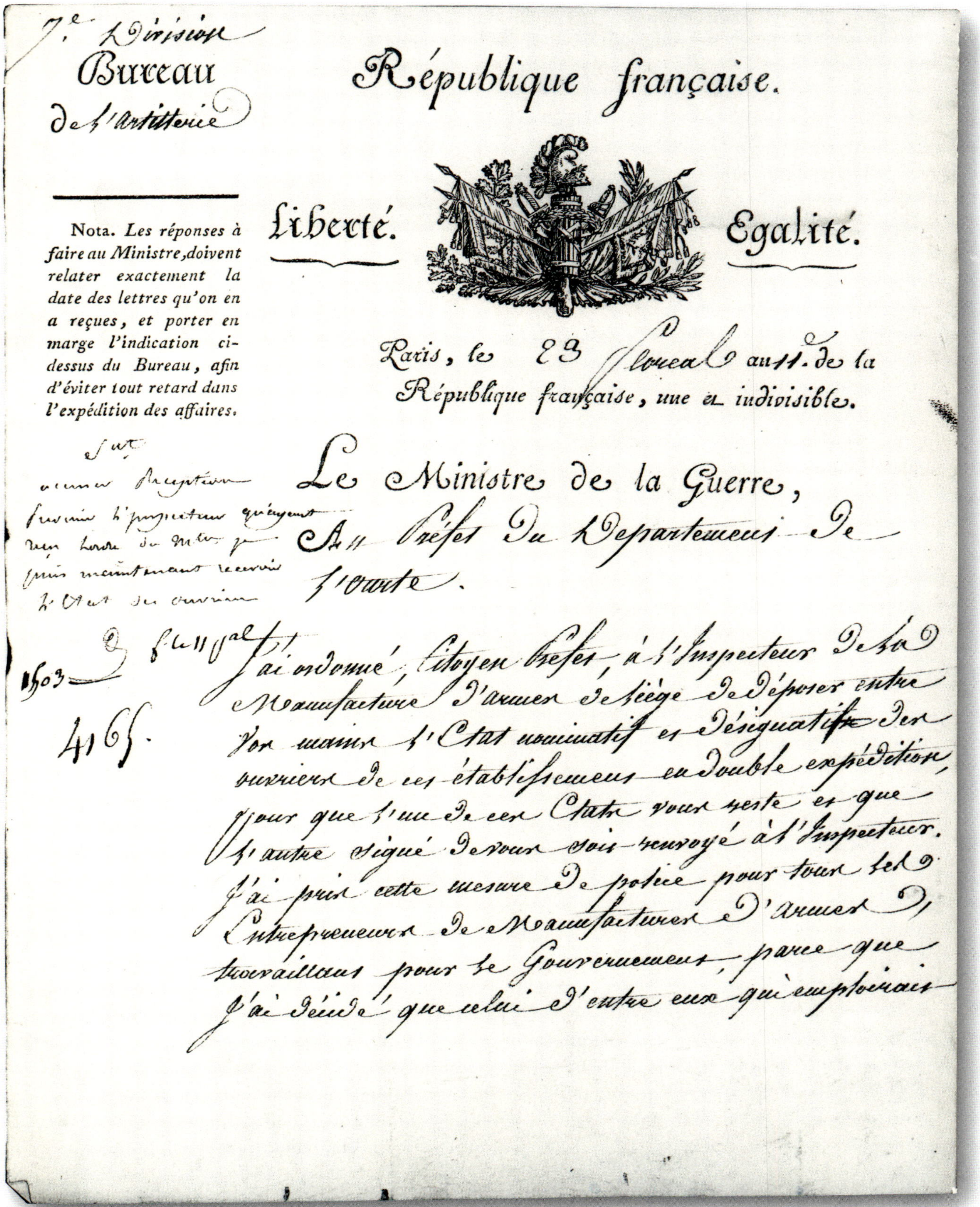
7e Division
Bureau
de l'Artillerie

République française.

Liberté. Egalité.

Nota. *Les réponses à faire au Ministre, doivent relater exactement la date des lettres qu'on en a reçues, et porter en marge l'indication ci-dessus du Bureau, afin d'éviter tout retard dans l'expédition des affaires.*

Paris, le 23 floréal an 11. de la République française, une et indivisible.

Le Ministre de la Guerre,
Au Préfet du Département de l'Ourthe.

4165

J'ai ordonné, Citoyen Préfet, à l'Inspecteur de la Manufacture d'armes de Liège de déposer entre vos mains l'Etat nominatif et désignatif des ouvriers de ces établissemens en double expédition, pour que l'un de ces Etats vous reste et que l'autre signé de vous soit renvoyé à l'Inspecteur.
J'ai pris cette mesure de police pour tous les Entrepreneurs de Manufactures d'armes, travaillant pour le Gouvernement, parce que j'ai décidé que celui d'entre eux qui emploierait

Provisions concerning arms manufacturers' workers, communicated by the Minister of War of the French Republic to the prefect of the department of the Ourthe (13 May 1803). The French regime (1794-1814) encouraged and modernized the production of armaments in Liège, but its nationalization measures were detrimental to the private arms industry.

The arms industry, a victim of the crisis

For the City it also marked the beginning of a period of political and economic turmoil. We shall not deal here with the political upheaval which shook France until the fall of the Directory. Suffice it to say that Liège and its industry were considerably affected and that the arms industry was to be one of the first victims of the crisis. Undoubtedly, during the months which followed the revolution, the manufacturers had to a certain extent benefited from supplying arms to the principality's troops and to those of the United Belgian States.

Brief respite

According to an account by a contemporary, the arms industry would benefit from that "invigorating breath of life which the First Consul is giving to the national industries, the restoration of which never ceases to occupy the solicitude of the government". Without giving the arms factories back their bygone prosperity, this concern provided them with both a favorable and prolonged development in two important areas, the improvement of the production methods and the refinement of taste and sense of quality.

Mass production

Further progress was made under Louis XVI. For the construction of guns, Gribeauval, who had become Inspector General of the Artillery, drew up tables establishing the exact measurements for each piece of artillery. This system, to quote the term used at the time, was gradually applied to infantry firearms and more particularly the Model 1777 musket which, with minor modifications, would remain in use until the last years of the July Monarchy (1840s) and was adopted by various countries including the United States of America.

But the disorder, the foreign invasions and eventually the ruin of the corporative system which, in spite of its disadvantages, had given professional life some kind of structure, plunged the arms industry into turmoil. In 1808, according to Thomassin, the number of manufacturers still in business had fallen to 14, while their annual production barely reached 8,000 rifles and pistols. The unemployment which followed this collapse had forced many craftsmen to emigrate. Some established themselves in St-Etienne, others in Paris and Versailles. The restoration of order and of the public economy which followed the introduction of the Consulate gave some respite to the arms industry.

The progress in production methods which Napoleon's administration brought about was of a decisive nature, and for that reason it requires some further explanation. Since the reign of Louis XIV, thanks to the reforms of Louvois, France succeeded Sweden as the principal player in the manufacturing of arms. In fact, the minister had made a reduction in the immense number of different weapons which were used by the army, and introduced discipline and reliability in the production of the models which he had carefully selected. Moreover, he saw to it that the weapons benefited from the improvements which had been conceived by the gunsmiths. So from 1680 onwards he had the matchlock musket gradually replaced by flintlock guns.

The Model 1777 musket is undoubtedly the first weapon, if not the first object in the world, to be manufactured according to a scientific method, the constituent parts of which could be assembled as they were or with slight alterations. This event, to which historians appear paradoxically not to have paid much attention, seems to us to be one of the most memorable in modern history. In this period too, the concept of mass production was born, as the principles of machine manufacturing and control, which had proved their value in the manufacturing of arms, were progressively extended to other sectors of industry. This phenomenon, with its economic, sociological and moral implications, so profoundly reshaped human society that the circumstances of its origin deserve, in our opinion, some attention.

Flintlock musket, 1777 system, corrected Year IX, dragoon model, manufactured by the Manufacture Impériale d'Armes de Liège in 1813, and close-up of the trademark. Produced according to the then revolutionary standards of interchangeability, this gun was long considered as a model of industrial perfection. Some 300,000 of them were manufactured in Liège under French rule. Liège Arms Museum 8680 ("Grand Curtius Museum").

Cased flintlock pistols, by J. Lambert, said Biron, Liège, presented by Emperor Napoleon I in 1813 to Field-Marshal the Count of Wurben-Littitz, ambassador of the Austrian Empire.
Musée d'Armes de Liège 8346 ("Grand Curtius").

Next page : view of the interior. The detachable shoulder stock allows for conversion of one of the pistols into a carbine.

Proof house for firearms

The manufacturing procedures developed by the military engineers went against the grain of all current practices and were used only in State industries. The Revolutionary wars - which incidentally were the first mass wars - induced the government to call on private manufacturers and, consequently, to impose perforce its own production methods. We say "per force" because in the course of these troubled years manufacturing discipline had declined sharply. It was only in 1801, under the Consulate that the rigorous regulations, which had proved their value under the Monarchy, were restored.

This process can be seen very clearly in Liège. From the moment of the annexation, the revolutionary Committee of Public Safety had set up a committee entrusted with the verification of firearms in the area (22 September 1794).

This committee, which was replaced the following year by a military inspectorate, had to check the arms purchased by the State according to gauges and calibers.

In Liège and elsewhere the severity of this inspection was increased after the coup of Brumaire. The foundation in the town of a National then Imperial Factory instructed local manufacturers in the new methods of production. An Imperial Decree of 14 December 1810 instituted a Proof House for Firearms.

Master Boutet

We have seen that under Napoleon refinement of taste was accompanied by technical perfection. For the first time in history, the gunsmiths of Liège were directly approached by the clientele of a large capital.

A great number of craftsmen, and even the sons of certain manufacturers, offered their services to the Imperial Factory in Versailles which was led by Nicolas Noël Boutet. They could not have found a better master than this famous gunsmith. Since the eighteenth century the Liège craftsmen had indeed been able to decorate a firearm with style and talent. Boutet taught them that in gun making the beauty of a decoration was worthless without the beauty of its composition and execution.

Portrait of Napoléon Bonaparte, First Consul, during his visit to Liège (1803), by Dominique Ingres.
The future emperor was interested in the industrial potential of Liège, particularly in the manufacturing of arms.
Grand Curtius museum complex, Liège (in trust for the Museum of Modern Art).

The golden era : 1815-1914

Liège, "one big factory"

"Liège is one big factory", said Michelet. This description could not have been more apt than for the period which extended from the end of the French Empire up to the outbreak of the First World War. In the course of these hundred years the old city of the Prince-Bishops, Guichardini's "paradise of the priests", had indeed become one of the foremost industrial cities in Europe.

This development started with the transformation of two of the region's traditional activities, the coal-mining and metallurgical industries. Certain shrewd and venturesome businessmen like Daniel Dony, William Cockerill, Michel Orban and Antoine Lamarche, gave an impetus to this transformation, and with the generalized implementation of machines as well as the liberation of substantial capital, gave the local industry the characteristics which were to stay with it for a long time.

The revolution of 1830 which separated Belgium from the Netherlands did not slow down the economic expansion for long. Only five years after the foundation of the Belgian State, the first locomotives and rails on the Continent were built in the workshops of John Cockerill in Seraing, near Liège.

During the whole of the last century more and more new factories were being built in the area, while the depression which afflicted Europe from 1873 to 1880 merely slowed down that progress. Moreover, the Liège craftsmen went on to export their know-how to other countries: in France, Germany, Spain, Russia, Africa and Asia we find them opening mines, setting up blast-furnaces and building factories.

Stock of a hunting shotgun, carved in "troubadour style". This masterpiece of decorative art, not at all functional, was made by Lienard and designed by Rambert, Paris, mid nineteenth century. It is one of the collection of plates which Charles Claesen of Liège published in 1856 to illustrate the best in arms manufacturing. Liège Arms Museum (the Grand Curtius Museum).

Manufacturing of luxury arms

This explosive creativity does confront the historian of ideas with some problems. Here we have a people whose vitality is obvious in industry and commerce, but who is not supposed to be reputed in the way of artistic creation. We invite all those who hold such a view to come and pay a visit to the Arms Museum of Liège, and in particular the halls housing the small arms which were manufactured in this city during the first half of the nineteenth century. The refined form, the profusion and fantasy in the motifs, the finish, all make these weapons more apt for delightful enjoyment than for usage, and therefore have the quality of works of art.

It would not be the first time in history that a people expressed the best part of its artistic skills in the production of utensils.

From that moment on, Liège became the European center for the manufacturing of luxury arms. While it had previously been in Paris that the decorations where made which inspired the gunsmiths, it is in Liège that in 1856 Charles Claessen published the most remarkable pattern-book of that time. The most skilled engravers and the most gifted decorators of the time had worked together on that collection, a collection whose baffling originality appears at times somewhat exaggerated.

The last years of the century saw a profound transformation in the art of gun making.

The appearance of various types of hammerless mechanisms led to a greater simplicity in the exterior design of the guns, while at the same time the opulent beauty of the Second Empire followed a more timid eclecticism. However, some manufacturers like Auguste Francotte & Co did not hesitate to apply to their art the ornamental designs of the 'Art Nouveau' style. It was in this spirit that some of the most luxurious guns to be produced in Liège were decorated.

Percussion shotgun, manufactured for the Prince of Orange by J. Devillers in Liège (1829). Checkered butts, basket-weave style, barrels in Damascus steel, coat of arms of the Orange family inlaid with gold. Liège Arms Museum 8402 (the Grand Curtius museum).

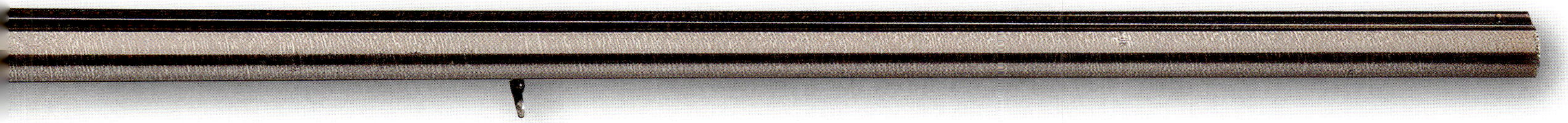

Pinfire shotgun, chased and inlaid by J. Boussart (Liège, 1865). This masterpiece of fine gunmaking appeared at the Universal Exhibition of Paris in 1867. Liège Arms Museum - 4719 (the Grand Curtius).

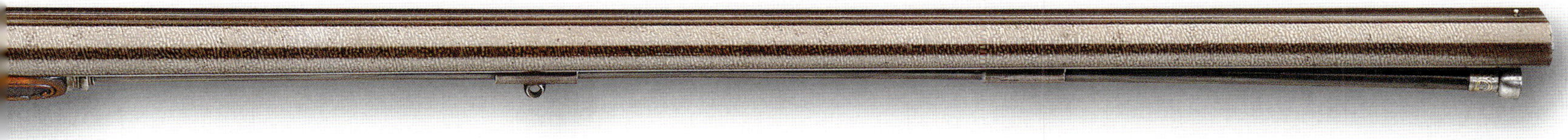

Richly decorated Liège percussion shotgun (about 1855). The stock was made by the Herstal arms sculptor Jean-Michel Tinlot. Liège Arms Museum 4717 ("Grand Curtius").

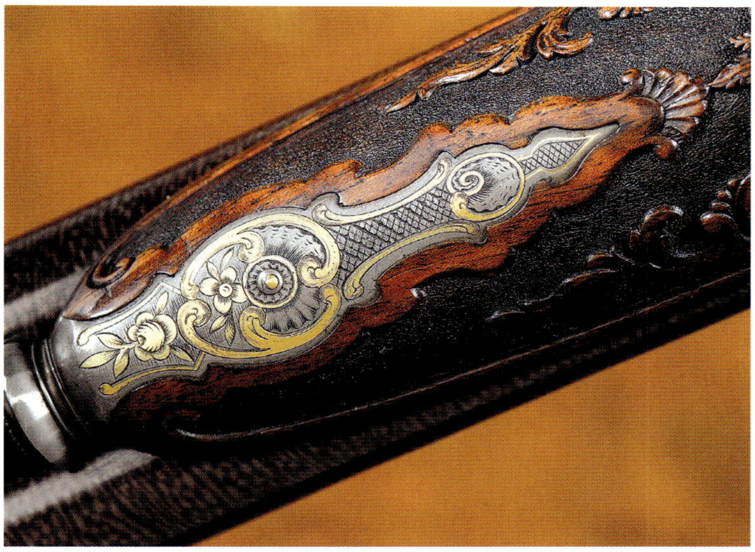

New inventions

Although the fact that luxury arms manufacturing is undoubtedly the most true expression of nineteenth-century artistry in Liège, we should not forget that the gunsmiths have also ventured into other, very different domains, in which they have often excelled.

One of these domains, the manufacturing of trade guns, does not really interest us. It concerns the production under rudimentary conditions of flintlock or percussion arms which were derived from the guns in use in the armies of Europe. They were destined for the native inhabitants of Africa and South America and were of mediocre quality.

On the other hand, the production and sale of other arms merits rather more attention. The period we will be considering was particularly rich in inventions which were to change the design and even the manufacturing of small arms. They all derive from the percussion principle. The first device which applied this principle was patented in 1807 by the Reverend Alexander Forsyth. It was adopted successively in the needle gun by Dreyse (1829), the rim-fire cartridge by Flobert (1843) and in the breech-loading needle gun by Chassepot (1866).

The progress made in percussion methods and in cartridge design in their turn allowed for the design of weapons which were capable of firing a reasonably large number of rounds. Thus we have successively the revolvers by Collier (c. 1820), Colt (1835-1837) and the Liège gunsmiths Mariette and Herman (1839), the repeating rifles by Spencer (1860), Henry (1862), Winchester (1866), Mauser (1884), Lebel (1886) and Mannlicher (1886), the machine gun by H. Maxim (1883) and finally the self-loaders by the Clair brothers (1889) and Browning (1890).
At the same time, advances in machine-tool techniques allowed the industry to make weapons of better quality.

With the first double-action steam engine, which James Watt had invented in 1786, engineers had a more powerful and reliable tool than the waterwheel. At the same time they diversified the existing models and increased their operating speed.

Between 1850 and 1856, Whitworth had designed a number of machines with which infinitely more accurate measurements could be made than ever before. Henceforth completely interchangeable pieces could be made. The first industrialist to make use of these new instruments was the revolver manufacturer Samuel Colt.

To complete the picture, we should add that since the invention of the dynamo by the Liège craftsman Zenobe Gramme (1869), the electric motor gradually replaced steam-engines in the factories, and that since 1888 production increased even more rapidly with the use of better quality steel in the machine-tools.

The Liège manufacturers progressively adopted the improvements in firearms design and production methods. At the same time, the government proof regulations which were instituted during the time of the French Empire were reinforced: a Royal Decree of 8 September 1846 abolished the contracting of the testing procedures and entrusted them to a public body, of which the manufacturers retained the ownership and management. The statute of the Proof House was defined in an Act of 24th May 1888.

The Zénobe Gramme dynamo. (Liège, Museum of Walloon Life).

The 'Electricity Fairy' pays homage to the inventor of the industrial dynamo. Monument to Zénobe Gramme (1826-1901) erected in Liège in 1905.

The "S.A. des Charbonnages et Hauts Fourneaux d'Ougrée" (Ougrée Collieries and Blast Furnaces Cy) around 1850, illustrated in "La Belgique Industrielle".

The first Belgian locomotive built in 1835 by John Cockerill (Cockerill photo archives).

10,000 hp Cockerill gas engines. In an attempt to recover the gas produced by the blast furnaces, Cockerill began to develop an engine run on lean gas in 1895. In the following period, Cockerill produced several hundred gas engines with capacities of up to ten thousand horse power. From 1898 onwards, these engines were even manufactured under license in France, England, the United States, Russia and Germany. (Cockerill photo archives).

Medal depicting the Liège gunsmiths, by Godefroid Devreese (1861-1941), made on the occasion of the Universal Exhibition of Liège in 1905.

Unprecedented prosperity

Thanks to all these circumstances, the arms industry drew maximum benefit from the current political and economic climate, and enjoyed an unprecedented period of prosperity. While in 1830 some 193,534 arms were hall-marked in Liège, the number reached 591,432 in 1850, 1,351,243 in 1900 and reached the all-time record figure of 1,579,479 in 1907. In 1896, the industrial census recorded more than 13,000 gunsmiths in the region, and we know also that the daily wages of the workforce varied between 2.50 and 5 gold francs.

Incidentally, in 1849 the arms manufacturers set up for their workers a Mutual Benefit Society which today is one of the oldest in the country.

To be able to understand the reason for such a development, we must take account of the fact that in the course of the 19th century, the chief arms manufacturers of Liège had been supplying the armies of Europe. In 1854, for example, during the Crimean War, the companies Ancion & Cie, Pirlot Bros, Renkin Bros and Auguste Francotte supplied the British government with 20,000 Enfield rifles. In the same year, Colt placed an order with Liège for a large number of revolvers destined for Russia. In 1867, the company Dresse-Laloux & Cie which was founded not long before that (1862) converted the rifles of the Belgian Army to the new system of Albini-Braendlin.

Belgian single shot rifle, for the Corps of Engineers, Albini Braendlin system 1868, 11 mm caliber, with bayonet. Liège Arms Museum - 4599 (the Grand Curtius).

Specimens of shotgun barrels made of Damascus steel, manufactured near Liège in the nineteenth century. Liège Arms Museum (the Grand Curtius).

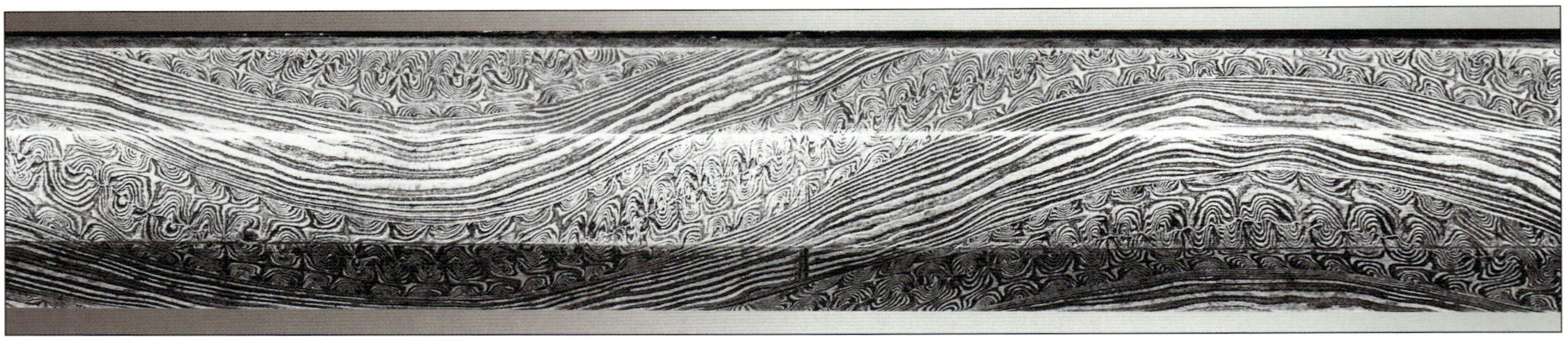

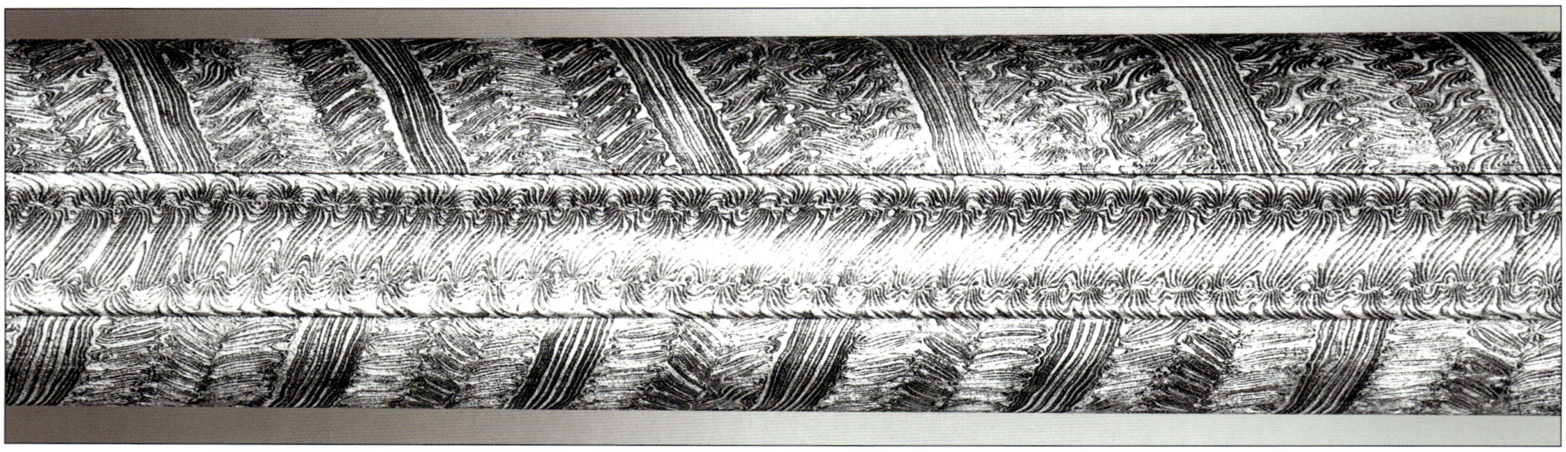

Decorative Damascus barrel. Liège, around 1850.

The "Fabricants d'Armes Réunis"

The increase in the number of weapons being ordered by the various nations, the short terms of delivery, the fear of foreign competition, especially from the German factories, compelled the arms manufacturers to unite. In 1870, the companies Ancion, Dresse-Laloux Cie, Auguste Francotte and Pirlot Fresart jointly set up a workshop in Herstal, called the Petit Syndicat, where the Comblain rifles were made which the Belgian government bought for the National Guard.

On 11 April of that same year – just a few months before the outbreak of the Franco-Prussian War – the same manufacturers joined together with other companies – the Société Liégeoise, Gulikers, Renkin Bros, Falisse, E. Malherbe, Beuret, Mordant, Benthin, P. Drissen – in a rather loose association with a chiefly commercial aim: obtaining orders and distributing them between the various members.

'Lienard' foliated scrolls. The Liège decorator Lienard, who had established himself in Paris, created a decorative style which imitated the old art styles. The Lienard foliated scroll, which is still used by engravers, goes back to the Renaissance. Liège Arms Museum ("Grand Curtius").

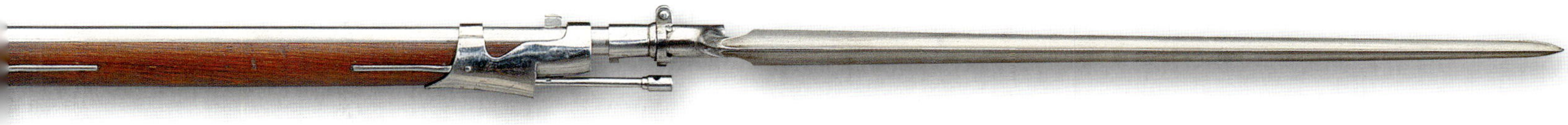

1868 Albini-Braendlin rifle : detail of the tip-up breech block.

This association, which was dissolved on 1 May 1876 but which actually survived for a few months longer, supplied several thousands of Chassepot rifles to the French government, 100,000 Snider mechanisms to the Turkish government, 8,000 rifles and 500 Comblain carbines to Greece, 4,000 revolvers to an unidentified state, 50,000 rifles and 10,000 Bouesco carbines to the Rumanian government.

The tendency towards solidarity and association was, as often happens, affected by the depression which lasted from 1873 until 1880. This solidarity would re-emerge in the next period of prosperity. On 10 November 1886, seven arms manufacturers, J. Ancion & Cie, Dumoulin Bros, Dresse-Laloux & Cie, J. Janssen, E. and L. Nagant Bros, Pirlot-Frésart, and A. Simonis (joined in January 1887 by H. Pieper) set up a joint association called Les Fabricants d'Armes Réunis (the Association of Arms Manufacturers) the aim of which was more or less the same as that of the 1870 alliance.

Apart from the conversion of 30,000 Beaumont rifles to the Vitali system which was ordered by the Dutch government, the association did not have much success with their venture. In Turkey, Austria, Hungary and Serbia, they had to compete with the factories of the Loewe group, the Mauser and the Steyr Waffenwerke which at that time were manufacturing repeaters of the highest degree of perfection, the Mauser and Mannlicher rifles.

Belgian soldier of the Civil Guard, holding the Comblain rifle. Cartoon engraving by B. Van der Gheynst (1903). Archives of the Liège Arms Museum (the Grand Curtius).

The inventor Hubert-Joseph Comblain with his single-shot breech-loading rifle (about 1876). This weapon was supplied to Belgium, Greece, Chile, Peru, Brazil and Persia. The manufacturing of the Comblain rifle by a group of Liège arms manufacturers, the 'Petit Syndicat' is the forerunner of the industrial association which gave birth to FN. Archives of the Liège Arms Museum ("Grand Curtius").

FN is born

The end of the shift at midday on 29 February 1912.

A government order

Even though the Fabricants d'Armes Réunis did not have much success, their venture nevertheless paved the way for a more lasting organization. For some time General Pontus, who was the War Minister in the Belgian government, had expressed his intention to equip the army with a repeater to replace the Albini-Braendlin regulation rifles, and the members of parliament from Liège had persuaded the minister not to buy these rifles abroad.

From that moment, that is to say, from March 1887, the Fabricants d'Armes Réunis considered the "setting up of an arms factory to meet the current requirements".

But the government had by no means intended to call on them to the exclusion of their competitors, Auguste Francotte and the Manufacture Liégeoise d'Armes à Feu. On the contrary, on 28 August 1888, after having officially notified its intention to give preference to manufacturers from Liège, the Inspectorate of Arms invited them all to state the price and the delivery terms at which they could supply the State with 150,000 repeating rifles of a model to be determined at a later date. (The War Ministry was undecided between the Mannlicher, Pieper, Nagant and Schulhof models).

So the War Ministry was undecided between four foreign models: Mauser, Mannlicher, Schulhof or Engh and three models of Belgian design: the Nagant, the Pieper and the Marga, the latter a late-comer to the competition, but not

without its merits. The debate was fierce. Reading the information published in the press at the time – certainly partial, possibly incomplete and in any case distorted - one gets a sense that the technical arguments took second place to nationalistic and economic considerations and internal politics. Above all, it is clear that personal contacts and business relationships were a determining factor, to the point of anticipating official decisions. The stakes were certainly high: amongst these were large investments to finance, the future of the Belgian – or Liège basin – arms industry, the divergent interests of the companies in the 'Syndicat' despite certain objective if not merely superficial understandings, the formidable expansionism of German industry.

A number of questions remain as to the exact involvement of those players who later founded FN. The role played by Henri Pieper, which will be covered later, is perhaps exemplary in this respect, but his is not a unique case.

Marga rifle, 7.5 mm caliber, five cartridge magazine.
Combined movement bolt.
Liège Arms Museum - 5663 (the Grand Curtius).

Nagant rifle, 8 mm caliber, five cartridge magazine.
Straight-pull bolt. Repeat stopping system. Model intended for tests in 1889.
Liège Arms Museum - 1926 (the Grand Curtius).

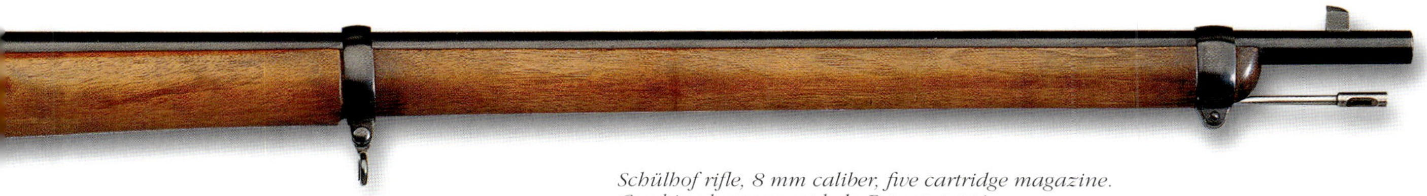

Schülhof rifle, 8 mm caliber, five cartridge magazine.
Combined movement bolt. Repeat stopping system.
Model intended for tests in 1889.
Liège Arms Museum - 7431 (the Grand Curtius).

Pieper-Mannlicher rifle, 8 mm caliber, five cartridge magazine.
Combined movement bolt. Repeat stopping system.
Liège Arms Museum - 5688 (the Grand Curtius).

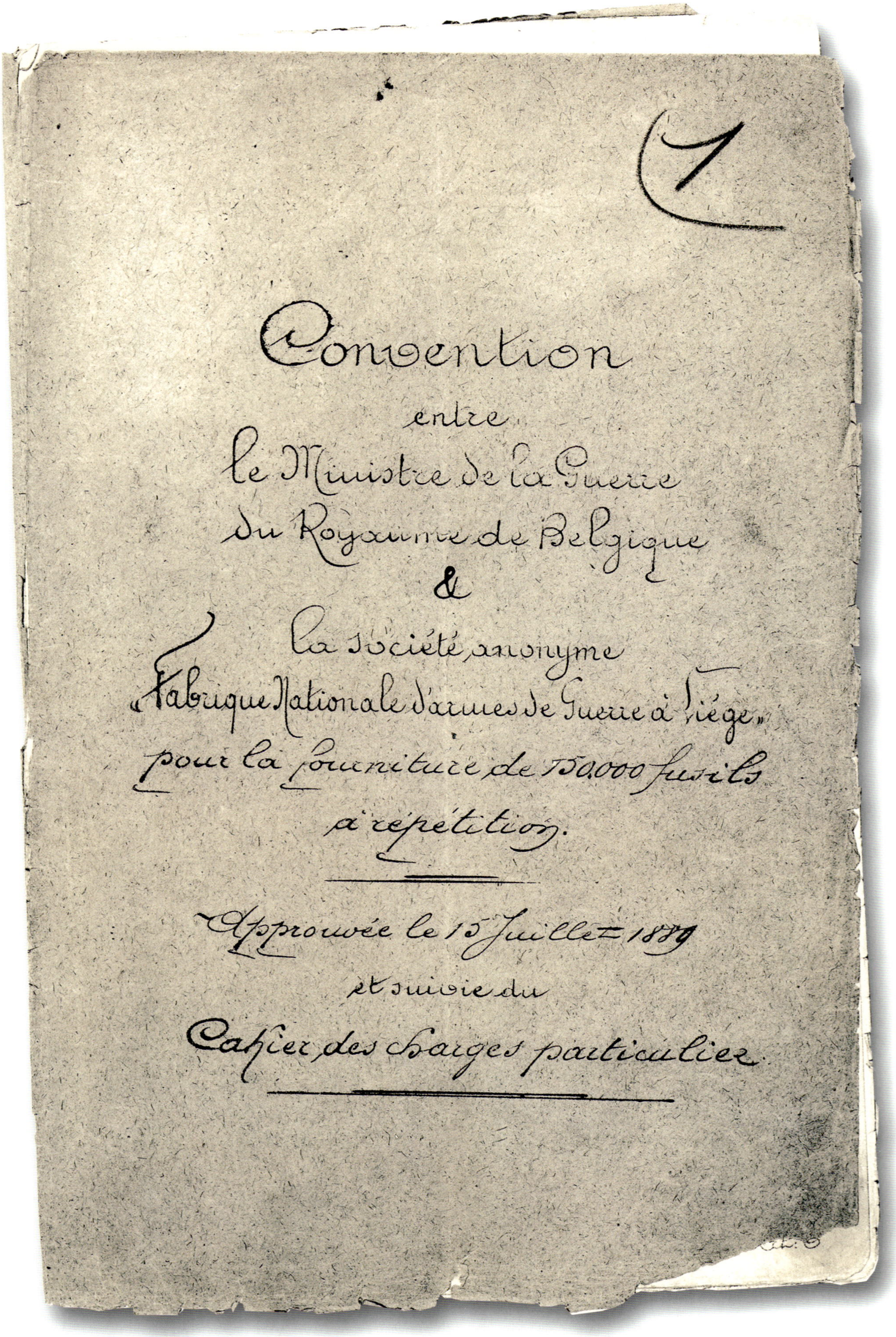
1

Convention
entre
le Ministre de la Guerre
du Royaume de Belgique
&
la société anonyme
«Fabrique Nationale d'armes de Guerre à Liége»
pour la fourniture de 150.000 fusils
à répétition.

Approuvée le 15 Juillet 1889
et suivie du
Cahier des charges particulier

Official order for 150,000 rifles issued by the Belgian government, which would give rise to the foundation of FN in 1889. The choice of a foreign rifle, the Mauser, obliged the new Belgian company to work under the license of the Mauser Waffenwerke, who in turn belonged to the financial group Ludwig Loewe. The latter would soon gain control over the Herstal company until its holding was bought up by the Société Générale de Belgique after the armistice of 1918.

A limited liability company with capital of 3,000,000 francs

The deal required considerable resources. Rather than engage in a conflict with their competitors, the Fabricants d'Armes Réunis did their best to win them over to their plan. This was achieved during autumn. On 15 October 1888, it was decided to set up a limited liability company with capital of 3,000,000 francs, which would then subscribe to the government order and to which the associates would contribute the rights which the State had implicitly credited.

The distribution of the shares and the assessment of the order gave rise to tough negotiations. In fact, some partners, like Joseph Janssen and Henri Pieper, attempted to increase their influence within the company by sponsoring moneylenders outside the arms business, the most important of which, the Crédit Général Liégeois, ended up with a participation of 900,000 francs. On the other hand, some manufacturers, like Emile and Léon Nagant, Auguste and Charles Francotte, who refused to make financial contributions to the company, demanded nevertheless to be rewarded for their contribution. Finally it was agreed that the contributions would be represented by shares without indication of value, so-called ordinary shares, the number of which would be equal to a third of the capital shares. It was also decided that the latter would be redeemed at par value and subsequently put on the same footing as the former.

The choice of the company's location was not much easier. Would the buildings of the National Arms Factory be leased; founded in 1838 in the Saint Léonard suburb of Liège, they were now languishing in inactivity? On the other hand, should a new factory be built which would then be the property of the new company? This solution, which was to have a decisive effect for the future, was not decided on until the spring of 1889 after many months of uncertainty. As far as equipment and machinery was concerned there was much less hesitation and, after contacts with Pratt and Whitney in the United States, it was decided to approach Ludwig Loewe & Co, who were considered to be specialists in the manufacturing of military arms.

The Fabrique Nationale d'Armes de Guerre

One problem seemed to have been solved relatively quickly, and that was the name of the company: this was agreed, apparently without too much discussion, on 8 December 1888. The name was to be 'Fabrique Nationale d'Armes de Guerre'.

Finally, on 3 July 1889, after many months of negotiation, the company was incorporated by a deed drawn up by Nicolas Biar LLM, notary in Liège. The table below gives the share distribution among the founders. The 2,000 ordinary shares represent the contribution to the company made by the government order. The 6,000 capital shares, so-called preference shares, which give a preferential right to the dividend, were issued at a unit rate of 500 francs.

The reason we have dwelt upon these events in detail is that they have profoundly marked the history of the company. We have seen that behind an appearance of unity there exists a fundamental division on certain principles between the various founders. Some, like Henri Pieper, Joseph Janssen, Allard Bormans, Managing Director of the limited partnership Dresse-Laloux & Cie, wanted to create a lasting company of the Fabrique Nationale. They contributed significantly to the company's capital and the decision to build an independent factory undoubtedly corresponded to their wishes. However, the intentions of Emile and Léon Nagant, Auguste Francotte, the Manufacture Liégeoise d'Armes a Feu, Pirlot and Fresart were very different. They saw FN as being merely an association with limited objectives and of limited duration. Moreover, these views emerged clearly in the statutory clause which in order to facilitate a progressive liquidation of the company, provided for a redemption at par value of the preference shares. The conflict was to explode fully as soon as the government order, for which the company had been set up, had been executed with the consequences which are dealt with below.

Founders	Preference shares	Ordinary shares
Albert Simonis, arms manufacturer	300	200
Jules Ancion & Cie, arms manufacturers	600	200
Dresse-Laloux & Cie, limited partnership, arms manufacturers	600	200
Manufacture Liégeoise d'Armes à Feu, S.A.	600	200
Dumoulin frères, arms manufacturers	600	200
Joseph Janssen, arms manufacturer	600	200
Henri Pieper, arms manufacturer	1.200	200
Pirlot et Frésart, arms manufacturers	600	200
Crédit Général Liégeois, limited company	600	–
Nicolas Vivario, proprietor	300	–
Auguste Francotte, arms manufacturer	–	200
Émile et Léon Nagant, arms manufacturers	–	200
Totals	6.000	2.000

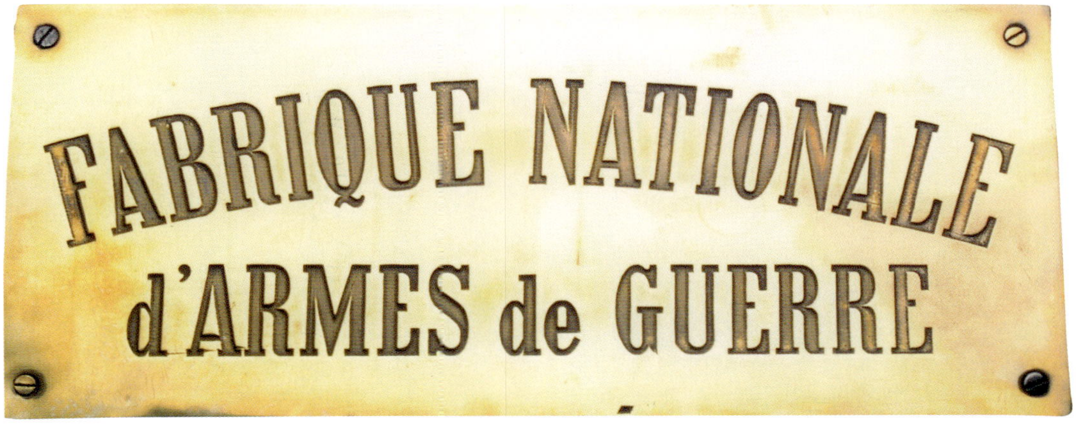

Old recruitment office plaque near the main entrance to the headquarters of Herstal SA. (Collection A. Vanderlinden, USA).

The first orders

Joseph Chantraine
Managing Director of FN from 1890 to 1895.
Archives Herstal SA.

Joseph Janssen
Chairman of the FN Board from 1892 to 1896.
Archives Herstal SA.

Once the company was founded, the general meeting of shareholders elected their first board of directors. The board members were Jules Ancion, Allard Bormans, Leon Collinet, Auguste Dumoulin, Joseph Janssen, Henri Pieper, Gustave Pirlot, Alban Poulet, Albert Simonis, Ernest Francotte and Leon Nagant. Allard Bormans was appointed Chairman, Joseph Janssen Vice Chairman. Henri Pieper became Managing Director.

150,000 Mauser rifles

From that moment, things would develop rapidly. On 12 July, Allard Bormans and Henri Pieper signed the contract committing FN – as the company shall be called henceforth – to deliver to the War Ministry, at a unit price of 79 francs, 150,000 repeating rifles of a model to be decided by the government. (The total value of the order amounted to 11,850,000 gold francs).
As might be expected, this model would not be one of those which had been considered the previous year, but one which Paul Mauser had recently developed and which he had just covered in Belgium with a half a dozen patents. Moreover, at the end of August, the company concluded in Berlin with Ludwig Loewe & Co – owners of the Mauser Waffenwerke – an agreement for the purchase of the necessary material for the production in large numbers of the above-mentioned rifle and for the technical assistance required for that purpose. The equipment – machinery, tools, mountings, testing equipment – cost 219,000 marks, or 240,276 gold francs.
The board of directors did not have to wait long for the ministerial decision. On 22 October, General Pontus obtained from Alexandre Resimont, authorized agent in Belgium for Mauser Waffenwerke, a license for the production of Mauser repeating rifles, 1889 model. The contract provided in explicit terms for the production of the guns by private industry.

Thus assured as to its future activity, the company went to look for production sites. The first had a surface area of 17 acres and was bought on 1 February 1890 from the limited company Charbonnages de Bonne Espérance et Batterie for the sum of 50,000 francs. It was a meadow called Terre de l'Evêque and was situated on the territory of the town of Herstal in a district called Laixhaut, where the main workshops of the arms section are now located. The plans for the factory were agreed in March. By April, the main works had been contracted and the construction of the buildings commenced.
The equipment for the buildings employed the latest technologies more particularly electric lighting and driving power.
In September 1889, the board of directors had appointed as manager an engineer, Léon Castermans. Fourteen months later, Léon Castermans resigned and was replaced forthwith by Joseph Chantraine, another engineer, whose know-how had a profound influence on the early years of the company.

30,000,000 rounds of ammunition

But it was not long before a new problem arose. In February 1891, the War Ministry had in fact informed the company of its intention to buy – in the form of separate components – 30,000,000 cartridges for the rifles it had ordered from FN. Even though the founders of FN had no experience whatsoever in explosives, they did not hesitate long before making arrangements for the construction of a cartridge factory next to the original factory. It was an inspired move. We now know that the fact of producing and designing weapons and their ammunition together has contributed enormously to the technical and commercial development of the company.

So in May an agreement was concluded with an old Belgian ammunition factory, the Société d'Anderlecht, and with the Deutsche Metallpatronenfabrik in Karlsruhe which, incidentally, was owned by L. Loewe & Co. The latter were requested on the 15th of the following month to supply FN with the necessary material for the daily manufacturing of 25,000 cartridge cases and as many bullets.

The links between the company and the Loewe group became ever closer. Already in March 1890, Loewe had offered the Board to buy a third of the company's shares and to conclude an agreement with him on an allocation of the orders, but the Board was not very impressed. In June 1891, acting on behalf of Mauser Waffenwerke which he had bought up several years earlier, he offered FN a direct license for the production of the repeating rifle. The offer was taken up forthwith and the contract was signed in Berlin on 26 November that year. Four days later, FN concluded an agreement with the Deutsche Metallpatronenfabrik whereby the latter would lend its technical assistance in the production of ammunition.

In the meantime, rapid progress was being made in the construction of the factory and the start of production. On 3 December 1891, the first three rifles were assembled at FN. One of them was presented to the War Minister. The official delivery of the first production rifle was made on 6 February 1892. Less than 9 months later, halfway through October 1892, the number of arms delivered daily varied between 100 and 150. The production of cartridges was no less satisfactory : by September, daily production had topped 15,000 to 16,000 bullets and similar figures were anticipated for the cases.

These results were not obtained without some difficulty. Construction of the buildings was not yet completed, the installation of the machinery had taken longer than expected and various technical problems had limited production.

The Société Pieper's famous logo with the Knight Bayard. Pieper was FN's main shareholder in 1889.

Henri Pieper and his family. (Collection Michel Druart).

Henri Pieper around 1895. (Collection Michel Druart).

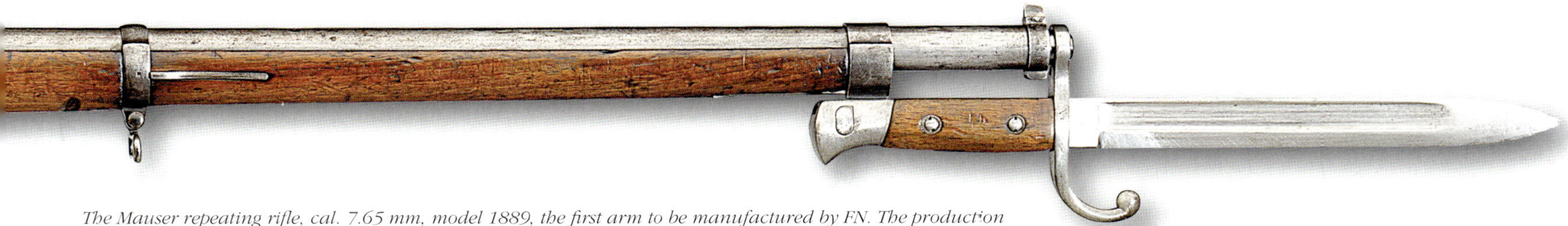

The Mauser repeating rifle, cal. 7.65 mm, model 1889, the first arm to be manufactured by FN. The production of the 150,000 rifles ordered by the Belgian government was started in December 1891 and was completed in December 1894. In the meantime, the government had placed an order with FN for 30,000,000 cartridges. They were manufactured in a new, dedicated department , with the technical assistance of the Deutsche Metallpatronenfabrik, which also belonged to the Loewe group. Liège Arms Museum 6205 (Grand Curtius).

Orders from foreign governments followed

Meanwhile the fame of the company started to spread. In the month of July 1892, the Russian government had asked for a batch of some 400,000 Berdan rifles to be adapted, and a considerable part of the work was carried out by Henri Pieper. Favorable prospects opened up in various countries, first of all in Belgium where, under pressure from the Liège members of parliament Xavier Neujean and Emile Dupont, the government foresaw new orders for rifles and cartridges from the Netherlands, Spain, Serbia, Brazil, Chile, and even from China.

So at first sight the position of the company was very favorable. During the general meeting of 10 October 1893, the Board was able to announce that on 30 June last, 42,000 rifles had been supplied to the Belgian government and that daily production amounted to 250 arms, 20,000 cases and 25,000 bullets. The profit and loss accounts showed a trading profit of 749,382.89 francs, which was entirely appropriated to amortization.

But even this success did not fail to create problems. Some partners viewed with apprehension how the company which they had set up for the execution of one order was becoming a permanent competitor. This anxiety soon became apparent. On 12 July 1894, two directors, Auguste Francotte and Léon Nagant, resigned and sold their shares to third parties.
The general meeting of 9 October that year made a double decision: in conformity with the articles of association, to convert the 6,000 preference shares of the company into ordinary shares and to remit to the previous holders of these shares 6,000 bonds of 500 francs each at 6% interest and redeemable within ten years : this in fact meant the long-term liquidation of the company.

Three months later, on 31 December 1894, the production of the 150,000 rifles ordered by the Belgian government was completed. The question was, would the company survive the occasion which had justified its setting up?

However, it soon transpired that in spite of the skills of FN's defense lawyers Dupont LLM and Mestreit LLM, the case took a bad turn for the company.
Conscious of the threat to the company's assets and existence, the Board of Directors decided on 21 January 1895 that the 1,200 bonds be redeemed. And on 1 February it decided as a desperate measure to ask the Minister of Employment and Industry for the forfeiture of the patents in question: this was in fact an implicit recognition of the fact that they did not wish to renew the license !

Committed as he was by the law, the minister hesitated to accede to such a demand. Nevertheless he wanted to come to FN's rescue by requesting Mauser Waffenfabrik to reach a compromise with their opponent. Sure as they were of their position, the representatives of Loewe - and more particularly their lawyer Victor Fris, Senator of Louvain - made draconian claims which FN rejected (June-July 1895).

This figure need not surprise us: while the Mauser rifle was sold at 79 francs, the manufacturing cost of the weapon was reduced to well below that figure and constant efforts were being made to bring down that cost even more. The care with which the rifles were made gave FN an even more beneficial reputation. This is shown clearly by the orders which were placed successively by Brazil (June 1894), China (October 1894), Norway (January 1895) and Costa Rica (May 1895).

FN's fate in the hands of Ludwig Loewe und Co.

A recent event had occurred which was not a good sign for the future. In the first week of December 1894, while Chile was preparing to buy from FN 60,000 rifles of the Spanish Mauser model, the Waffenfabrik Mauser warned the company that it would take legal action to prevent it from executing the order. Taken aback, Chile broke off negotiations and decided to buy its arms in Germany. At the same time, Mauser summoned FN to forego using the patents covering its Spanish model. FN took the case to court, a procedure provided for in the license contract of 26 November 1891.

The case continued while the company continued to press for the forfeiture of the patents through parliamentary questions and press campaigns. But all these efforts were in vain. Rightly alarmed by the events Joseph Chantraine, who had led the factory for five years, handed in his resignation (18 November 1895). After having stirred up a minor precedence quarrel, the Company Chairman Joseph Janssen followed suit on 17 January the following year. On that same day, the Board was informed that the Court of Arbitration had found against FN. The Board promptly decided to appeal. But it soon became clear that there was very little enthusiasm left.

On 6 February 1896, five of the eight remaining directors - Bormans, Pirlot, Simonis, Dumoulin and Ancion - had also resigned, while Ludwig Loewe und Co. had bought more than half the company's shares, doubtless at favorable conditions.

The Board of Directors was replaced at once. Isidore Loewe and Alexis Riese, both directors of Ludwig Loewe und Co., were therefore appointed, also the engineer J. Dallemagne, G. de Laveleye and Charles del Marmol. Alexis Riese and Henri Pieper were also appointed Managing Directors. Loewe's lawyer, Victor Fris, was appointed as consultant to the company. The first concern of the new Board was to retract the request for the forfeiture of the Mauser patents which had been submitted the year before, to revoke the appeal lodged against the decision of the Court of Arbitration and to start negotiations with the Waffenfabrik Mauser for the revision of the license contract of 1891. Thus the fate of FN was henceforth linked to that of a foreign group which was at that time the most powerful in the field of small arms and ammunition.

The scarcity of sources makes it difficult to assess the precise – but clearly key – role played by Henri Pieper (1840-1898) in the creation of FN. A brilliant mechanical engineer of German origin and a leading industrialist in the field of armaments, he had emigrated to Liège (where the Mauser brothers had also worked), where he founded the only factory capable of competing with the future FN. However, he avoided becoming obsessed with the specter of competition, as so many of his colleagues did, but subscribed a considerable share (20%) in the new company's preferential shares, and became its Managing Director. A negotiator during the Loewe Group takeover, he remained in this post which he shared with a co-director of the principal shareholder, until his death in 1898. It is also noteworthy that another German mechanical engineer, Jean-Baptiste Goebel, joined FN in 1896 from Loewe's construction department to take the helm of FN's construction activity and was soon appointed Deputy Manager. Then, in 1904 he became Manager of Etablissements Pieper. Finally, the German cartel owned holdings in the Compagnie Internationale d'Electricité, a company founded by Henri Pieper in 1889 and managed after his death by Henri junior (1867-1952). In fact, it was this electric company which supplied FN with its dynamo generator, the biggest built in Europe at that time. In any event, it is hard to avoid the conclusion that the Pieper-FN-Loewe triangle represented a deliberate strategy on the part of the German group.

Portrait of Isidore Loewe. 1911.

Mauser repeating rifle, 7 mm caliber 1893 Spanish model, manufactured by the Mauser company in Germany. Strictly controlled by the Loewe group, FN failed to gain a number of orders. Liège Arms Museum 6773 (the Grand Curtius).

General view of the arms manufacturer "Waffenfabrik Mauser" in Oberndorf/Neckar, near Stuttgart, circa 1908. From the special edition (sixth year, n° 2) of "Deutsche Industrie- Deutsche Kultur". Archives of the Liège Arms Museum ("Grand Curtius").

The German school

Joining the company to the Loewe group was very beneficial to the development of the factory. In contrast to Belgium, and more particularly Wallonia, which were among the first in Europe to follow England down the route of industrial revolution, Germany was a late-comer to the process. She applied herself to it so diligently and effectively however, that by the end of the nineteenth century she had developed a potential for creativity and productivity which was second to none… and FN was to reap the benefits. From now on, FN was to abandon empiricism and adopt scientific methods in its technology and organization of labor. This transition was to a great extent facilitated by the fact that in Herstal the new managers found a workforce which was by their own confession remarkably skilled, as well as the production methods which they themselves had inspired, and a production apparatus which they themselves had supplied less than five years earlier.

It is worth mentioning that of all the sections it was the cartridge factory which

View of FN in 1911. According to the custom of that time, this view greatly exaggerates the size of the installations, especially in the depth. The administrative offices of the management are situated to the left of the main entrance.

made the most conspicuous progress. True to say, it had been feeling the consequences of the founders' inexperience, so that more than the other services it had much to learn from the German school.

On the other hand, the fact of belonging to a cartel did much to restrict the company in its freedom to maintain relations with the military authorities of foreign countries. In fact, this limitation could have been detrimental to FN if it had not compensated for it in those areas where its independence was total. We will see that the company immediately went about this in an increasingly successful way.

The influence exerted by the new directors made itself felt favorably in a totally different domain. Steeped as they were in the principles which had already prevailed for more than thirty years in their country of origin, they showed great openness and consideration for the social problems of the day. This undoubtedly explains why the company's relations with the workforce were generally better than was the case in other enterprises in the area.

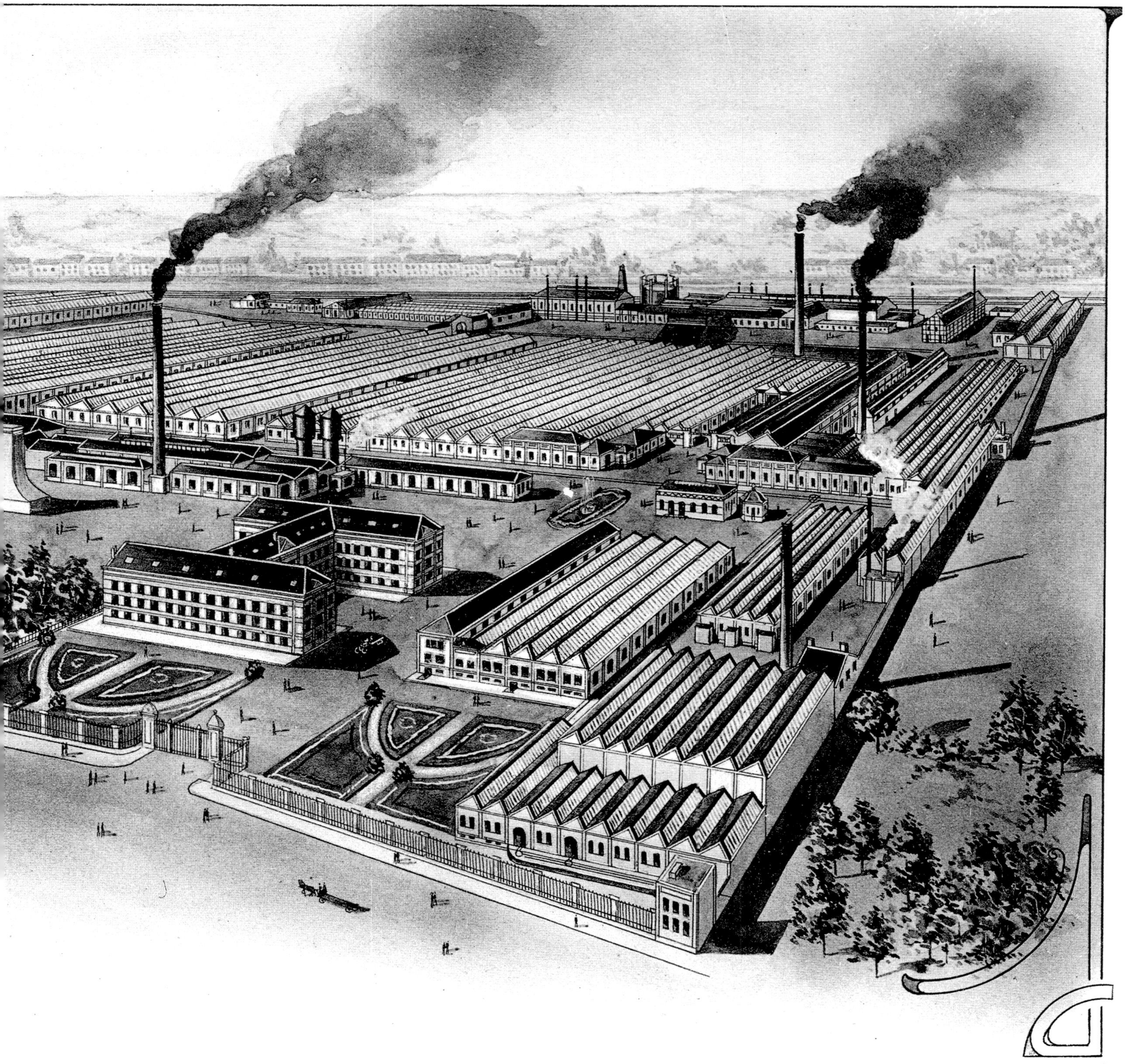

The workshops were destined for the manufacturing of arms, cartriages and two-wheeled and four-wheeled vehicles. At the time, the Herstal company employed approximately 3,500 people.

The main entrance to FN in 1907. The first buildings had been constructed at a distance from the main road to Liège.

New buildings which were added progressively eventually filled the empty space.

Steam-driven electric generator at FN, turn of the century. A compound Corliss 450 hp steam-engine (supplied by Vanden Kerckhove, Gent), powered by four 'Piedboeuf' boilers, drove a dynamo-generator, the largest one ever built at that time in Europe, from the Compagnie Internationale d'Electricité of Liège.

This supplied the electrical engines which controlled the driving shafts of the machine-tools. This installation also supplied the buildings with lighting and steam for heating. The generalized use of electricity was considered very innovative at that time.

Increase of capital

The first concern of the new Board was to restore the financial situation of the company: remember that FN was a debtor of bonds for a total amount which was equivalent to the initial capital. Thanks to loans from Belgian and German banks, the board was able to prepay this enormous debt, while in October 1869 the capital was increased from 3,000,000 to 5,500,000 francs and the number of shares from 6,000 to 11,000. Three years later, FN freed itself from all its commitments to its Belgian financial backers by agreeing with the Schaaffensen'scher Bankverein of Cologne a loan of 2,000,000 marks, guaranteed by the Deutsche Waffen- und Munitionsfabriken (D.W.M.), a company which already managed the holdings of Ludwig Loewe und Co. One of the directors of the Schaaffensen'scher Bankverein, Dr. Ludwig Hagen, was appointed on that occasion Director and Vice-Chairman, thus reinforcing German predominance within the company. Finally, on 26 November 1907, when the situation of the company had improved, the general meeting decided to increase the capital for a second time - this time by 1,100,000 francs - and to issue 1,641,500 francs worth of bonds.

Technical reorganization

The technical reorganization came about at the same time as the restoration of the financial situation. The first significant step taken in that respect by the Board was the appointment of a manager, Henri Frenay (27 March 1896). This proved to be a very wise move, and was soon followed by a considerable enlargement of the engineering staff. The expansion of the company led to the creation in 1899 of a Technical Board and a Commercial and Administrative Department, while Henri Frenay was promoted to the function of General Manager. This management reshuffle would remain virtually unchanged until 1914 apart from the unexpected death of Frenay in November 1906 and his replacement by Alfred Andri on 13 December 1906. As for the executive staff, these were divided into a certain number of departments led by university-trained managers.

The increasing reliance on technical staff who were graduates of the higher schools of engineering, some of whom had in addition served their apprenticeship in the factories of D.W.M. or Mauser Waffenfabrik, was to lead to a considerable improvement in production methods. This trend gained even more impetus with a decision which was taken on 30 May 1904 to implement a plan which Alexis Riese had already been working on since the spring of 1896 and which was to have a long-term beneficial effect. This concerned the allocation of 20,000 francs to the construction of a laboratory where the raw materials would be delivered and where the metals would be tested. Thus FN became one of the first factories in Belgium to have a scientific research center. Also a young engineer of considerable merit was taken on to head the laboratory: Alexandre Galopin.

Restrictive commercial agreements

While belonging to a foreign group was particularly beneficial on the technical and financial front, the effect was extremely restrictive in matters of trade and commerce. In June 1896, the company was forced to accept strict rules regarding the allocation of orders for arms. Not only were some markets closed to them, but in most countries where it was permitted to sell its products it was obliged to share the work with the other members of the cartel. Moreover, the quota allocated to FN was in comparison very small, as the following figures, which were laid down in an agreement concluded in January 1897 and renewed without significant changes, clearly show :

Deutsche Waffen- und Munitionsfabriken (Berlin)	32.5%
Mauser Waffenfabrik (Oberndorf)	20.0%
Steyr Waffenwerke (Steyr)	32.5%
Fabrique Nationale d'Armes de Guerre (Herstal)	15.0%

Similar agreements were concluded with the factories belonging to the ammunition production cartel, that is to say, D.W.M., the Deutsche Metallpatronenfabrik in Karlsruhe and the Vereinigte Köln Rottweil Pulverfabrik.

Thus tied up in a web of restrictive agreements, FN had little opportunity to execute armament orders of any consequence. Some business which had originally been allocated to them was even taken away again, such as the order for Krag-Jorgensen rifles which Norway had placed with them and which was then promptly transferred to Steyr Waffenwerke (7 August 1896) Strictly speaking, only one order of any importance came to them, and that was an order for 45,000,000 cartridges for Serbia (22 June 1899), and which induced the management to develop the cartridge-manufacturing section, thereby increasing production capacity to 200,000 rounds per day. Unfortunately, this deal was to have very unfortunate consequences, because Serbia had suspended its payments by the end of 1900. In March 1896 there was also an order for 14,000 rifles and carbines and 4,500,000 cartridges for Uruguay and another one for 5,000,000 cartridges for Spain, and in June 1903 and November 1906 respectively, two orders for 20,000 rifles for the Belgian Army, etc.

Bicycles and rifles

As FN was now restricted in its original domain, it made substantial efforts to try and find compensations in other areas, so as not to leave its installations permanently idle.

The first product which the management considered was the bicycle. The previous management had already looked into the manufacturing of the most delicate components of this vehicle. The study was made so enthusiastically that by April 1896 production could commence. In May 1898, the company had already invented a model without a transmission chain, the 'Acatene Chainless' or 'FN chainless', which was then immediately put into production.
But soon another area of activity would catch the management's attention, that of hunting and sporting arms. In fact, this field, which the founders of the company had always refused to consider for fear of creating a dangerous competitor to their own enterprises, was to arouse the interest of the new management.

In principle, the equipment used for the manufacturing of armaments could also be used for the production of hunting weapons, while the production methods which had proved successful for the former should guarantee comparable results for the latter. In November 1896, the Board decided to produce 50,000 sporting rifles of .22 caliber. A little later, an agreement was concluded between FN and a group of Liège manufacturers, the Association for the Pro-duction of Standard Gun Parts, whereby the former committed itself to supplying to the latter large batches of hunting shotgun parts on a regular basis. To give an idea of the volumes involved, in 1903 FN delivered in the form of separate parts the equivalent of 25,000 hunting guns.
However beneficial these deals may have been in the short run, for FN they did not bring in orders for models of their own nor specific customers they could directly deal with.
A fortuitous event, though, did serve to achieve both.

A series of postcards illustrating the great Belgian companies, published at the beginning of the twentieth century, highlights the existence of a dining hall for FN's female workers, a social initiative which was unusual for the time.

Frontage of the new laboratories (1912).
In 1905 FN was one of the first Belgian companies to have an integrated laboratory for testing, controlling and receiving raw materials and products. Inspired by the theories of Henry Le Chatelier, a professor at the French College, and the example of the car manufacturer de Dion-Bouton, the central laboratory of FN was installed in 1912 in this large building. For over 80 years, young engineers served their apprenticeship here before taking up responsible positions in the company. During the last years of the twentieth century, in view of the complexity and growing specialization of techniques, the tasks of the laboratory have been for the most part decentralized.

FN chainless bicycle (1912). The production of this bicycle equipped with a "multi-coneshaped transmission" started at Herstal in 1898 (Herstal Group Collection).

Workshop for the assembly of bicycles (1904). FN, which was seeking to branch out into other areas, embarked in 1896 on the manufacturing of bicycles. This line of production would last for thirty years, and produced the famous chainless bicycle, where the transmission chain was replaced by a system of conical pinions.

FN chainless bicycle (1914). This military version was used by the Belgian army troups at the borders. It was equipped with a Mauser rifle fixed to the frame (Herstal Group collection).

Detail on the crankset showing the letters F and N.

Poster by Emile Berchmans (1867 - 1947).

Details of the leather saddle, lightened brake handgrips, FN frame logo, and of the two brackets on the rear wheel. One was used for uphills, the other for downhills.
To change gear, the driver "just had to" stop, get off the bicycle, turn the wheel the other way round, and go...

FN race bicycle (1920).
Herstal Group collection.

The finished arms inspection office in 1914. Shown here are Mauser rifles 1889.

Acatene bicycle for the Belgian Gendarmerie, with mount for the Mauser rifle (1908).

Belgian Gendarmerie acceptance board for FN bicycles (1908).

John M. Browning in FN's yard on 28 June 1910.

A certain Mister Browning

On 2 April 1897, the board sent its Sales Manager, Hart O. Berg, to the United States to inquire about the advances which had been made there in the manufacturing of bicycles. On this journey, Berg met two American arms manufacturers, the brothers John Moses and Matthew Browning. John, whose ingenuity and creative spirit had earned him some celebrity in his country, had just taken out a patent for a 7.65 mm blowback operated automatic pistol. The manufacturing license was offered to FN. After it had been submitted to the board on 19 June and strongly defended by Berg, the contract was signed on 17 July by the latter and by the Company Chairman, Baron del Marmol.

Production required several months of preparatory work, and in January 1899 the first pistols were delivered. The superiority of the weapon soon became apparent. In 1900, the Belgian State armed its officers with it. After Switzerland, it was the second country to equip its regular army with such an automatic pistol.

This was only the start. In February 1902, John M. Browning came to Herstal with the prototype of an automatic shotgun which he had unsuccessfully offered to Winchester and Remington. More far-seeing than these two companies, FN soon expressed an interest in the weapon.
A license contract was signed on 24 March and on that same day Browning ordered from this licensee a batch of 10,000 guns. Most of the relations which FN would maintain with the Browning family originated from this dual contract.

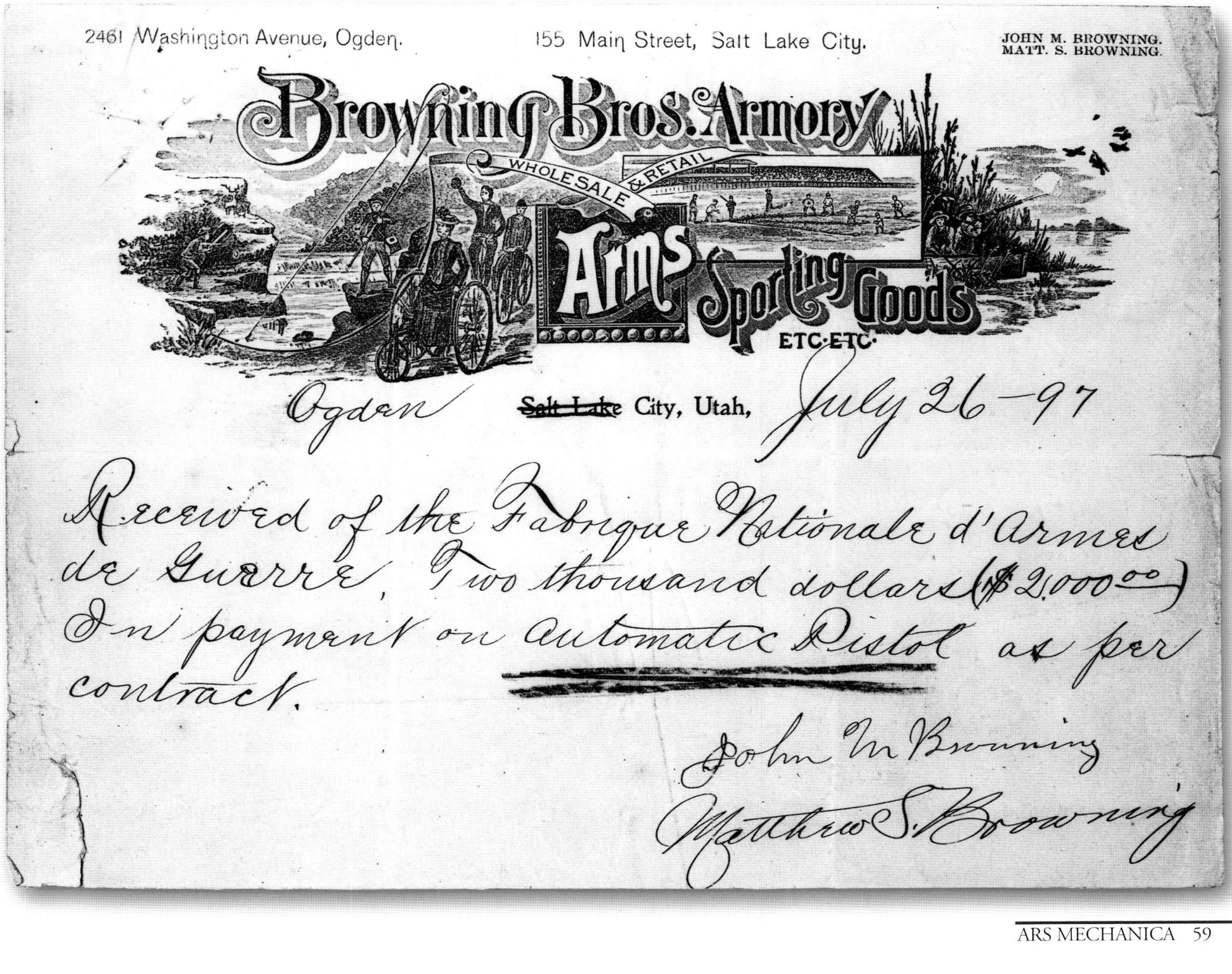

2461 Washington Avenue, Ogden. 155 Main Street, Salt Lake City. JOHN M. BROWNING. MATT. S. BROWNING.

Browning Bros. Armory

Wholesale & Retail

Arms Sporting Goods ETC. ETC.

Ogden ~~Salt Lake~~ City, Utah, July 26–97

Received of the Fabrique Nationale d'Armes de Guerre, Two thousand dollars ($2,000.00) In payment on Automatic Pistol as per contract.

John M Browning

Matthew S. Browning

Receipt signed by John M. Browning and his brother for the payment of his rights on the first automatic 7.65 mm pistol (1897). The production of this weapon marked the beginning of FN's relations with the great American inventor and his family, whose company was taken over by the Herstal company in 1977. Herstal Group archives.

Prototype for the 1900 Browning automatic pistol (1898).
FN Herstal's R&D Department.

Browning automatic pistol, model 1900, inlaid with gold in 'Art Nouveau' style.
Liège Arms Museum 12,589 ("Grand Curtius").

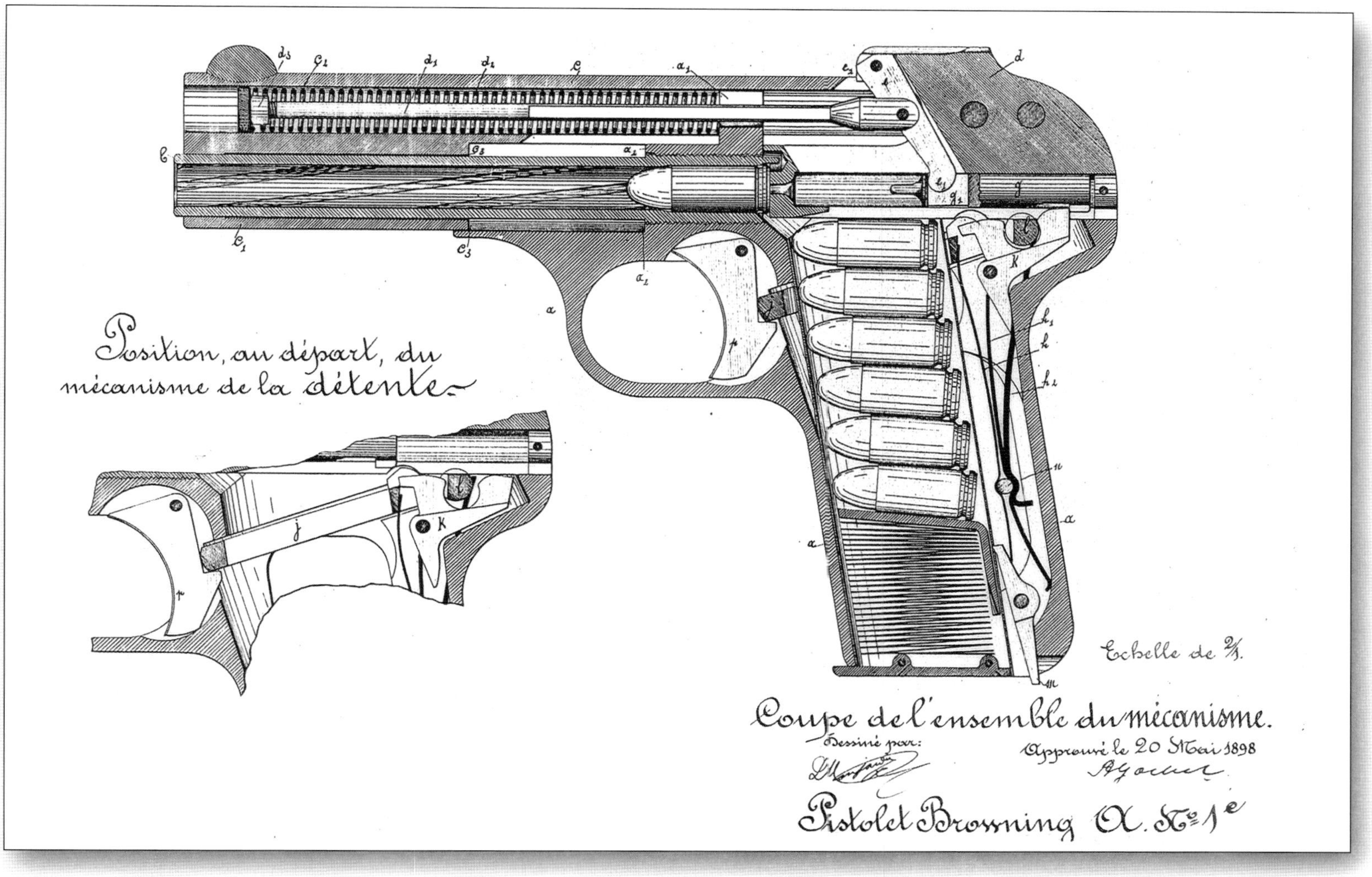

Prototype for the 1900 Browning automatic pistol.
Sectional drawings dating from 1898.
FN Herstal R&D Department.

The 'Belle Epoque' being modernized : firing from the shoulder the Large Model 9 mm pistol using a holster butt (1911). The word Browning was very soon to become a household name for an automatic pistol, compact in design and with high firing capacity. One year after this picture was taken, FN had already produced a million of them in various models.

Fig. I.

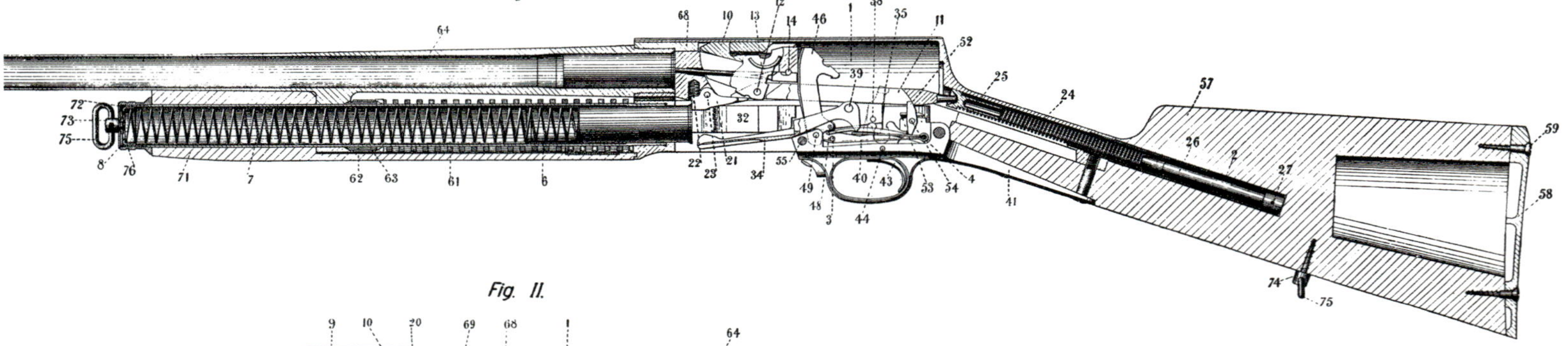

Fig. II.

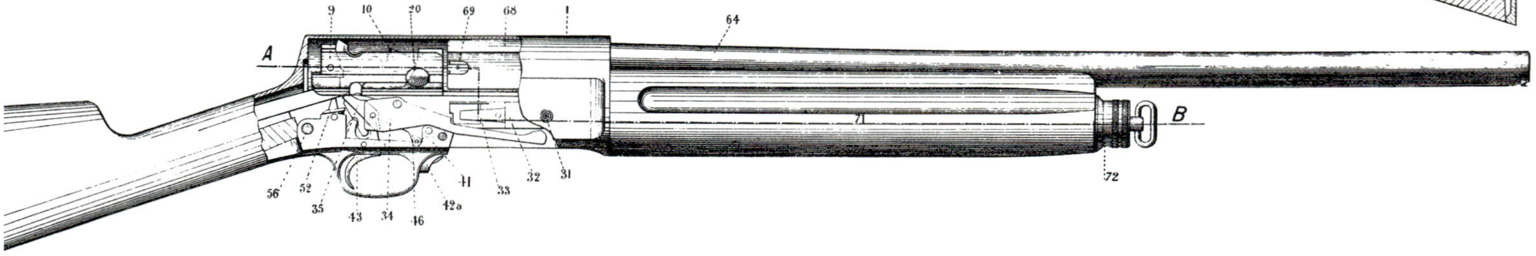

Fig. III.

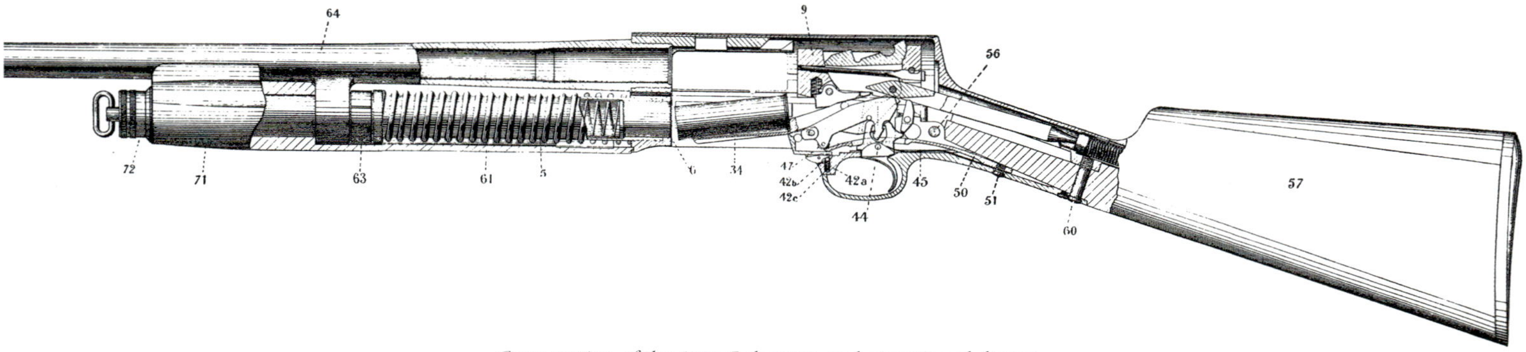

Cross-section of the Auto-5 shotgun and operational diagram.
The automatic operation of this shotgun is based on the principle of the long barrel recoil.
It was the first practical automatic shotgun which was produced in large numbers and had considerable commercial success.
The presentation of the prototype in 1902 was also the first personal visit by J. M. Browning to FN. Herstal Group archives.

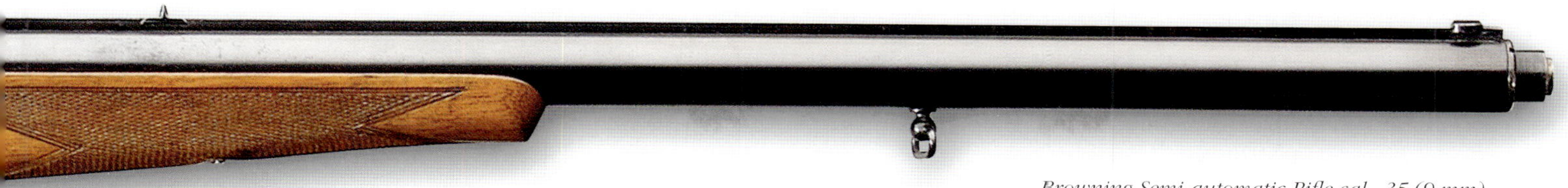

Browning Semi-automatic Rifle cal. .35 (9 mm). Produced from 1910 to 1931.

The oldest example of the Auto-5 shotgun kept by Herstal SA.

"Classical" portrait of John M. Browning with his Auto-5 shotgun. The photograph was taken at FN during his last visit, during which in 1926 he died of a heart attack in his office.

The millionth pistol !

To the historian of arms manufacturing, the inventions of John M. Browning are characterized by their abundance and diversity. This dual quality, which was accompanied by a very keen intuition of the clients desires, presented from the businessman's point of view very considerable advantages. Conscious as they were of these advantages, the directors of FN wasted no time in obtaining the right to manufacture most of Browning's new arms as he invented them.

It concerned various models of pistol, one of which was a 9 mm caliber which was bought in considerable quantity by the Swedish army (1907), another pistol – known as the '1910 model' - which could fire 7.65 mm as well as 9 mm cartridges when the barrel was changed, (29 July 1909), an automatic .22 caliber rifle (4 January 1913), 50,000 units of which the inventor immediately ordered , and a 20 gauge pump action shotgun of which he bought 25,000 units (20 December 1913), but unfortunately the war prevented its production

In the meantime, on 24 May 1907, following FN's request, John M. Browning granted them the right to use his surname as a trademark, thereby emphasizing the common interests which linked his family to the company.

It is an interesting fact that the inventor never gave such permission to the large American companies which had commercialized his first inventions. The full importance of this arrangement would become obvious over the years. The popularity of weapons developed by Browning and manufactured by FN would continue to increase, both in Europe and America. The sales figures reflect this trend, especially as concerns pistols. From 1899 to 1906, the number of pistols sold amounted to 250,000; it reached 500,000 in 1908, and in July 1912 the millionth pistol was assembled in Herstal.

This event was celebrated on the 31st January of 1914 with a magnificent party to which were invited, apart from John Browning himself, two ministers in office and some 500 people from the world of business and politics.

The first semi-automatic 7.65 caliber Browning pistol.
From top to bottom, the inventor's original model, the 100,000th 1900 model pistol produced at FN (1904), the 250,000th (1907) and the 500,000th (1909).
Browning Firearms Museum, Ogden, Utah, USA.

Browning and his pistol, cartoon by Jacques Ochs (1914).
A resident of Liège, the artist Jacques Ochs (1883-1971) was a marvellous cartoonist of native popular types and of personalities in Belgian politics.
Very well-known in the area-the workers of FN called him the "master"- J. M. Browning made 61 visits to Herstal to develop and commercialize his inventions.
He had obtained a total of 128 patents for 80 different weapons.

Photograph taken on 18 June 1910 on the occasion of J. M. Browning's departure for the United States.
From left to right : M. Christiaens, FN's representative in Western Flanders, a member of the acceptance board;
A. Galopin, Chief Engineer;
E. Michels, Management Secretary;
M. Chesney, permanent representative and receiving agent for Aerators Ltd (Sparklets);
M. Darrier, Head of the Automobile Department;
M. Englebert, FN Holland agent;
M. Israel, Head of Department, armaments design office;
T. Böll, Chief Inspector for armaments and cartridge manufacturing, a member of the Russian Commission;
J. M. Browning;
M. Braüning, inventor of a semi-automatic military rifle;
A. Andri, Managing Director;
M. Hansgraeve, Sales Director; an FN representative;
J. Leroy, Head of Industrial Accounting, future Company Secretary of the Manufacture d'Armes de Paris;
G. Joassart, Secretary and Head of the legal department, future Managing Director;
M. Tritschel, Armaments Production Manager;
M. Stassart, Arms Sales Manager.

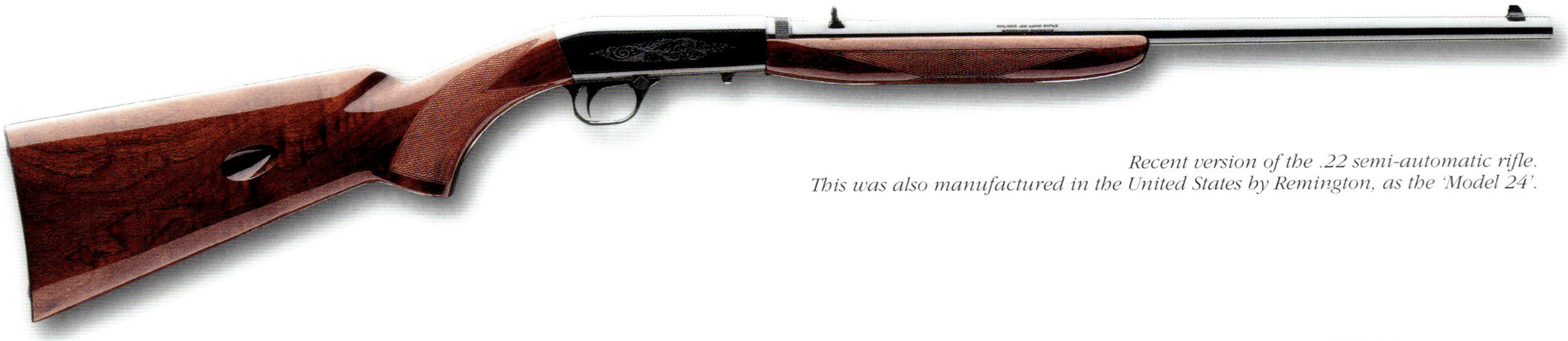

Recent version of the .22 semi-automatic rifle.
This was also manufactured in the United States by Remington, as the 'Model 24'.

Advertisement picture from 1913 : the new Browning automatic .22 rifle, balanced on the index finger of a hand.
As a precaution, a stem had been introduced into the barrel to hold the weapon still.

John M. Browning with his new automatic .22 rifle (Herstal Group collection).

A banquet held at the FN headquarters in honor of John Moses Browning (underneath the red dot), on 31 January 1914. The occasion marked the production at Herstal of the millionth pistol of his invention, actually produced in 1912.

Machining workshop for arms parts in 1912. The 'Grand Hall', with its forest of transmission belts and its high number of

The cartridge factory in 1908. Set up in 1891, in cooperation with the Deutsche Metallpatronenfabrik, which also belonged to the Loewe group, the FN cartridge factory was extended in 1929 with a workshop for the manufacturing of cartridges for hunting shotguns.

female workers, remained the epitome of the Fabrique Nationale de Herstal's work until the 1970s.

FN motorcar 2100. (1909-1910).

FN motorcar 1500-1560. (1909-1911).

FN motorcar 2400. (1911-1913).

The celebrated FN

A few months after the start of their relations with the Browning family, FN embarked on a road which appeared to be just as promising: the manufacturing of motor vehicles. In October 1897, the board had bought a car to have it studied by its engineers. Two years later, it decided to build a first series of 100 vehicles equipped with a two-cylinder engine with side-valves and belt gear.

In 1902, a 14 hp car – or rather a chassis, because in those days the coachwork was built outside the factory – with shaft-drive was released.

From that moment on, successive models appeared rapidly. For instance, the FN 6900 with two two-cylinder engine blocks developing 30 to 40 hp (1905); the FN 2000 (1906), several versions of which, the 2000 A and the 2100, were equipped with a throttle-chamber carburetor (1907); the FN 1400 (1907) and derived models ; the FN 1600 (1911), for which a spring system with shackles had been developed, which is still being used; the FN 2400 (1913), equipped with a four-cylinder engine block, a disc clutch and shaft with ball bearings, and, last but not least, the FN 1250 (1913), undoubtedly the first of the smaller cars, which fascinated the connoisseurs by its extremely simple engine and leather-cone clutch. Three hundred of these cars were sold at the 1913 Paris Motor Show.

Contrary to the usual practice, the car was supplied with an open body in wood, which was made in Lyon, France.

FN motorcar model 1900.

The first FN car.
From a series of stamps depicting early automobiles, issued by Monaco in 1961.

First FN motorcar model 1900.

The FN 1200 car belonging to the Managing Director, A. Andri, photographed on 26 November 1910.

FN
FABRIQUE NATIONALE D'AR

FN poster by Raymond Tournon (1907). Set in "Art Nouveau" style, and in our opinion a nice touch of "retro", it represents the essential of FN's fine civilian production : hunting shotguns, bicycles, motorcycles, motorcars. (Herstal Group archives).

Workshop for the assembly of motorcars, 1912.

FN motorcar 2000 (1910).

FN motorcar "Landaulet" 2700, n° 694 (1913).

An FN 1500 motorcar in front of the Cathedral of Saint Isaac and the statue of Tsar Nicholas I in Saint Petersburg (around 1910).

FN's stand at the Paris Motor Show in 1908. In the foreground, the 1400 model automobile which had been released the same year.

FN motorcar belonging to Crown Prince Frederic Willhelm of Prussia, son of the German Emperor Willhelm II (1911).

It was to decorate this wing that the painting depicting 'Vulcan's Furnaces', illustrated on pages 550-551, was made.

In 1912 the company "Ets Deschryver et Blanchart", based at 15 rue Veydt in Brussels, was FN' s agent for the Brabant province.

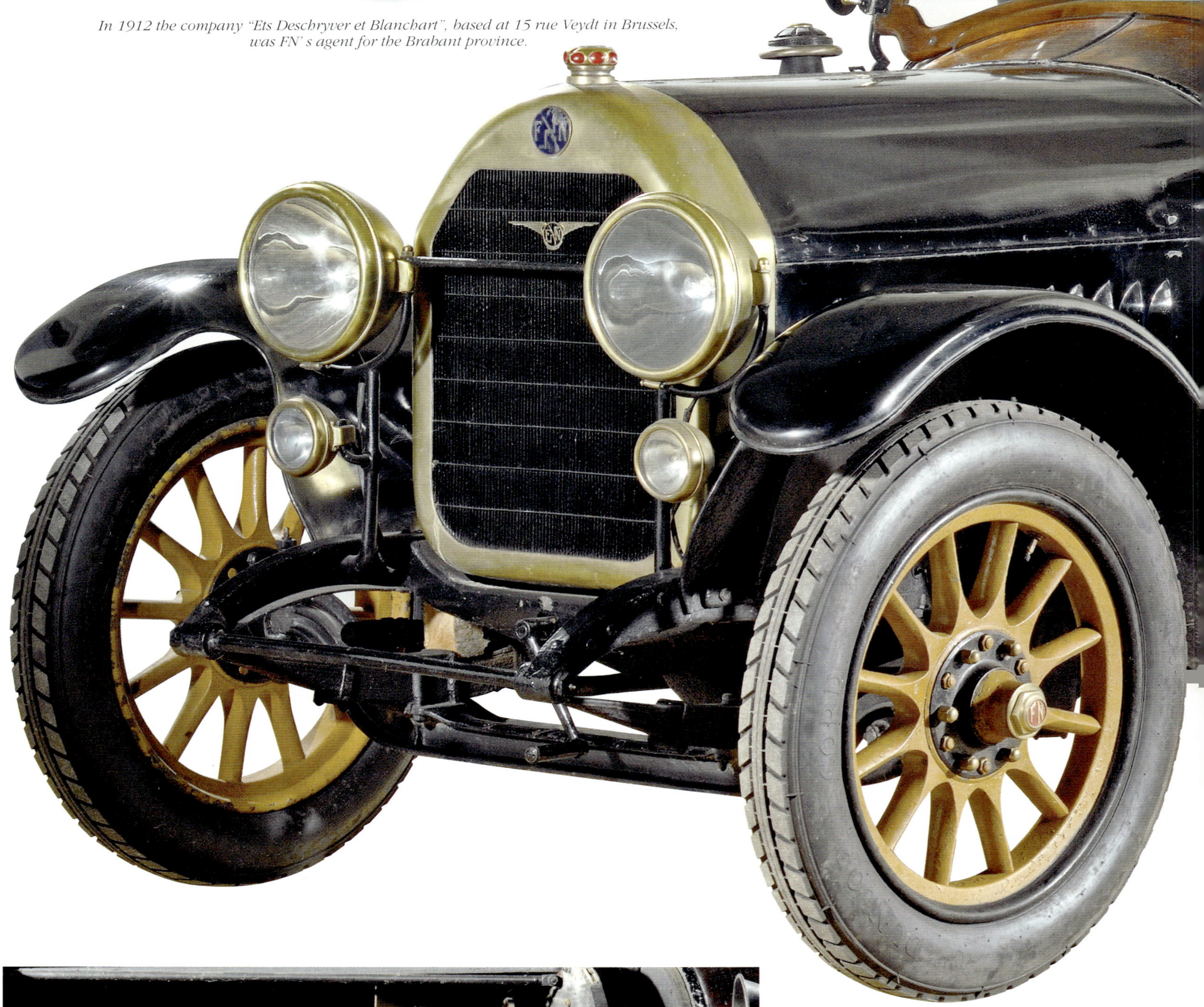

6,000 cc engine, 4 cylinder, 30 hp.

On the dashboard, a St. Christopher medallion : "Look upon St Christopher, then drive on with confidence".

FN motorcycle 1903 - 188 cc.

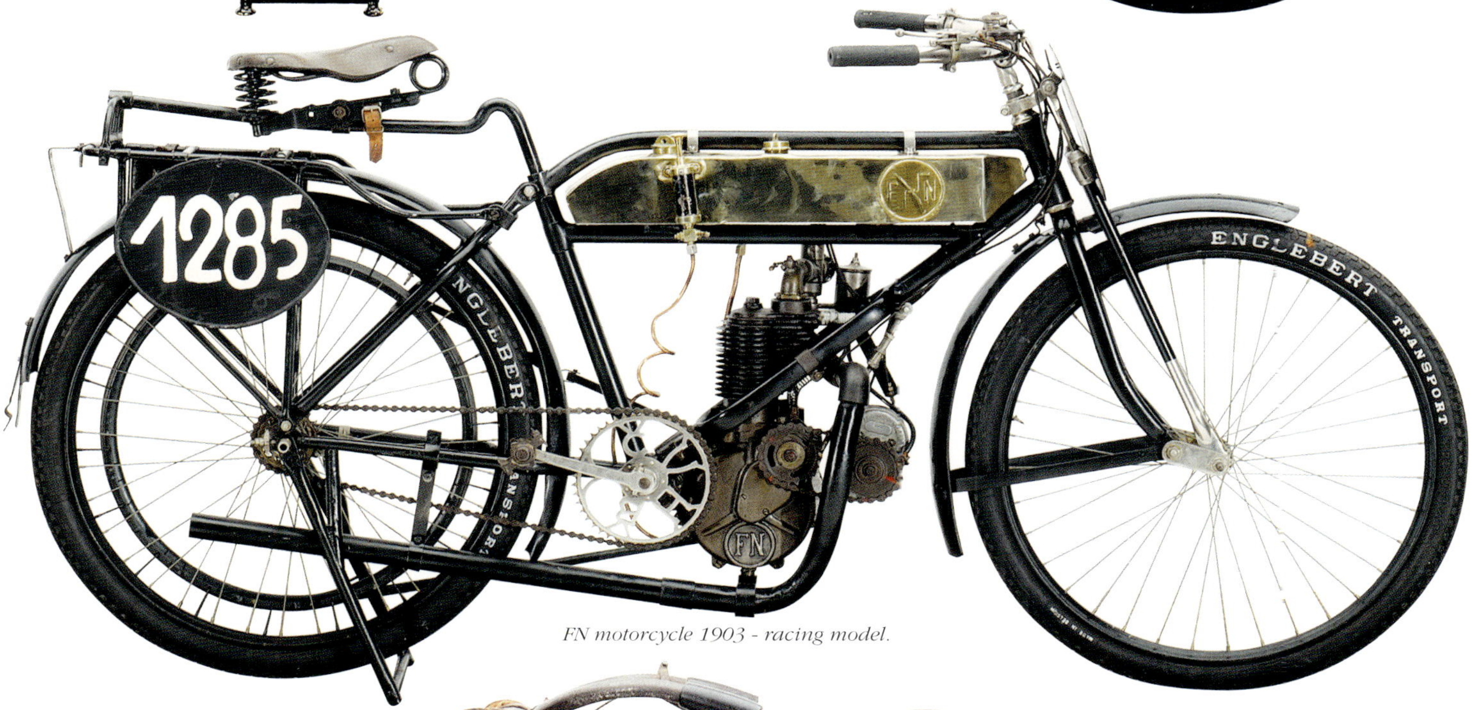

FN motorcycle 1903 - racing model.

FN motorcycle 1904 - 300 cc.

FN motorcycle 1901. In December 1901, FN began the production of motorcycles. The first models were little more than bicycles fitted with a single cylinder engine.

The model underwent a series of improvements; the cylinder capacity was increased from 133 cc to 300 cc and the power from 1.25 hp to 2 hp. Six thousand machines of this type were sold between July 1902 and July 1904.

Four cylinder FN motorcycle (1906 model). This was the first FN motorcycle in the proper sense of the term. The engine contained four cylinders with a total cubic capacity of 362 cc, later increased to 410 cc, as on this model.

The inlet valves worked automatically while the outlet valves were operated manually. The merits of this motorcycle soon earned it a substantial reputation, especially in Great Britain, where the public gave it the descriptive title of "car on two wheels".

Belgian soldiers equipped with FN motorcycles at the large-scale manoeuvres of 1908. The Belgian army, which had always been a client of FN's for its small arms and ammunition, has for many years also purchased its vehicles: bicycles, motorcycles, lorries, tricars, AS 24, etc.

The famous Italian composer Giacomo Puccini in the sidecar of a 4 cylinder FN motorcycle (1913).

The Imperial Russian army equipped with FN motorcycles (1907). Before the 1917 revolution, the Russian government placed an order with FN for the conversion of a batch of 400,000 Berdan rifles and also ordered a number of army motorcycles.

Assembling motorcycles in 1910. On the left, a worker is on his knees working on a 285cc motorcycle.

Four-cylinder FN motorcycle, 1910.

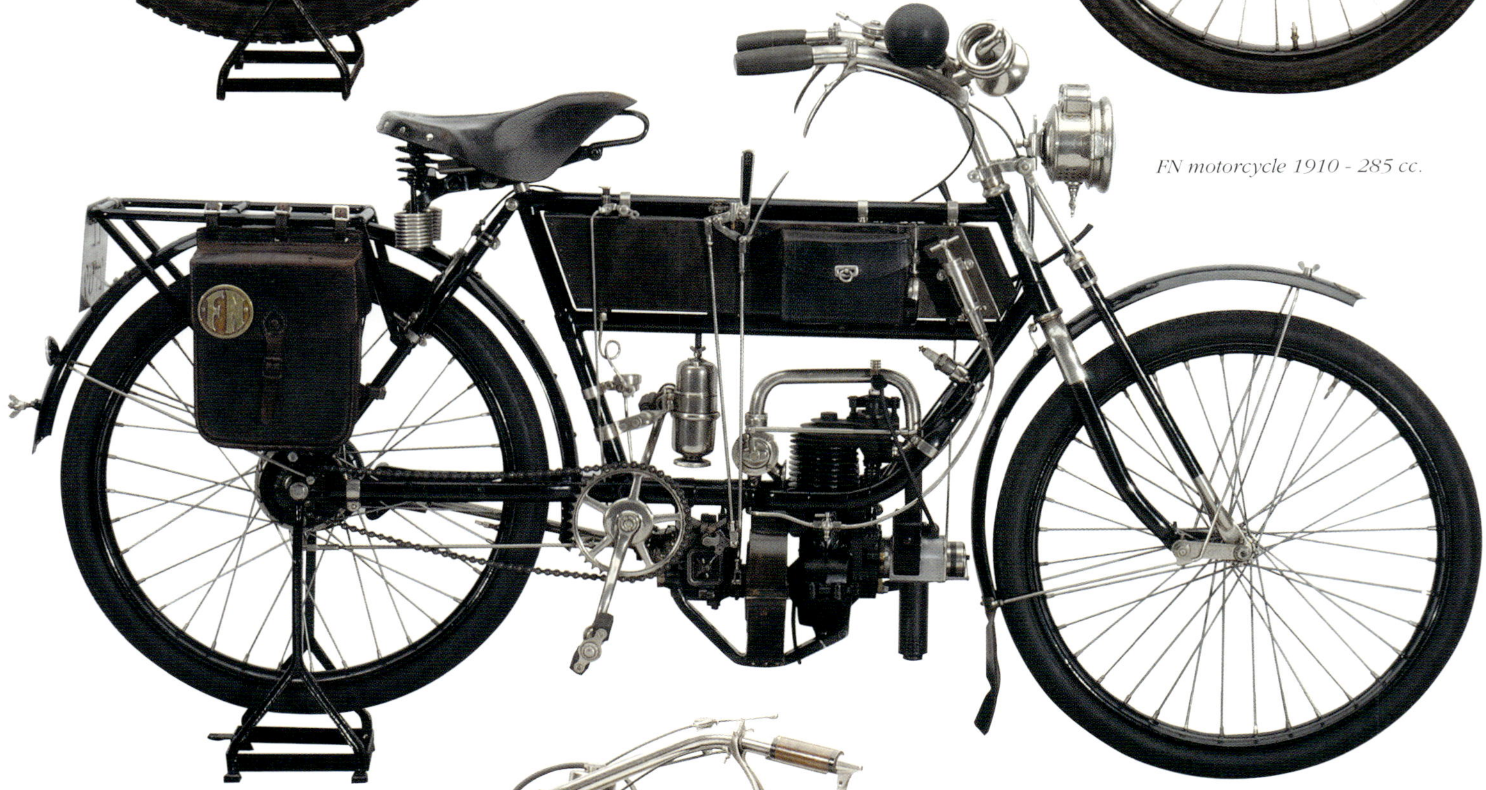

FN motorcycle 1910 - 285 cc.

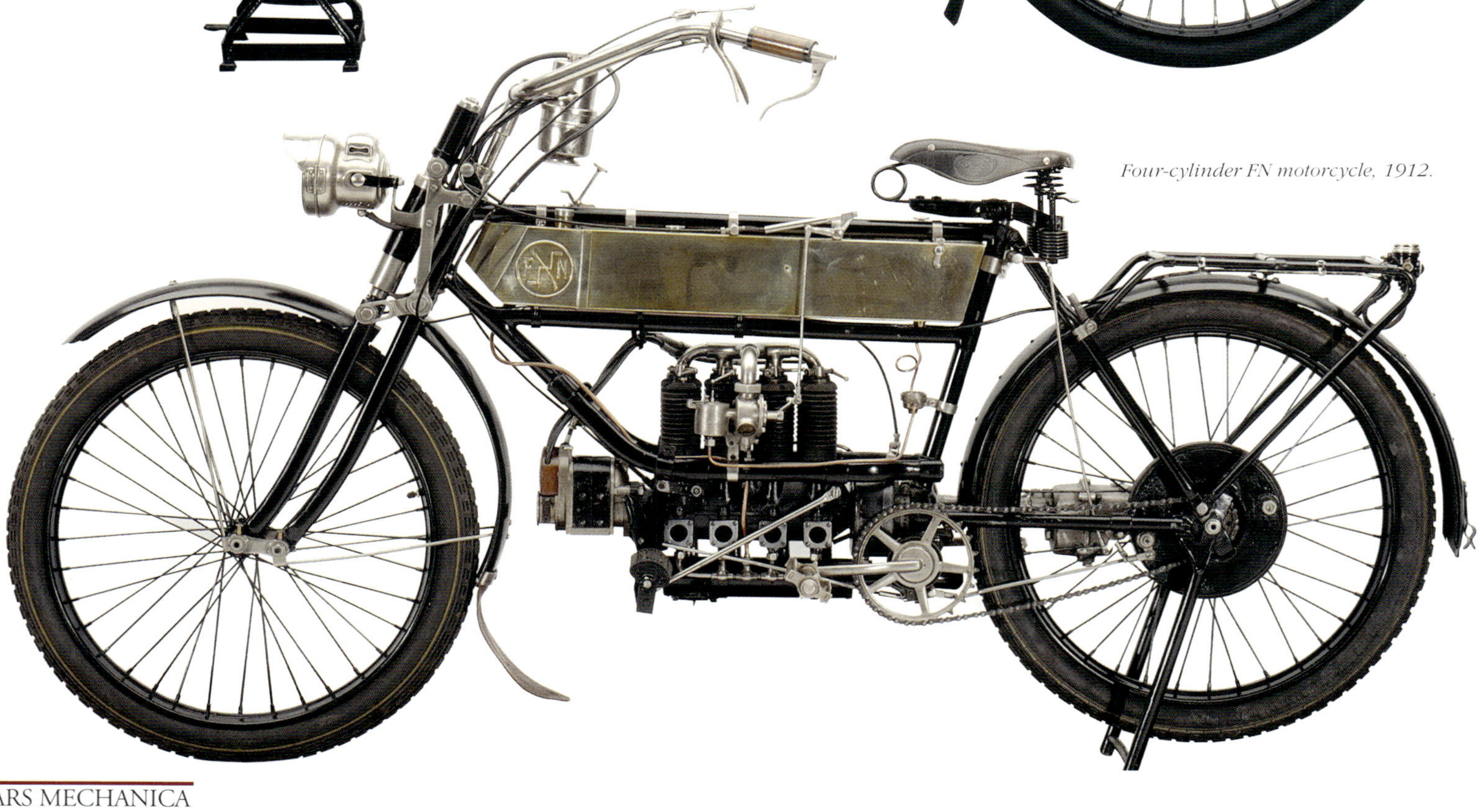

Four-cylinder FN motorcycle, 1912.

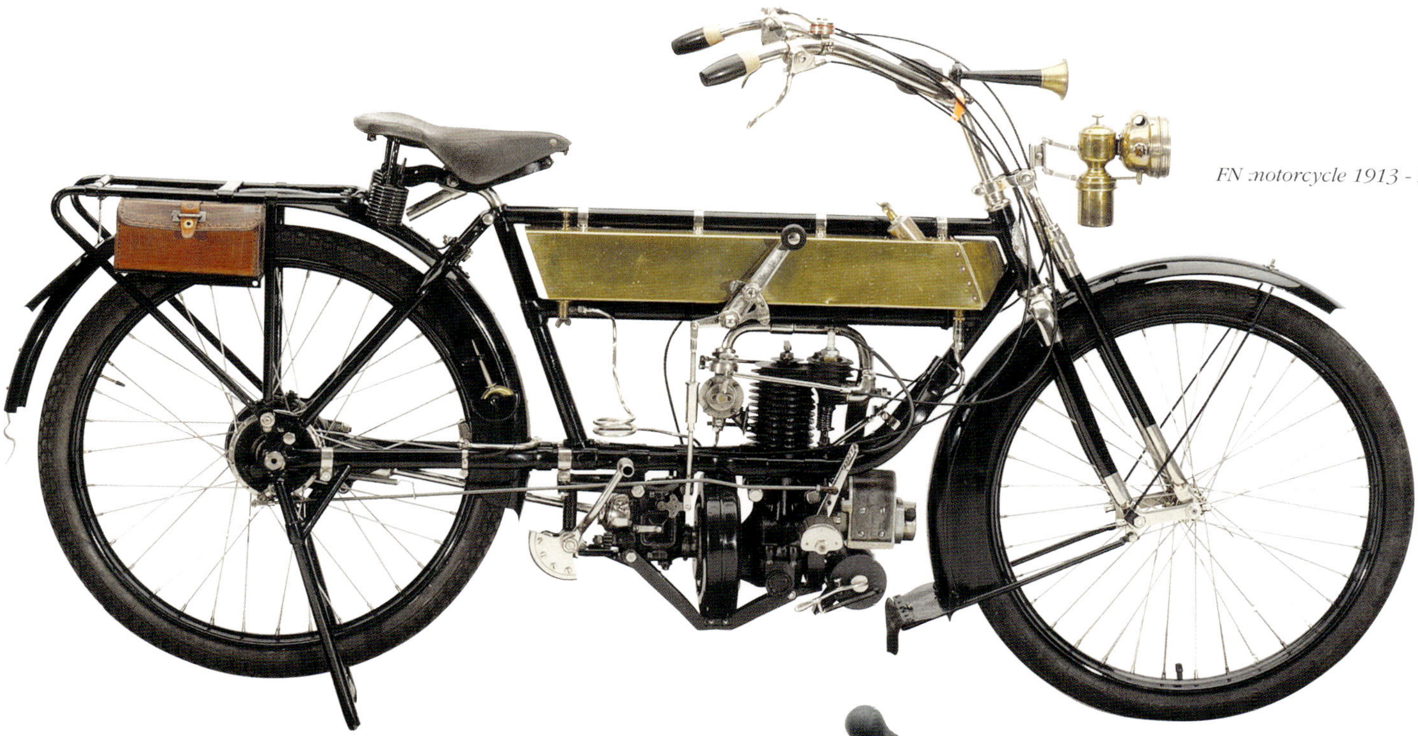

FN motorcycle 1913 - 285 cc.

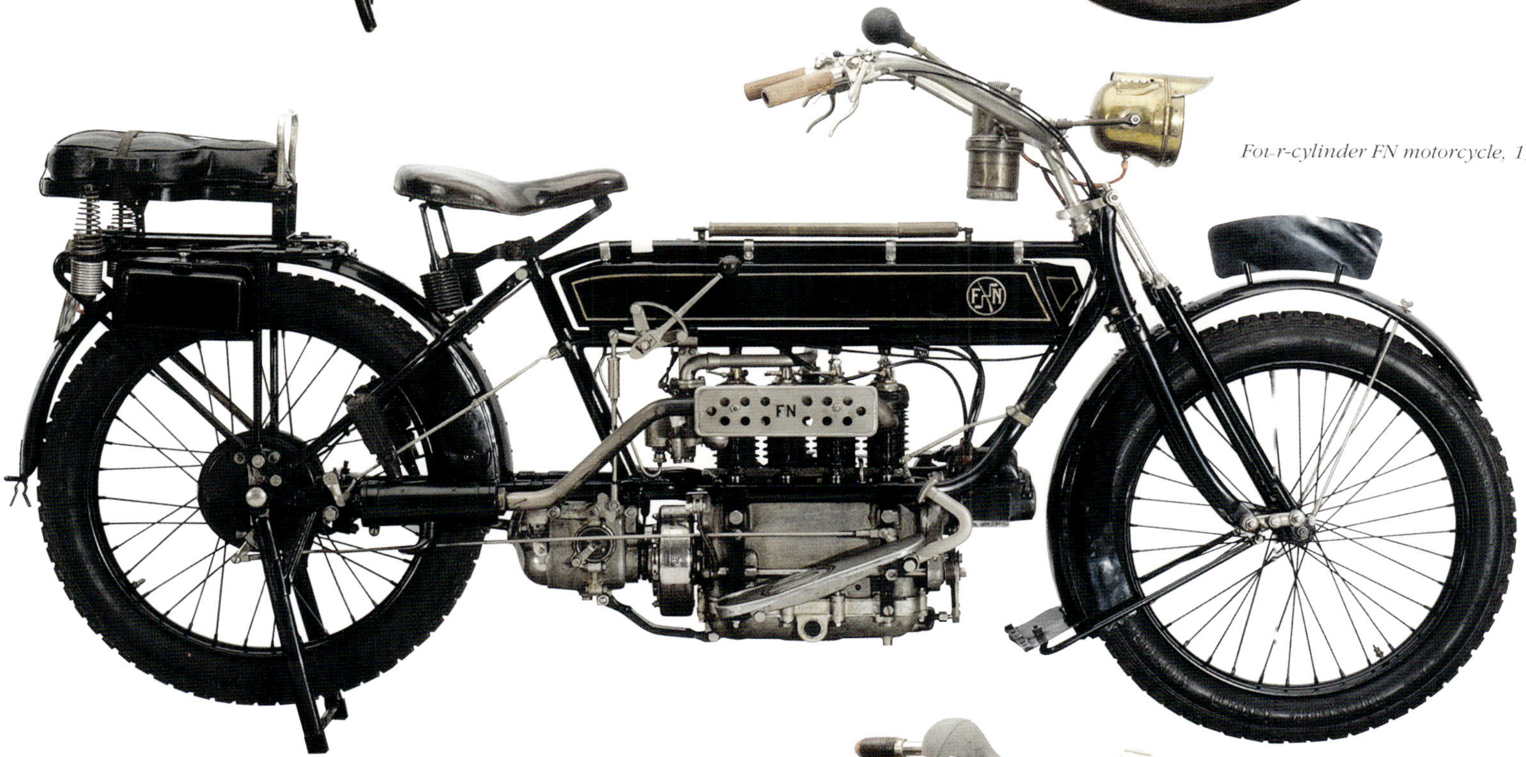

Four-cylinder FN motorcycle, 1914.

FN motorcycle 1914 - 285 cc.

FABRIQUE NATIONALE D'ARMES DE GUERRE
(SOCIETE ANONYME)
HERSTAL-LIÈGE
BELGIQUE
20
100
FN

Probably the work of Emile Dupuis, a French artist working in Liège since 1902, this poster depicts an innovation, the four cylinder FN motorcycle which came out in 1905. FN's production of motorcycles, from 1901 to 1964, is still admired in Belgium and internationally.

FN ahead of its time in labor relations

As we have already seen, the management of FN was very much concerned with its social responsibilities. This was shown clearly in November 1896, when the company, which was then in full crisis, set up for its workforce a medical and pharmaceutical service, the costs of which were borne by the company, and also had the workers insured against industrial injuries. It is worth noting in this respect that it was not until 24 December 1903 that compensation for industrial injuries in Belgium became the employer's legal responsibility. As soon as the situation of the company had improved, the board set up a mutual pension fund for the workers and employees (1906), while the management instituted savings banks and leisure societies. Financial recovery, administrative reorganization, improvement in production methods, extending the range of products, favorable industrial relations, all of these factors helped FN to acquire a level of prosperity which seems quite extraordinary today. After having repaid first the debenture creditors and subsequently the Schaaffensen'scher Bankverein, the company was able to distribute to its shareholders a dividend which had risen from 40 francs in the financial year 1903-1904 to 150 francs in the financial year 1907-1908, and which stayed at that high level until October 1913. Correspondingly, the market value of the shares, which had reached 2,750 francs on 30 June 1912, had risen to 3,550 francs in the summer of 1913.

Manufacturing of rifle parts (1912)

Visit to FN by a group of Australian cadets on 17 June 1914. Since the turn of the century, FN has always attracted many military and civilian visitors because of the large scale of installations and its range of products. It is a Belgian company of international repute.

Manufacturing of arms (around 1908).

Machining workshop producing parts for the Auto five shotgun (around 1910).

Small Browning, massive impact the assassination in Sarajevo (28 June 1914)

Although an enormous simplification, it could be said that two shots fired from a Model 1910 9 mm (.380ACP) Browning pistol triggered the First World War. This weapon, sold together with two identical weapons by FN on 3 December 1913 to the gunsmith Doucet of Koksijde near Ostend, found its way into the hands of members of the Black Hand terrorist organization as the result of an everyday retail transaction. At that time, in Belgium as in a number of other countries, the purchase and possession of handguns was not subject to any legal restriction whatsoever.

Secretly supported by the Serbian Chief of Police, the Black Hand was attempting, through acts of violence, to separate Bosnia-Herzegovina from the Austro-Hungarian Empire, to which it had been annexed in 1908, and join it with Serbia. Backed by Russia and indirectly by France and England, Southern Slavic nationalism opposed the policy of Austria and its ally, Germany. Which is to say that at the beginning of the twentieth century, Bosnia had become the powder keg of Europe.

In June 1914 the heir to the Imperial Austro-Hungarian throne, the Archduke Franz Ferdinand, accompanied by his wife, Sophia of Hohenberg, came to Bosnia to observe military manoeuvres by the Austrian army. The nationalists viewed their presence as an extreme provocation and hatched a plot to assassinate the royal couple. Seven people were indirectly involved.

On 28 June, the Archduke and Archduchess had arrived in the Bosnian capital, Sarajevo, where an official reception was to be given in their honor at the Town Hall. To travel to the reception, they were seated in the Graef & Stift 'Double Phaeton', put at their disposal for the duration of the manoeuvres by Count Franz Harrach, who had acquired it in 1910. Their route was known and marked out, lined with a crush of onlookers along its entire length. In the light of subsequent events, the organizers' lack of foresight or naivety seems astonishing; they could not fail to be aware of the hostility among a section of the local population, or of the risk of an assassination attempt by foreign agents : an open car, a large crowd, the inadequacy of the personal protection, a pre-announced route, the casual approach to the risk after the first incident... so many elements which suggest a serious disregard for the most elementary security rules.

At 10.15 am a bomb was thrown at the Archduke's car. He merely picked it up and threw it out, and it exploded against the car behind him, causing only light injuries among his aides de camp. The attacker, a young typographer, was arrested.

The procession continued along its route and the Town Hall reception went ahead as planned. On its return along the same route, the royal car had to slow down in front of a brasserie, precisely where one of the conspirators, 19-year-old student Gavrilo Prinzip, had happened to take refuge after the failed attack.

He immediately seized his chance, leaping towards the vehicle and firing two shots at the royal couple from his 9 mm Browning, at virtually point-blank range. The first shot hit the Prince in the neck, went through his jugular vein and entered his windpipe. The second pierced the car's bodywork and hit the Archduchess in the lower abdomen. Prinzip was immediately surrounded and seized. The royal couple was transferred in a critical condition to the Governor's residence. The doctors could do nothing, however, and the victims died shortly afterwards.

The assassin was to die in captivity of tuberculosis in 1918.
The sudden deaths of the heir to the Imperial throne and his wife unleashed, slowly and inexorably, a terrible chain reaction in the region's chancelleries.

Confident of Germany's support, on 23 July Austro-Hungary presented Serbia – indirectly responsible for the double murder – with an ultimatum. For obvious reasons, the latter refused to allow the Austrian police to conduct an investigation on its soil. On 28 July, the Austro-Hungarian Empire declared war on Serbia, which Russia was committed to defending. Loyal to its undertakings, Germany then entered the war on the side of its principal ally. Then followed France and England...
Little by little, Europe and the world mobilized. And so began the First World War, the most terrible in history up to that time.

Page on the right:
Top : an illustration of the Sarajevo assassination which appeared in the 'Petit Journal'.
Center : an illustration of the arrest of Gavrilo Prinzip.
Bottom : extracts from the export register for Browning pistols. Encircled are the pistols sold by Doucet and used by the Black Hand terrorists in 1914.

FN Browning semi-automatic pistol model 1910.

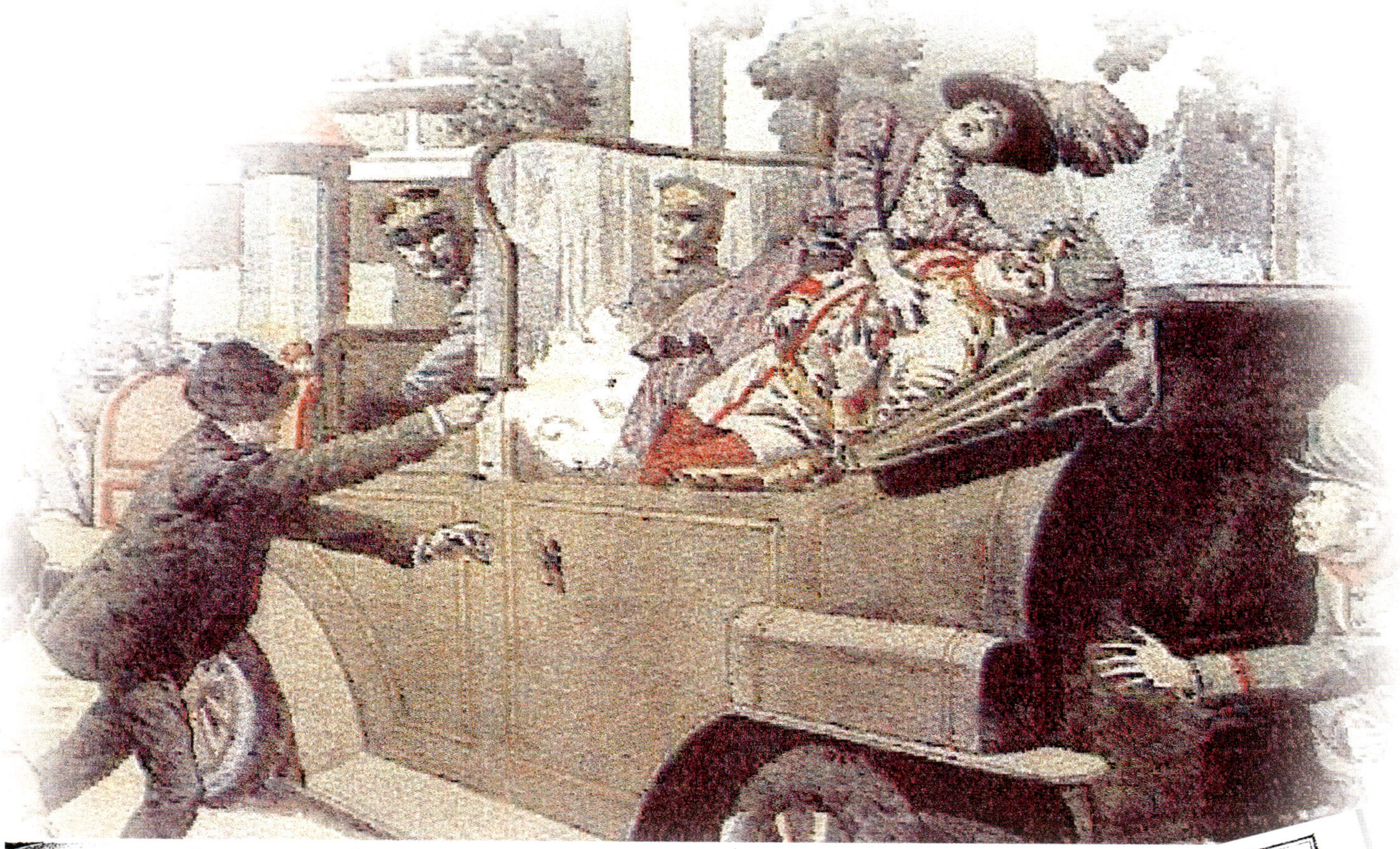

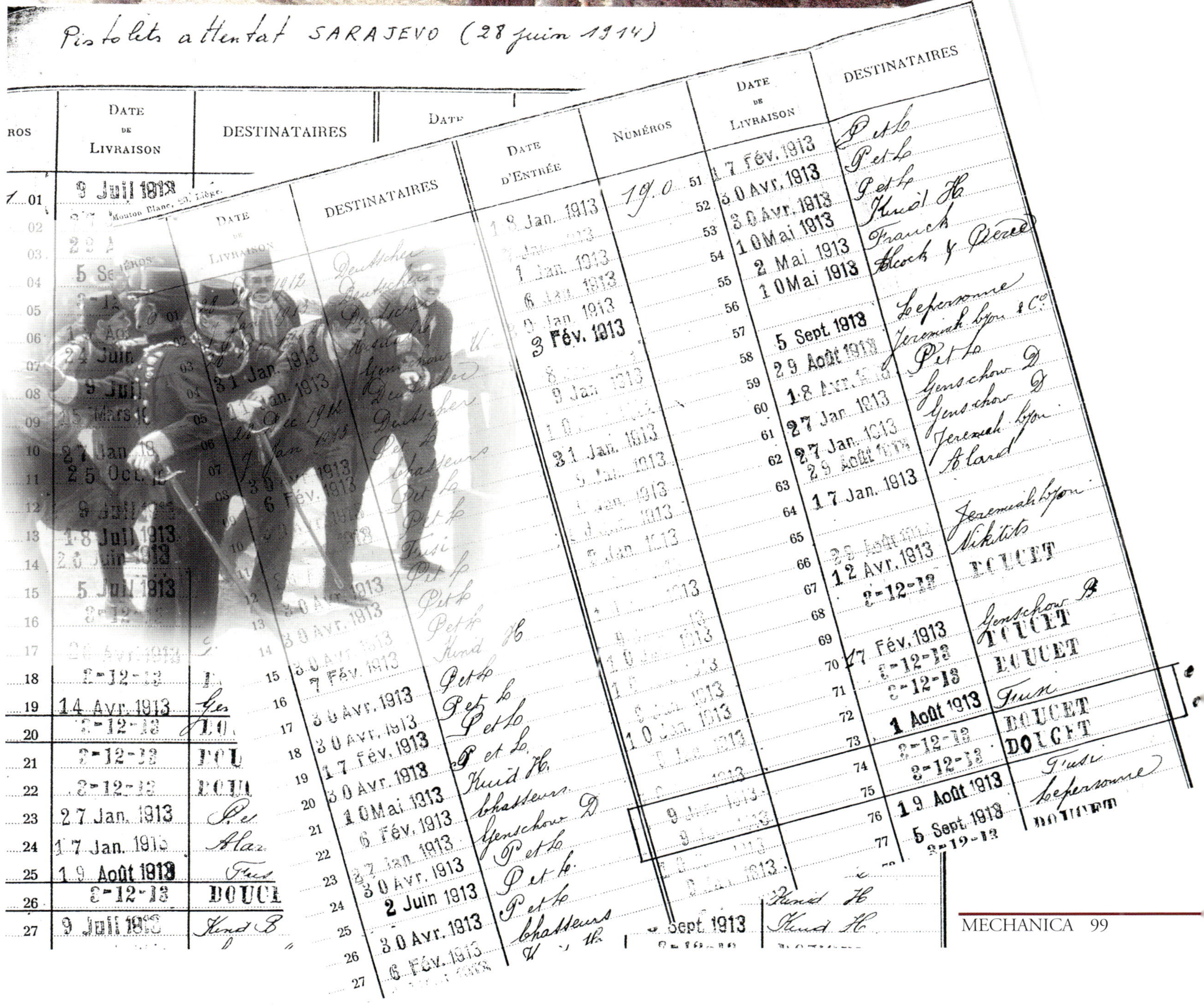
Pistolets attentat SARAJEVO (28 juin 1914)
DATE DE LIVRAISON
DESTINATAIRES
DATE D'ENTRÉE
NUMÉROS

FN requisitioned

Following the outbreak of war, we know that after a heroic resistance, Liège fell to the German army on 16 August 1914. At an early date, the Kaiser's government expressed its desire to have FN continue industrial activity. FN being mainly an arms manufacturer, the board, although German directors were in the majority, responded by closing down the factory and granting the staff advance payments and subsidies (October 1914). The German government then promptly requisitioned various armaments and machinery.
The hostilities continued. In January 1915, the board decided to fabricate new machinery and tools to replace those which had been requisitioned, on condition that the German government guaranteed that the new machines would not be requisitioned in their turn. Work started in 1916.

Machine-tool hall at FN, converted into a hospital for the army of occupation (1914). After Liège was taken by the Germans in August 1914, FN ceased its activities. In 1917 it was sequestrated by the Germans, but it resumed work normally from November 1918 and in 1919 severed its links with the Loewe group.

In the meantime, during a general meeting which was held on 20 February 1915 to approve the results of the financial year 1913-1914, a violent incident divided the Belgian and German shareholders.

As the results showed a profit, the Germans suggested that a percentage of the profits should go to the directors. The Belgian minority, however, whowere of the opinion that in view of the circumstances no dividend should be distributed, especially in consideration of the distress being suffered by the staff, unsuccessfully tried to block such a decision.

Meanwhile the occupying authorities pressed the management to restart other manufacturing sections. Supported by the unanimous decision of the board, Alfred Andri refused. He was taken away and imprisoned in a prisoner of war camp in Gutersloh, Westphalia (May 1915). In the same year one of the Belgian directors, Georges Laloux, who was caught smuggling correspondence to the Netherlands, was sentenced to one year imprisonment in a German prison (winter of 1915-1916). Finally, having failed to secure the cooperation of the company, the military authorities sequestrated the factory (1917) and themselves installed a vehicle-repair workshop on the premises.

The "Société Générale de Belgique" appears on the scene

But events took a turn. After the failure of the offensives launched by Ludendorff in the spring and summer of 1918, there was no more doubting the victory of Belgium and its allies. It was then that a number of Belgian banks, more particularly the "Société Générale de Belgique", set up a limited company, the Union Financière et Industrielle Liégeoise, the objects of which were to be the acquisition of the FN shares which were still in the hands of the Deutsche Waffen- und Munitionsfabriken. In the general climate created by the war, the fact of having an arms factory controlled by German nationals was felt to be quite unacceptable.
On the eve of armistice, a Decree provided for the sequestration of all property abandoned in Belgium by nationals of hostile countries. This measure also pertained to the 5, 700 FN shares belonging to D.W.M., and was applied from 30 November 1918, when G. Joassart, general secretary of the company, was legally appointed trustee of the shares. A few weeks later, at the end of December, the three German directors handed in their resignation. Meanwhile G. Joassart, who was acting on behalf of the Union Financière et Industrielle Liégeoise, negotiated the purchase of the sequestrated shares with their holders.
These negotiations were brought to a successful conclusion on 12 March 1919. On that day, G. Joassart signed in Cologne the agreement whereby the D.W.M. surrendered their entire participation to the U.F.I., who kept 3,900 shares and sold the others to various people who had assisted him in the operation.

Thus another era came to a close.

Alfred ANDRI, Managing Director of FN from 1906 to 1919. During the war, he was imprisoned for refusing to collaborate with the occupying authorities.

FN's Board Room during the German military occupation (1914-1918).

The armaments design office in 1916.

The canteen at FN during the German occupation (1914-1918). The price list is framed by a portrait of the Kaiser and the flags of the nations allied to the German Empire.

After Liège fell to the Germans in August 1914, FN ceased its activities. Sequestrated by the occupier in 1917, normal work was resumed in November 1918 and the company left the control of the Loewe group in 1919. FN personnel on 4 August 1916 during the German occupation. They were principally employed in the manufacturing of machine tools.

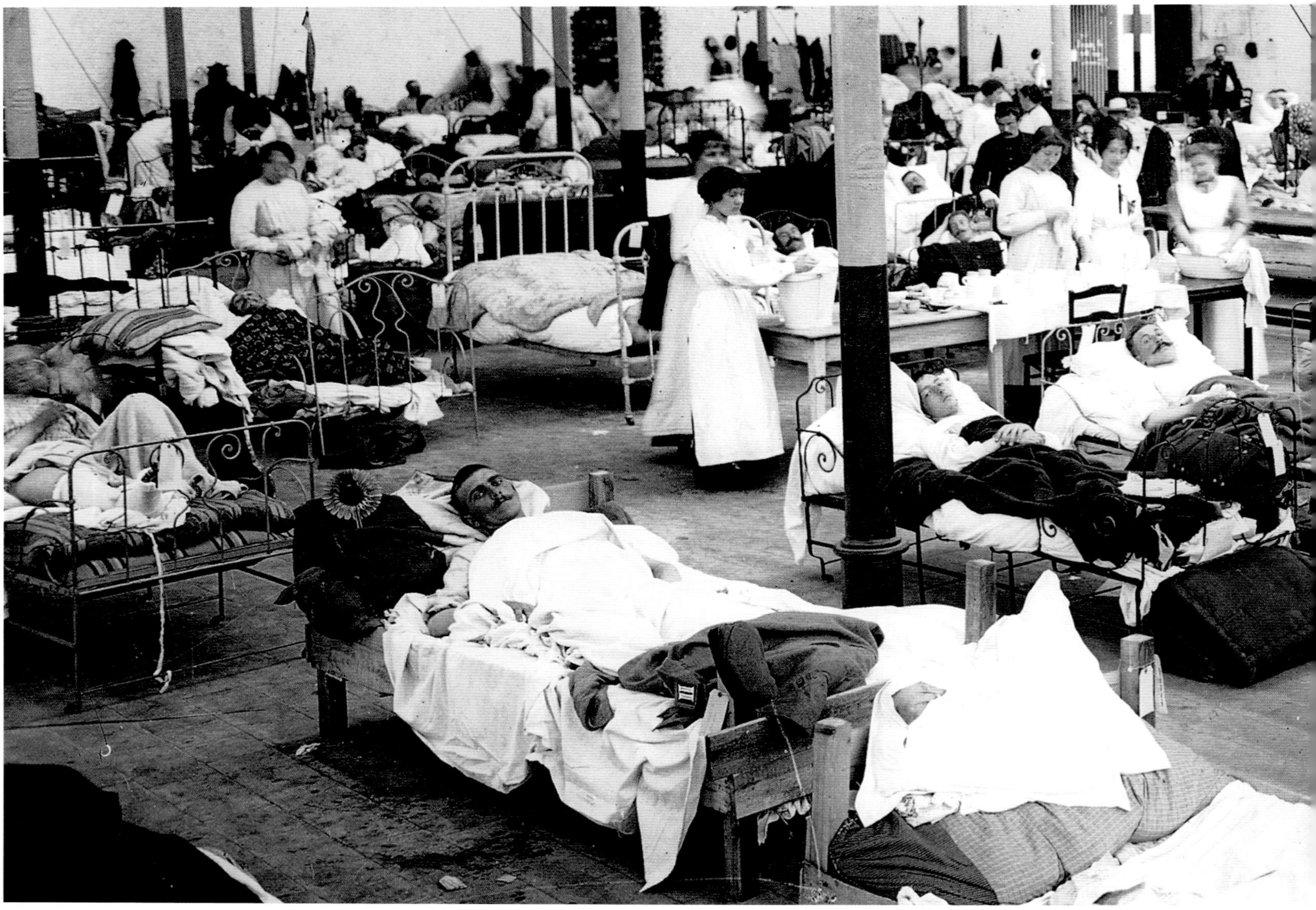

FN's main entrance. The company's mission of vehicle repair can clearly be seen here.

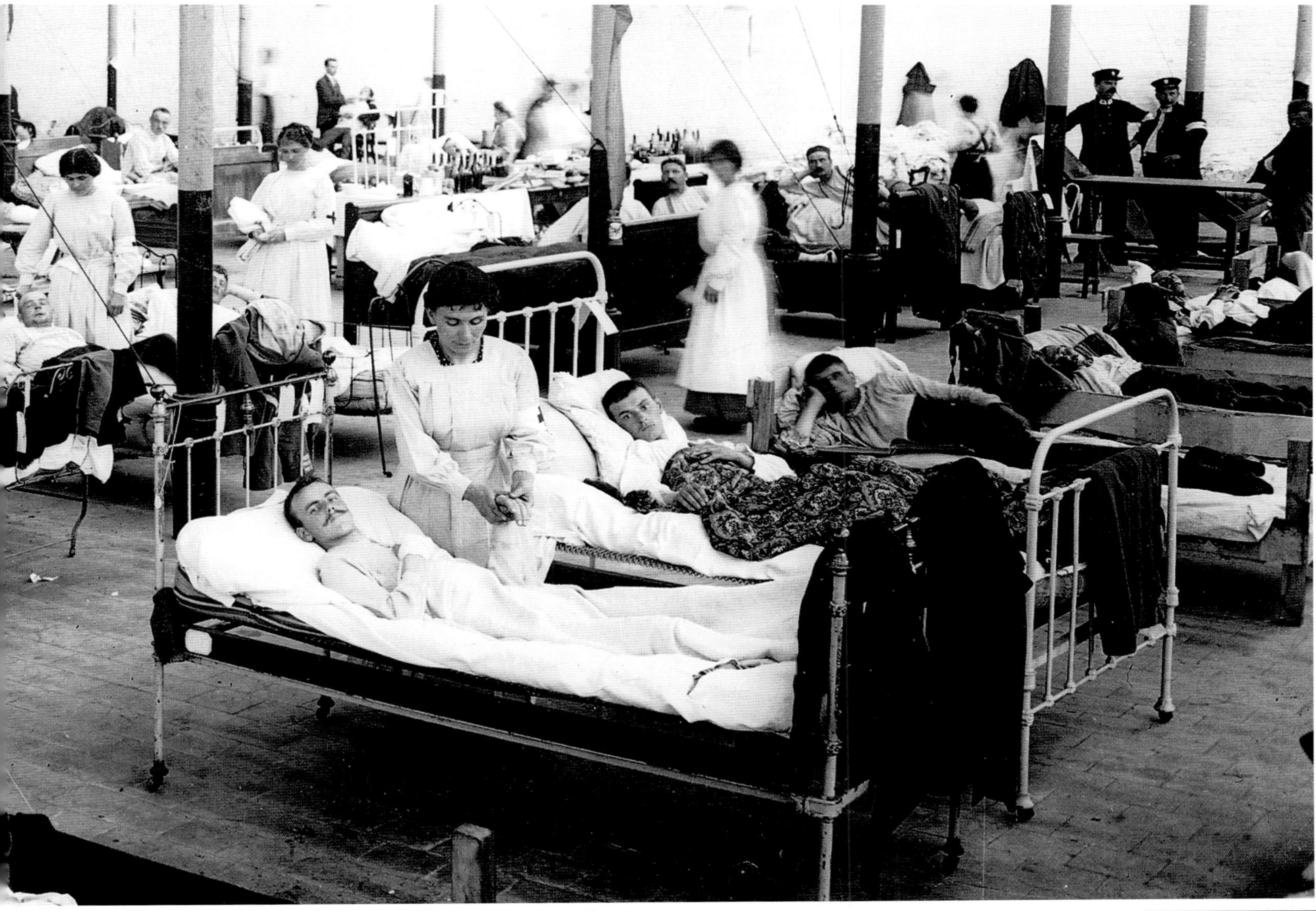

FN's machining hall converted into a hospital for the occupying troops (1914).

Assembly and repair of motorcycles.
Note the predominance of German military personnel, both in uniform and work clothing.

One of the FN's yards during the German occupation.

Regaining peace after the war

Alexandre Galopin, an outstanding personality in the Belgian business world. Alexandre Galopin, who had been an engineer from the University of Liège, came to FN in 1904 and was entrusted with the organization of the central laboratory. He became a manager in 1913 and was promoted to General Manager in 1919, Managing Director in 1923, and Chairman of the Board of Directors from 1932 until his death. He became Governor of the Société Générale, but was assassinated by collaborators in 1944.

The Board of Directors, which was now presided over by Jean Jadot, Governor of the Société Générale de Belgique and Chairman of the Union Financière et Industrielle Liégeoise – soon joined by Alfred Andri who had recently returned from captivity, was soon faced with a great many difficulties.

What had been ruined by the great war

In the euphoria of victory, it was thought that once the effects of the war had been wiped out, the company could purely and simply carry on its traditional activities and in that way regain its former prosperity. The management became quickly disillusioned.

The war had shattered the freedom of exchange and the monetary stability upon which the European economy had been based. The Treaties of Versailles and Sèvres, which hallowed the principle of national sovereignty in all its aspects, merely served to aggravate the situation.

As FN was geared to export, it was affected very severely by the state of turmoil after the war. Not only was the currency constantly being devalued - in 1926, when a law had stabilized it for some time, it was only 14.41 % of its value in 1914 - but all kinds of obstacles came to restrict sales on neighboring markets. For instance, in the period concerned, the customs duties imposed on imported motor vehicles increased from 10% to 45% in France, from 5% to 65% in Italy, from 5% to 16.5% in Germany and from 5 % to 12% in the Netherlands. The duties on sporting arms were similarly increased : on entering the United States, a duty of 82.5 % was imposed on the automatic Browning shotgun's sale price.

Reorganization

The internal reorganization of the company also set some serious problems. We have seen that before the war, FN regularly received technical assistance from DWM, while many of the higher positions in the company were taken up by German nationals. The change of management on 19 March 1919 put an end to that situation. A large number of Belgian engineers and technicians had to be taken on, some of whom were soon entrusted with considerable responsibilities. The company's management was replaced too. Alexandre Galopin, who since 1913 bore the title of Manager, was appointed General Manager (January 1919), while Gustave Joassart became his second in command (March 1919).

On the other hand, the reconstitution of the manufacturing and other personnel after more than four years of inactivity was not an easy task. Qualified workers, still easily available a quarter of a century ago, had become very scarce. To remedy this situation, the management opened a vocational training school within the precincts of the factory. A little later, in 1926 to be exact, Henri Pommerenke, who was then Deputy Manager, set up a special staff selection center to improve recruitment methods.

The psychotechnical laboratory at FN. Machines for testing the aptitudes of candidates applying for work in the company.
Left: speed test in curvilinear movements.
Right: Hand movement dissociation test (April 1928).

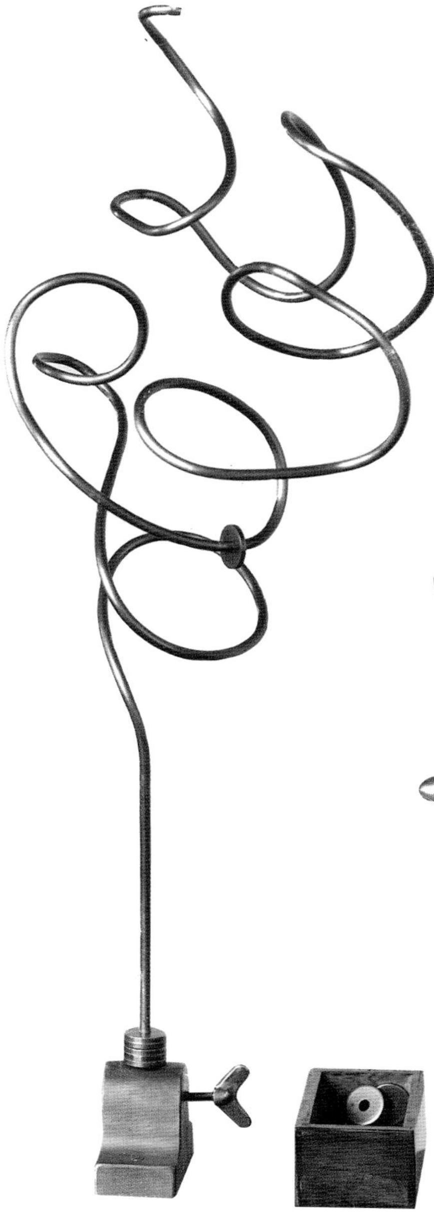

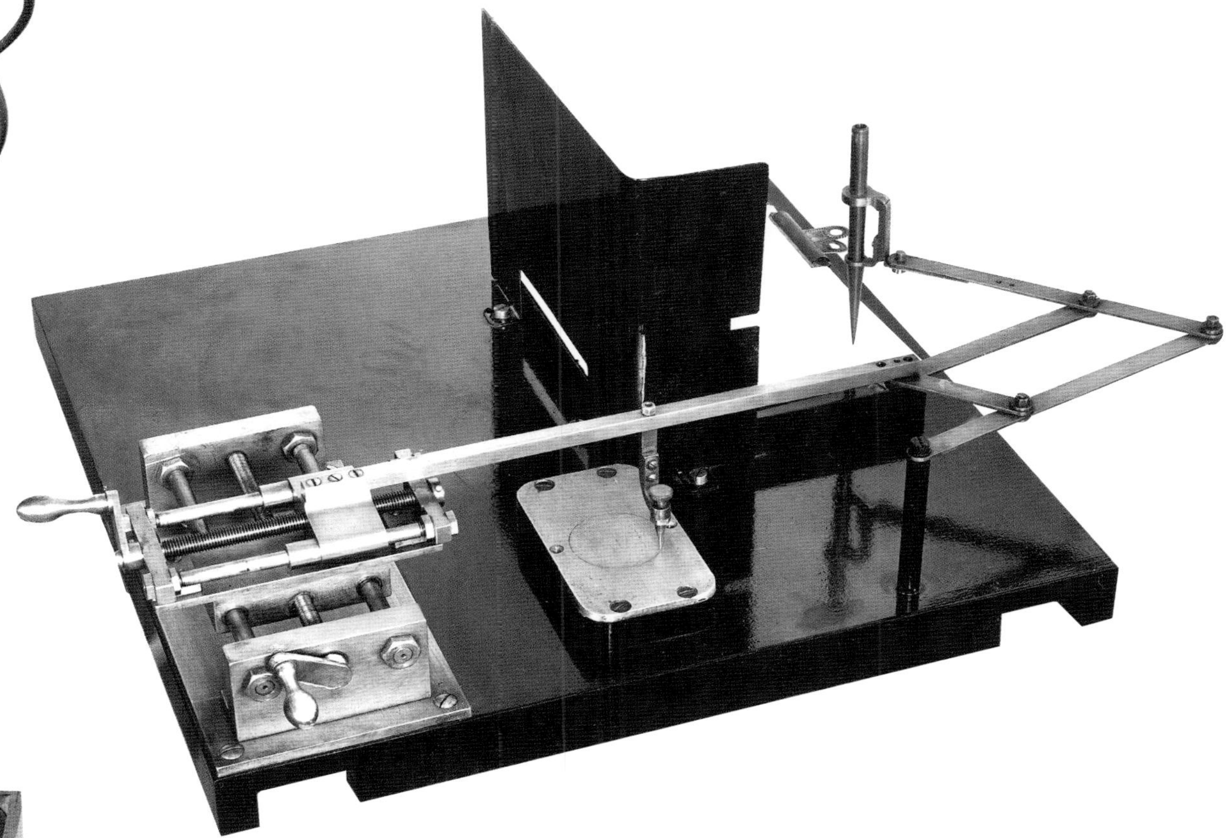

A class in the FN school in 1924. FN decided in 1921 to set up a vocational training school with a view to training qualified and specialised staff for its workshops. The number of disciplines and categories of workers which were covered by this training expanded constantly, and in this way it contributed, together with the public educational establishments, to making the Liège region a center of excellence in the field of advanced mechanics.

Radiator chassis for the 1925 FN 10hp Sport.

Meanwhile, in contrast with the fifteen years of peace which preceded the war, the decade which followed was marked by several social conflicts, the worst of which broke out in 1926 after the management had reorganized the health insurance fund.
This deterioration was certainly due to the general economic recession and to the advent of new ideologies and new aspirations among the working classes.

The material restoration of the factory and its installations went far more smoothly. Legislation for war damage reparations allowed the company to obtain from the Société Nationale de Crédit à l'Industrie a considerable loan pending the time when a judgement by the Court for War Damages in Liège would award FN 59,081,300 francs compensation (22 November 1922). Alexandre Galopin who had been Managing Director of the company since October 1922 , seized the opportunity to modernise the equipment, particularly in the section of motor vehicles, on which he counted to compensate the loss of arms production which in his opinion was doomed as a result of the new international situation. In his report to the general meeting of shareholders on 13 October 1925, he says :
"It is in this division (that of motor vehicles) that we are placing all our hopes for the future".

Car chassis assembly (1921).

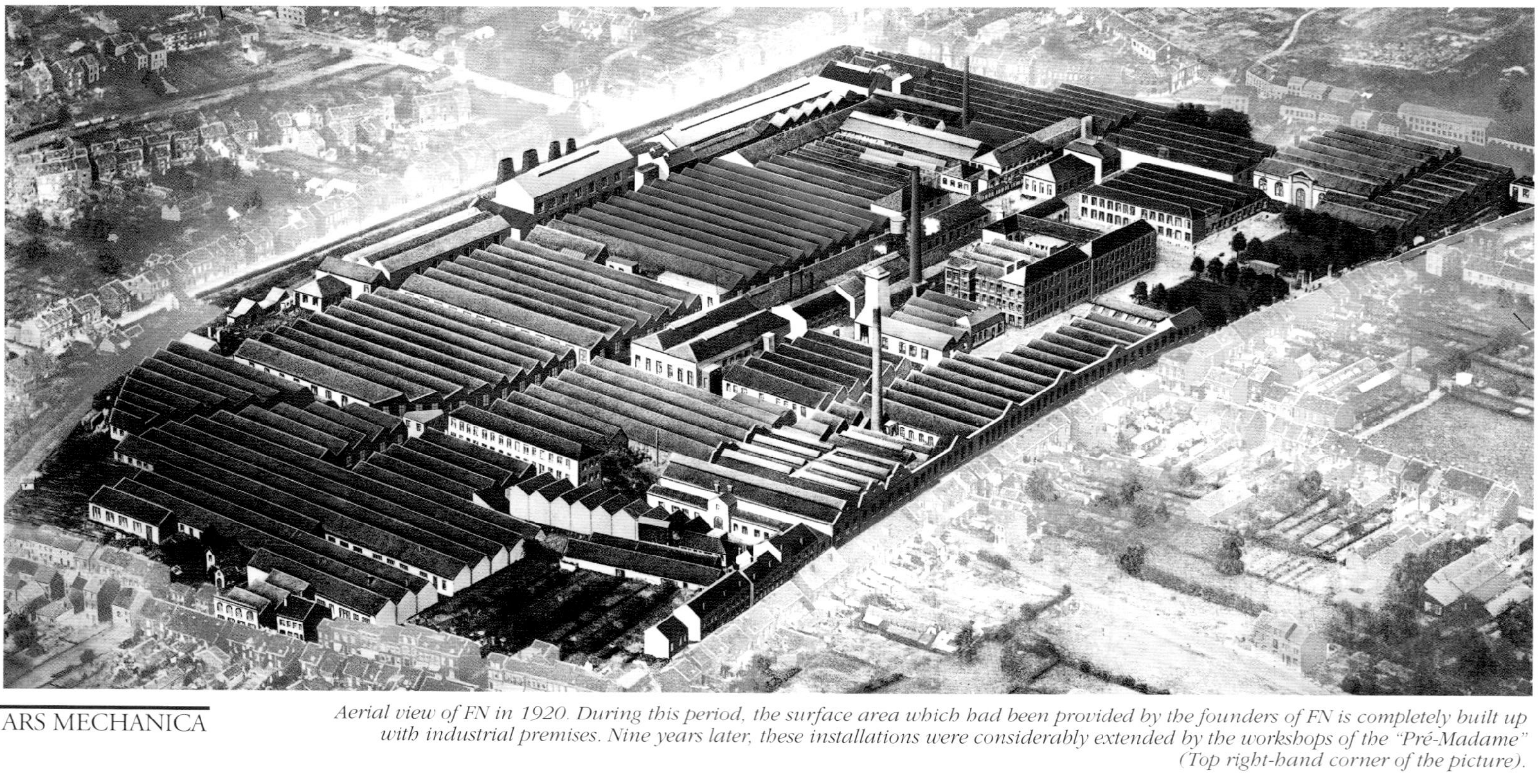

Aerial view of FN in 1920. During this period, the surface area which had been provided by the founders of FN is completely built up with industrial premises. Nine years later, these installations were considerably extended by the workshops of the "Pré-Madame" (Top right-hand corner of the picture).

FN 1250 A n°489 motorcar with boot lid raised (1920).

Skiff bodied FN 1300 Torpedo.

123 Group. Assembly of rear drive axles (1921).

FN laboratory, Photographic section. Top : light room. Bottom : darkroom towards the dryers (1921). The majority of FN's photographic archives, from which many of the illustrations for this book have been drawn, originated from this section of the former laboratory.

Mechanical machining hall for military arms parts (1927).

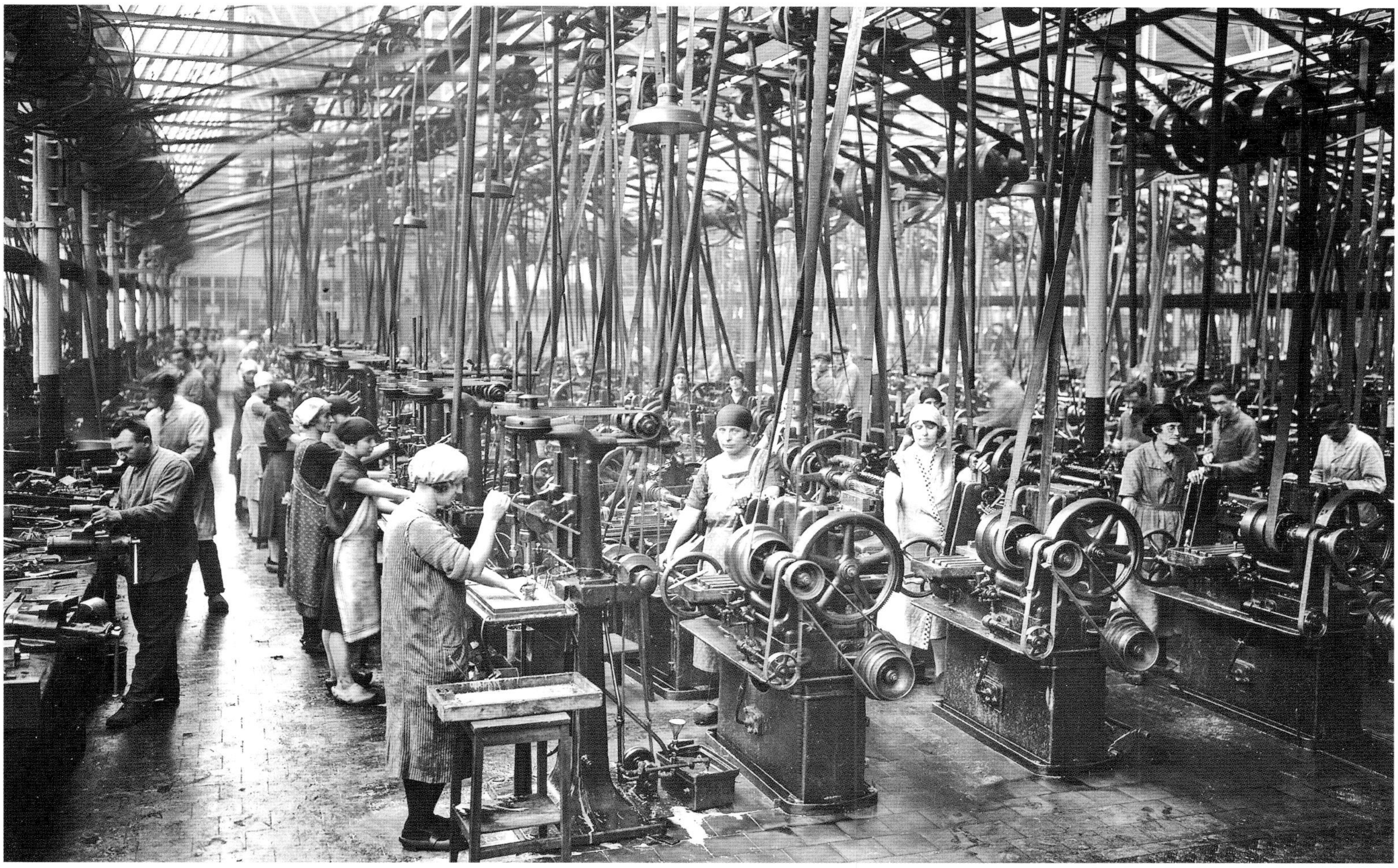

FN laboratory. Dividing machine (1922).

FN laboratory, Chemicals section: research room (1922).

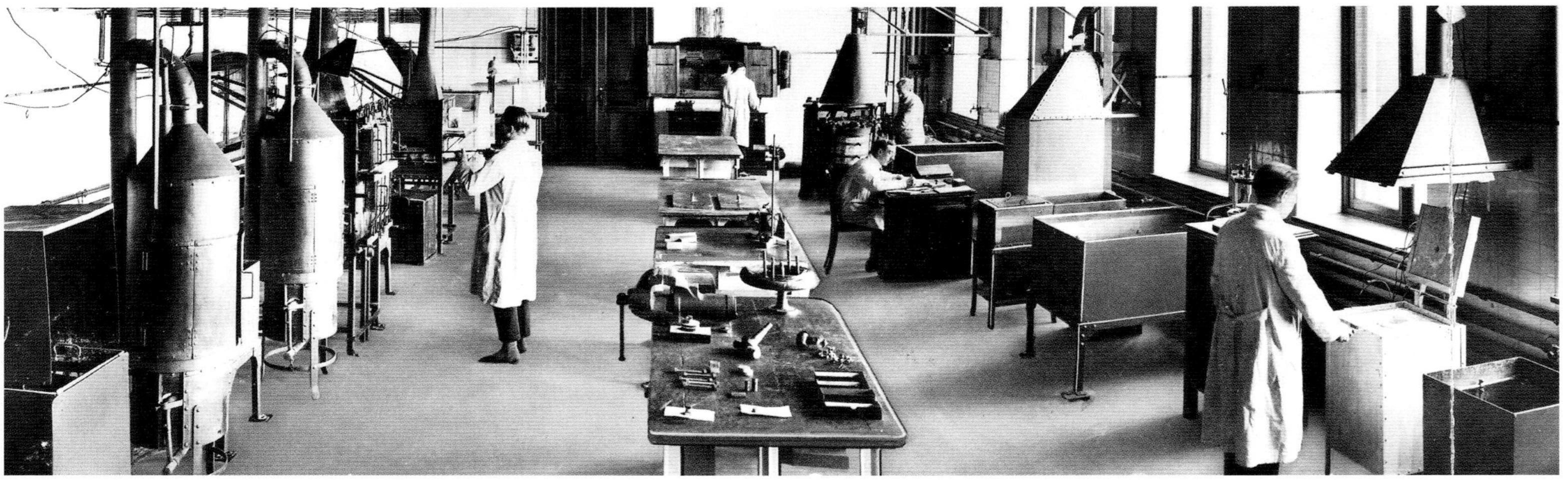

FN laboratory, Physics section (1929).

The foundry laboratory. 75 kg Charpy Izod machine and a Brinnel hardness tester (1922).

FN laboratory, Chemicals section. Right, behind the glass panes, the weighing room (1929).

FN laboratory, Physics section (1929).

Central laboratory office (1922).

FN SPORT in the race

For some years events confirmed these. Moreover, thanks to their intrinsic quality, the vehicles manufactured by FN enjoyed considerable commercial success. Apart from the chassis and the engine, the company also began construction of the coach work. This was made in a workshop in Liège, in the Rue du Vieux-Mayeur.

There were a great number of models, of which we shall simply mention those which have gained some distinction in sporting contests and which are the most remarkable. In 1924, a 1300 model car won the Klausen Hill Climb, while vehicles of the same type won the Royal Cup at the Twenty-four Hours of Francorchamps of 1925 and 1926, while a 2200 G 16 hp car ranked top in the Tour of France circuit race of 1926.

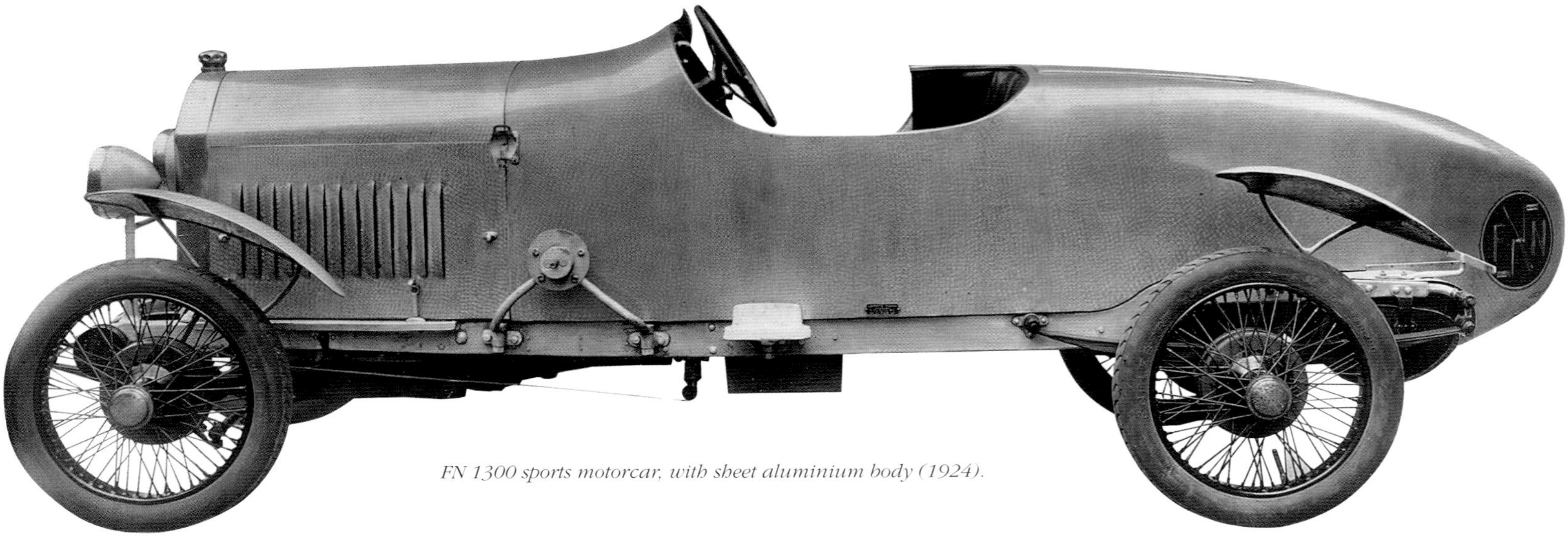

FN 1300 sports motorcar, with sheet aluminium body (1924).

Car belonging to Alexandre Galopin, Managing Director of FN (1922).

1300 Town Coupé, photographed on the 'Terraces' in Liège (1924).

FN 1300 Sports car, bodywork by Labaye in Liège (1925). During the inter-war years, most European car makers had their cars bodied by subcontractors. The Belgian body builders were numerous and well-respected.

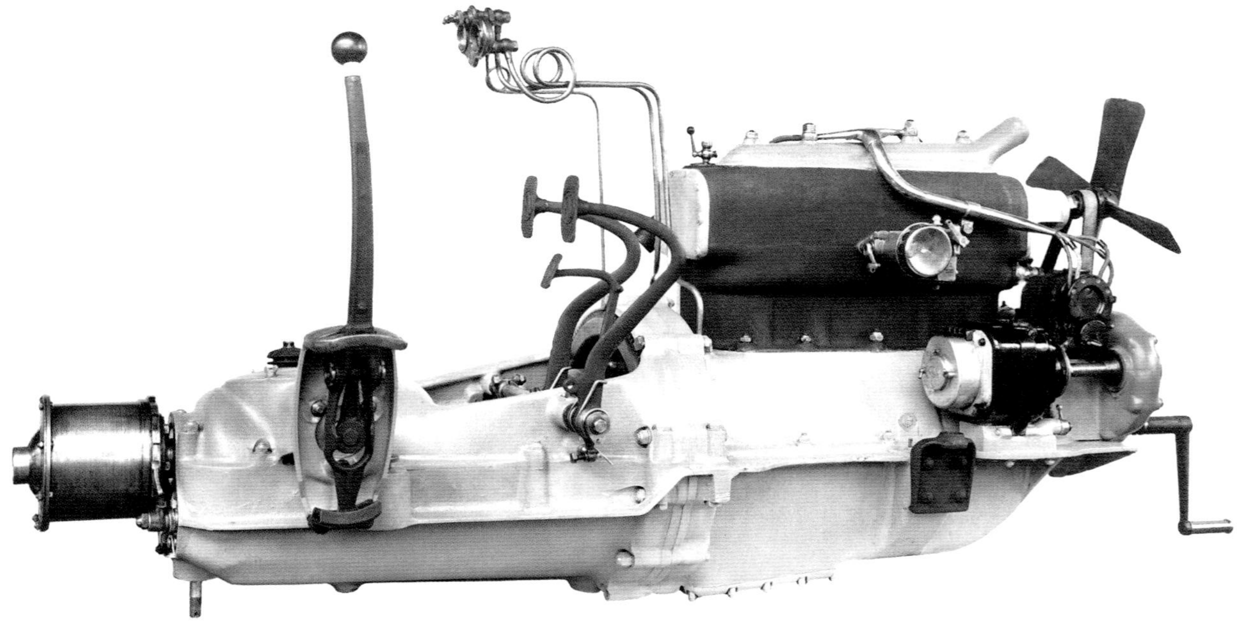

*Engine unit for the FN 3800 car.
Up : carburetor side.
Bottom : exhaust side.
(February 1922).*

The FN 1300 cars were Royal Cup winners in 1925 and 1926.

Brussels Car Show (1925).

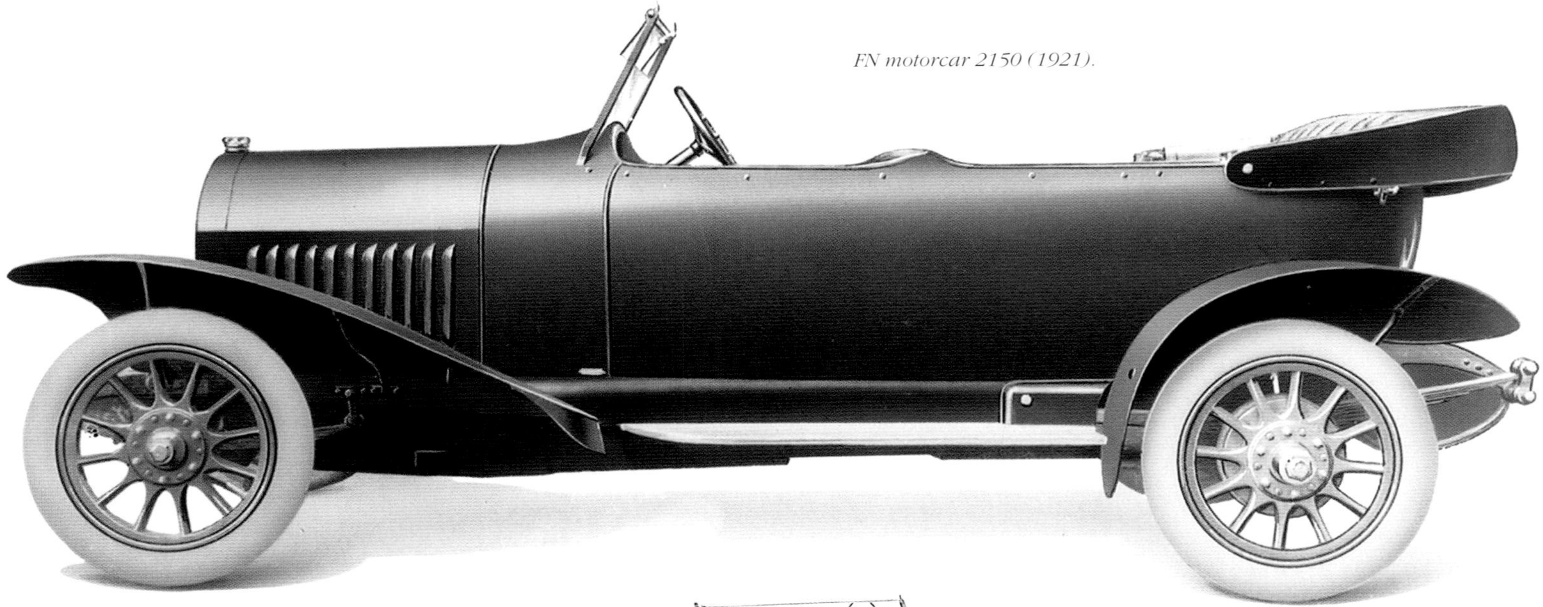

FN motorcar 2150 (1921).

Roadster car on an FN 11 hp chassis (1929).

FN 1300 'Princesse Astrid' (1927). This car was named after Princess Astrid of Sweden, wife of the future King of Belgium, Leopold III.

FN 1400 A sedan (1929).

Passenger compartment of the FN 1400A sedan (1929).

The race to the Cape inspired the Liège artist Auguste Mambour (1896-1968) to draw this remarkable poster.

In 1928, two FN 1,400 cc motorcars, driven by four Belgians, the journalist Roger Crouquet, the aviators Jacques Lamarche and Robert Fabry and

One of two FN cars returning from the Race to the Cape (1928).

the young diplomat Hubert Carton de Wiart, arrived at the Cape within 105 days after a memorable cross-country race over the African continent.

FN's motorcycles too proved their quality : in 1927, three motorcyclists successfully crossed the Sahara with their M70 motorcycles (350 cc), and the following year the same vehicles carried off the Rabat Grand Prix and won four world records for FN. The popularity of these motorcycles induced the company to drive the daily output up to a hundred units and to set up assembly plants abroad. In contrast with the motor vehicles, the sale of bicycles and their accessories declined steadily. Production had to be ceased in 1926.

In the meantime, on 28 March to be exact, Alexandre Galopin resigned in favor of Gustave Joassart after having been invited to take up a management function with the Société Générale de Belgique. As general manager Joassart pursued the policy of his predecessor. Seeing the development in the area of motorcar and motorcycle production, he ordered the construction on a site called "Pré-Madame" and situated to the Northeast of the existing installations, of a new factory where the production of motor vehicles and coachwork would be concentrated. Work started in 1928 and was completed in the summer of 1929.

At the same time, FN bought from the "Anciens Etablissements Nagant" large premises on the embankment of Coronmeuse and installed its repair and maintenance services there.

FN motorcycle M70 "Sahara" (1927).

The first Sahara crossing by motorcycle (1927). This was performed with FN-M70 350 cc motorcycles by the Belgian sportsman Joseph Weerens and two French army radio-telegraphists Gimié and his superior Bruneteau. The latter had been an assistant to Georges Estienne, the pioneer of trans-Saharan motor-travels. The 1927 7,000 kilometer trek linked Paris with Dakar, as the famous rally that had become very popular with the last years of the twentieth century.

World record for the 12 hour endurance motorcycle race, on an FN machine, by Lovinfosse and Flinterman (August 1926).

FN motorcycle M60 (1925).

Flinterman, Sbaiz and Lovinfosse establish a new world record for the 24 hour endurance motorcycle race on an FN 350 cc motorcycle, averaging 105.253 km/hr (1926).

Upholstery for the FN 1400A car. Cutting the fabric using an 'Eastman Universal Cutter', capable of cutting through up to 24 thicknesses of fabric (1929).

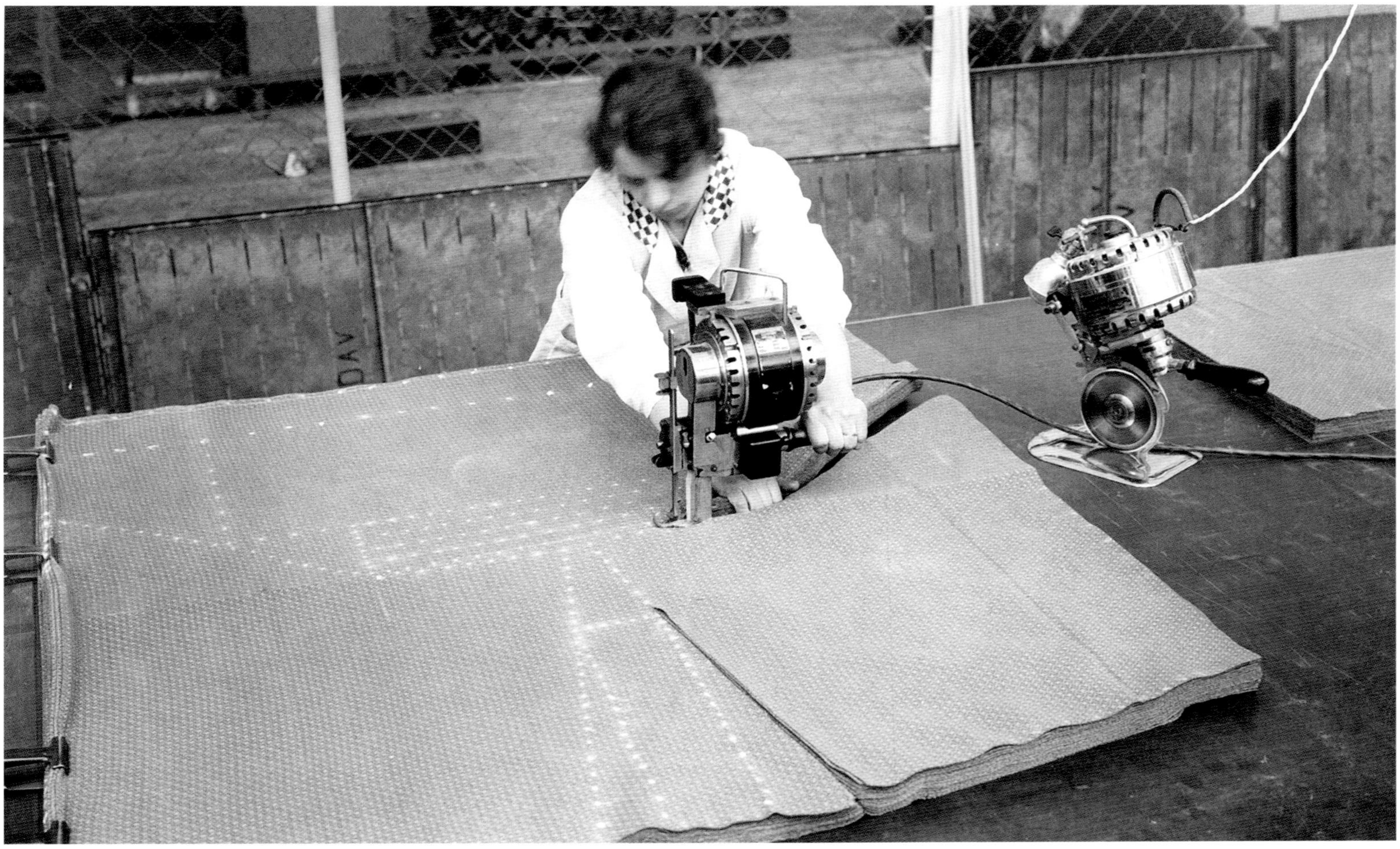

FN M70 motorcycle assembly line (1930).

Motorcycle assembly workshop (1921).

Riders on FN motorcycles at the Spring Circuit race (17 May 1921).

Liège Post Office sidecars, on FN M67 motorcycles (1929). In the center background, the former headquarters of the Liège Grand-Poste.

FN motorcycle 285T (1922).

Four-cylinder FN motorcycle (1924).

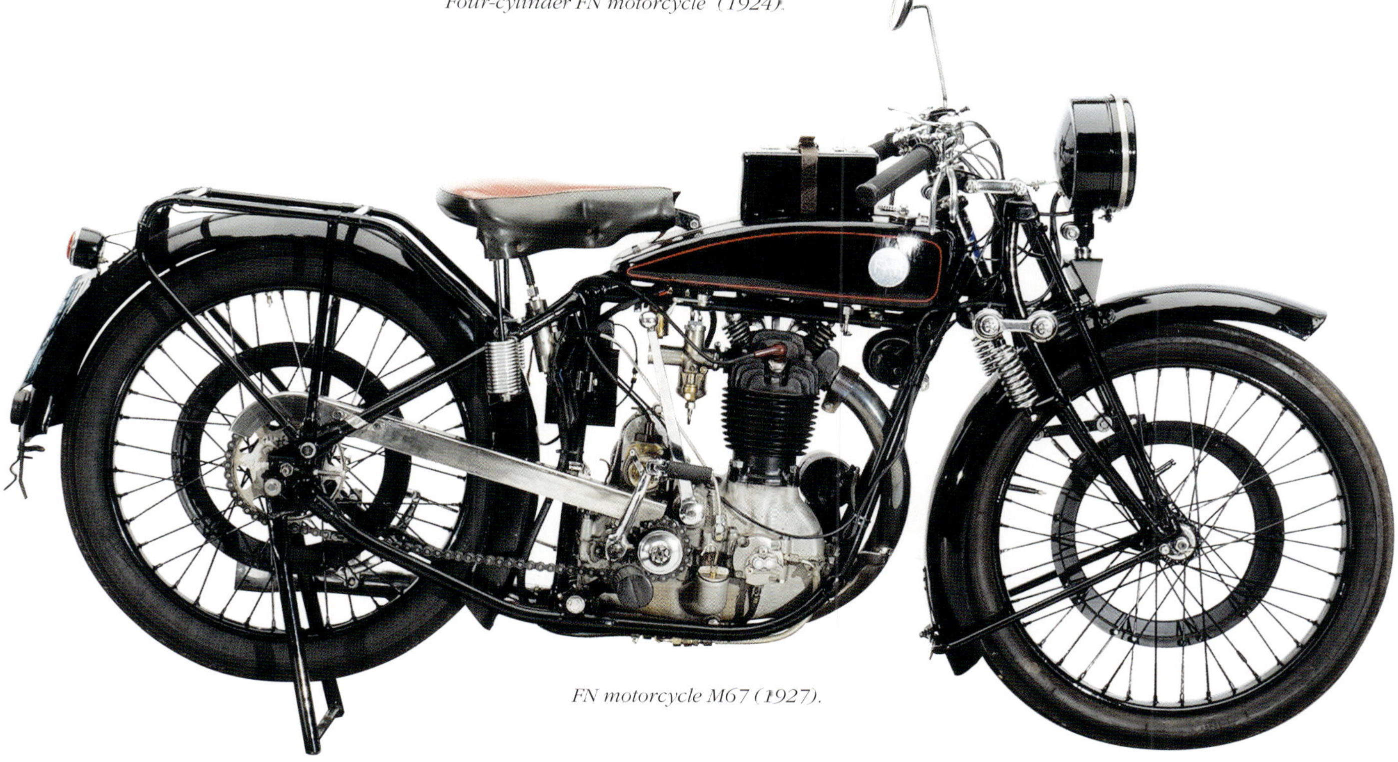

FN motorcycle M67 (1927).

Motorcycle poster by Albert Chavepeyer, a graphic designer from Châtelet (near Charleroi) and active in Liège (1928).

A large advertising poster for the FN M70 motorcycle, one of Herstal's best known products (Herstal Group Collection).

Manual packaging at the Cartoucherie Belge (1927). This company, located in Liège and taken over by FN, was the start-off for FN's production of hunting ammunition.

A policy of participation

The desire to enhance the importance of the company's civilian products and to ensure a regular outlet for them, in spite of the restrictions imposed on international trade, prompted FN to adopt an ambitious policy of participation and expansion.

During the war, Alexandre Galopin and Gustave Joassart had managed to escape to France where, together with a number of French industrialists, they had founded the Manufacture d'Armes de Paris (M.A.P.). The relations thus established lasted, and in 1921 FN took a share in the capital of that company at a time when it was being converted into a typewriter manufacturing plant. In the following years, M. A. P. was invited to assemble motorcycles for the French market. In 1919, on the other hand, FN participated in the incorporation of the company Société Anonyme Belge de Constructions Aéronautiques (S.A.B.C.A.) and in 1920, together with the Russo-Belgian cartridge factory, in the foundation of the Cartoucherie Belge, which was to produce sporting ammunition and artillery shells.

In 1922, the company took a holding interest in the Société des Machines à Sténographier Grandjean, and in 1924 in the Cartoucherie Française, which sold its hunting and sporting equipment in France, and in 1925 in the company Achille Fusi in Milan, which had acted as its Italian agency since the beginning of the century. In 1928, FN set up a branch in Sweden, the FN Aktiebolaget, with a view to selling its motor vehicles in that country, and in the following year the company built a factory in Aachen for the assembly of motorcycles for the German market. In addition, a number of licenses extended the company's range of civilian products. From 1920 to 1923, FN studied the Ellis calculating machines, which in the end they never produced, as the license was granted to other companies.

Grandjean shorthand typewriter. For a short period, between 1922 and 1924, FN manufactured this shorthand typewriter in Herstal under a license contract from the French inventor, Marc Granjean.

M.A.P. typewriter (1922). The Manufacture d'Armes de Paris, which had been founded during the Great War in co-operation with the directors of FN was subsequently converted into a factory of civilian products. This company belonged to FN from 1921 until 1936. It manufactured this type of typewriter, among other products.

Progress in the manufacturing of arms and cartridges

It is clear that FN was moving rashly, perhaps even too rashly, into activities which were foreign to its proper domain. If for instance we look at the production figures for the post war years, we can see clearly that apart from the model 10 pistols, FN only produced small quantities of hunting and defense weapons. Two almost simultaneous events, G. Joassart's accession to the post of General Manager and the orders placed by the new kingdom of Yugoslavia, induced FN to resume its former activity.

Indeed, on 28 February 1923, the government of the Kingdom of Serbs, Croats and Slovenes placed an order with the company for 60,000 automatic 9 mm short pistols of a modified model, the 10/22, along with 6,000,000 cartridges.

An even more important contract was signed on 14 July that year. It concerned the delivery by FN of 50,000 7.9 mm Mauser rifles and of 50,000,000 cartridges of the same caliber. It also provided for the installation by FN of an arsenal in Kragujevac which could produce 200 Mauser rifles and 200,000 cartridges every day. This contract, which was worth 53,750,000 francs, was followed on 16 February 1926 by an order for 40,000 Mauser rifles and 60,000,000 cartridges.

The execution of all these various orders – to which were added orders for the supply of rifles to Brazil (1923-1924) and Mexico (1926-1927) – progressively led FN to renew its installations and working methods in the arms manufacturing section. The machine-tools which had been set up in Herstal during and after the war had completed the equipment of most of the workshops. In addition, the use of cam fixtures and high-speed steel became general. The measurements of the parts were subjected to stricter rules, thus making filing for the most part unnecessary. Henceforth calibration was carried out as and when the parts were machine-finished, while at the same time the central inspection department was closed down. These improvements allowed in their turn for complete interchangeability of the parts, without a preliminary selection using operating gauges being necessary. Similar progress was made in the cartridge factory. The sequence of operations was simplified and rationalized.

The use of chromium-plated chasing-chisels became widespread. The power supply of the machines and the inspection of the dimension figures became progressively automatic, while the application of electric tempering – whether partially or completely – allowed a more effective action on the structure of the metallic elements of the cartridge.
The laboratory also developed a machine equipped with a double perforator, with which the solidity of both the inner and outer wall of the shells could be measured.
The ballistics section was furnished with a Mann cannon tester.

Auto 5 16 caliber shotgun. Leopard hunting (1924).

10/22 9 mm caliber pistol in "Modern Egyptian style" (1930).

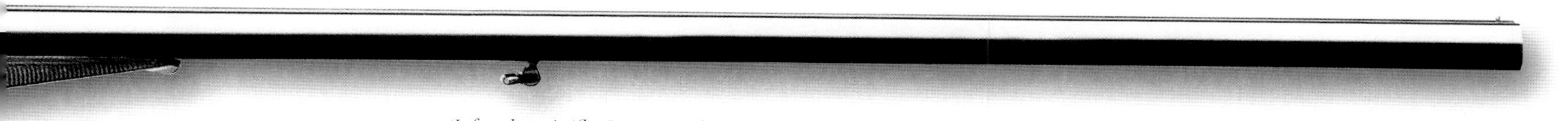

'Lefaucheux' rifle. Inaccurately called 'Lefaucheux' because its external hammers gave it the appearance of a pinfire shotgun, this central percussion arm was produced by FN until 1927, both as a finished product and as sets of parts supplied to other armaments manufacturers in the region (Herstal Group Collection).

Tableau showing duck hunting on the Korean coast with the Browning Auto 5 (1928).

Overall view of the arsenal at Kragujevac (1932). For the procurement of its military equipment, in which FN took a large share in the 1920s, the young kingdom of the Serbs, Croats and Slovenes (to become Yugoslavia) entrusted the company with the installation of an arsenal in Serbia. It was one of FN's first engineering contracts, soon to be followed by many others, be they military or civilian, right up to the present day.

Plan with measurements of the breech-frame of the Mauser rifle, 1924 model. During the twenties, FN became the most important manufacturer throughout the world of Mauser rifles and carbines. This supremacy necessitated a considerable modernization of the equipment and production methods, which was carried out forthwith.

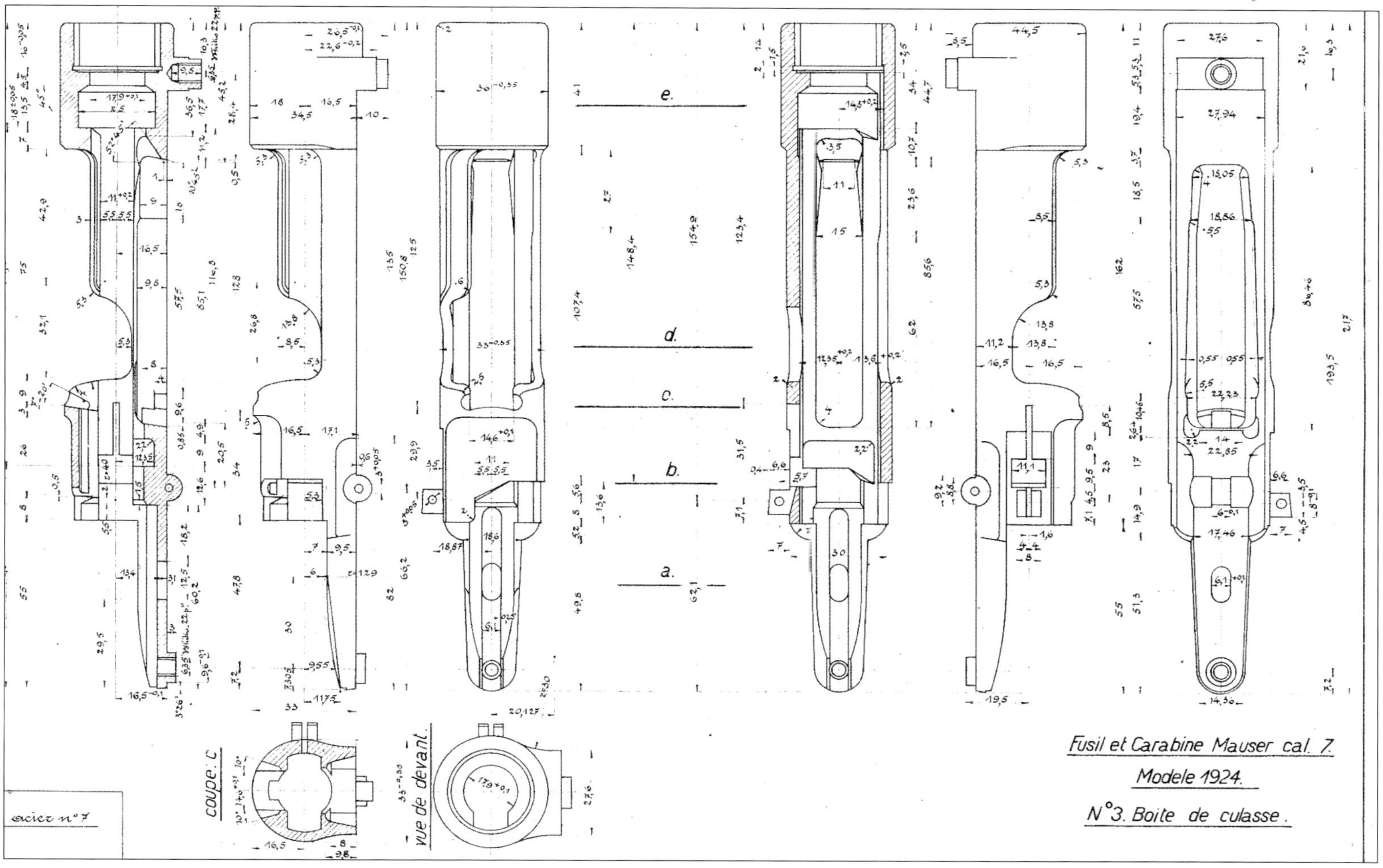

Assembly of the Browning Auto 5 shotgun (1921).

The return of Mister Browning

In the meantime, John M. Browning had come back to Herstal where in collaboration with staff from FN he made more studies and inventions. Among those were a .22 caliber "trombone" rifle, a high-efficiency automatic pistol and a hunting shotgun with over-and-under barrels – the future renowned B25 – which he was still perfecting when he suddenly died in 1926, in his office of the company with which he had collaborated so fruitfully over the past thirty years.

The three types of .22 cartridge can be mixed in the tubular magazine. They are counterbored and fired with no jamming.

.22 'Trombone' rifle.
Its mechanism enables it to fire Short, Long or Long Rifle .22 cartridges, with a loading capacity of 15, 12 or 11 cartridges.

Recent, modernized example, inlaid with gold (prairie dogs), of this firearm invented by J. M. Browning and produced by FN since 1922, principally for the European market.

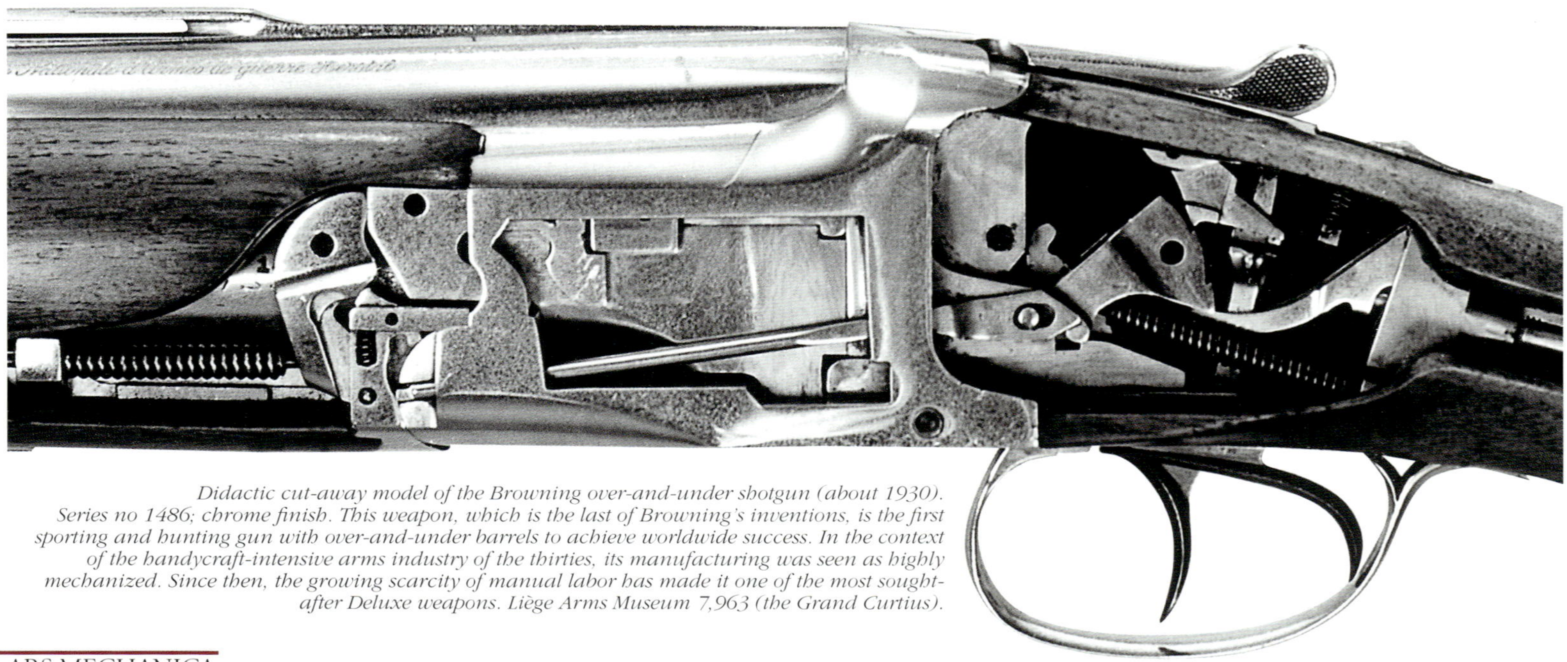

Didactic cut-away model of the Browning over-and-under shotgun (about 1930). Series no 1486; chrome finish. This weapon, which is the last of Browning's inventions, is the first sporting and hunting gun with over-and-under barrels to achieve worldwide success. In the context of the handycraft-intensive arms industry of the thirties, its manufacturing was seen as highly mechanized. Since then, the growing scarcity of manual labor has made it one of the most sought-after Deluxe weapons. Liège Arms Museum 7,963 (the Grand Curtius).

Browning over-and-under B25. The engraver has taken advantage of the reflection from the ducks on the surface of the water playing with the colors. Shielded from reeds among water lilies, birds and plants are inlaid with gold, silver, palladium and copper. Side plates represent a green winged teal and a common golden eye. On the bottom there is a mallard and on the trigger guard a duck pilet, even some mosquitoes participate in the quietude of moment.

At the time, the inventor was perfecting the prototype for his "over-and-under" model which he had designed three years earlier. He wanted both to improve its operation and to adapt it – as far as the techniques of the time would allow – to mechanical production at a competitive price. Its completion was to be the task of his son Val Allen Browning (1895-1995), whose eminent role both as an arms technician, having inherited his father's creative genius, and as a manager of the family business should not be underestimated.

The fruit of the combined efforts of father and son – and of the Herstal technicians –, the over-and-under Browning shotgun was to become a twentieth century world-class benchmark in the manufacturing of hunting and sports arms.

Almost 400,000 of them were produced in Belgium over a 45 year period, 65% of which were destined for the North American market. Considered at its beginnings as a highly mechanized product (although still highly labor-intensive), the scarcity of skilled craftsmanship later made it a luxury item reserved for special models which went under the name Browning B25.

Frequently envied, sometimes imitated by a number of competitors, the Browning "over-and-under", followed more recently by other models designed in response to inflationary cost price pressure, contributed more than any other product to FN's establishment in the high-class civilian arms market, especially after the Second World War.

John Moses Browning's death, however hard it was felt – work was interrupted as a sign of mourning and all the staff paid their last respects to the inventor's mortal remains, which for the occasion were placed in the board room – did not put an end to the creative activities of the company.

Its research department developed a sporting shotgun, one version of which was an Anson type model and another was fitted with side-locks.

FN hunting shotgun, Anson system.
The double-barrelled shotgun, Anson system, is one of the rare civilian arms formerly produced by FN not invented by Browning. Manufactured at Herstal since the turn of the century until 1972, in three successive models, this arm is a luxury edition, chased by the master engraver Louis Vrancken, around 1962.

Hammerless de Luxe sidelock rifle. Detailed English bouquet engraving, Birmingham type, tempered and marbled (1930).

The body of John Moses Browning, lying in state in the chapel of rest adapted in the FN Board Room (1926).

For the Herstal group, history will recall two distinct phases in John M. Browning's inventive activity. The first, the longest and most prolific, lasted from 1897 until 1914; the second only lasted seven years, from his return to Herstal in 1919 until his death in 1926.
The first period allowed FN to establish itself in a masterly fashion in the civilian arms sector, in particular in the sectors of semi-automatic shoulder and handheld arms. It also opened up international horizons which were closed to it in the military sector.
The second, whilst consolidating FN's position in the above trend, also gave it access to the fields of automatic military arms and the then-innovative over-and-under hunting shotguns.

But that was not all. Prior to the First World War, the Herstal company, to some degree obsessed by the immense stature of the American inventor, had generally been content to manufacture the products he offered them, only too happy to compensate in this way for the reduced profit from its German shareholder's quotas. Freed from this tutelage, FN saw Browning return in 1919 with new patents which had been taken out during the conflict and which were valid for European commercialization : for the automatic rifle and the .50 caliber machine gun. By launching the manufacturing of these arms, the company crossed a technological threshold which enabled it, in one go, to perfect its own invention and development services. It is true that these services worked in close collaboration with Browning during the last years of his life, but they progressively became independent of the master to the point of refining, even finishing, his inventions and making them viable.

And so were created, at Herstal, the 10/22 pistol for Serbia, the HP pistol (based on the 1923 American prototype), the 'over-and-under' – which materialized the last John Moses firearm creation completed by his son Val Allen - and later the 'Baby' pistol. The automatic rifle with firing rate reducer and quick-change barrel, the .30 caliber machine gun version for aircraft and the .50 versions adaptable for vehicles and various mountings, all saw the light of day in the same conditions. At the same time the range of ammunition was diversified.

For promotional reasons the contribution of the Herstal technicians to these advances were minimized by the company's management in favour of the prestigious Browning name. This is one of the characteristic features of the Liège arms industry, whose own brand image was often sacrificed in the name of immediate commercial imperatives. However, with the passage of time we are able in FN's case to render unto Caesar the things that are Caesar's.

The transportation of the automatic rifle and its magazines is taking place here in a rather unsoldierly fashion! (1926).

During the 1914-1918 war, John M. Browning, who was at that time in the United States, had invented two weapons: a gas operated automatic rifle and a water-cooled machine gun with short barrel recoil. They were both used by the American Army and at the French front.
The Colt company, who manufactured them in Hartford (Connecticut), had obtained the worldwide exploitation rights for these arms. In 1924, the Polish government expressed its interest in the first of these arms, the automatic rifle, and decided to buy several thousand of them and to have others manufactured locally. For that purpose, it contacted Colt and its agent FN, which had acquired a very good reputation with the installation of the arsenal in Kragujevac.
The two companies swiftly reached an agreement and on 26 December 1924 they signed with the Warsaw company K. Rudzki i Ska a license contract and a technical assistance agreement, which would come into effect once the Polish government had placed an order. The latter waited until August 1927 to make a decision. But the conditions it demanded were of such draconian severity that Colt, alarmed, equivocated and finally withdrew on 27 October 1927.

Even though FN had no experience in manufacturing such a complicated weapon as was the Browning automatic rifle, they immediately accepted to take up Colt's commitments and, in consultation with that company, signed on 10 December 1927 the contract which committed them to supplying to the Polish government the 10,000 automatic rifles which they had asked for.

Looking back, this audacious decision, which was due mainly to Gustave Joassart and his Technical Director F. Nepper, appears to have had a far-reaching effect. In the eyes of the military authorities, it placed FN among the largest companies in Europe and the United States, and we shall see how much benefit this reputation would bring them in the course of the following years. In addition, the efforts which a swift execution of that order had required from the company were to have a lasting and fruitful effect. More than 600 new machines were bought, the engineers in the factory became familiarized with the more delicate problems of measurements and acquired an extensive knowledge of automatic infantry weapons.

Browning automatic rifle. Firing exercise by a unit of the United States Marines (1926).

A teaching model of the Browning automatic rifle, made for export to Poland. It bears the series n° 10. Liège Arms Museum - 12,822 (the Grand Curtius).

The branch in Bruges

In the meantime, FN, in response to a request by the Belgian government who was anxious about Herstal's proximity to the German border, had built a factory near Bruges for the manufacturing of small and medium caliber military ammunition and artillery shells. At the same time, the company bought up two firms who were long-established specialists in the manufacturing of ammunition for hunting: the Cartoucherie Belge, in which FN had already acquired a capital share, and Etablissements Bachmann (3 May 1929).

The equipment thus acquired was installed in the workshops of the new factory, which was later to be complemented by a section for the production of primers. The Bruges branch, which was headed by an eminent explosives expert, Colonel Quintin, enjoyed an almost independent status and would remain operative until 1940.

The extension of the cartridge manufacturing section led as an indirect consequence to the setting up of a ballistics laboratory in Herstal and of a shooting range which was built initially in Houthalen and subsequently transferred the short distance to Zutendaal in the heart of Flemish Limburg shortly before the outbreak of the Second World War.

Black Friday

A desire to expand in all directions and with tremendous vigor constituted, as the reader will have undoubtedly noticed, the hallmark of the ten years which we have just been discussing. This ambitious policy, which had been undertaken under very adverse economic conditions at a time when currencies were being devalued very rapidly, would soon create serious financial problems. From 15 February 1921 to 29 June 1929, the capital had to be increased three times in a row from 6,600,000 francs to 105,000,000 francs, while the number of shares rose from 13,200 to 210,000. In spite of these measures and the massive issue of bonds – 20,000,000 in 1924, 40,000,000 in 1929 – the company sank deeper and deeper into debt with its bankers. The development of the factory, however brilliant it might have appeared on the outside, could only be fragile, in view of the cost at which this progress had been achieved.

On 18 October 1929, on Black Friday, memories of which still continue to haunt economists, the New York stock market crashed. World recession set in at a moment when FN was financially most vulnerable.

Publicity for the 'Metal Star' (1936).
The Metal Star cartridge was one of the most advanced of its time. The use of a metal tube eliminated the risks of deformation and swelling through humidity, while the star-shaped crimping ensured the regular dispersion of the shot.

The Bruges plant (1929). In 1929, FN acquired a factory for hunting cartridges by taking over the Cartoucherie Belge in Liège and the Etablissements Bachmann in Brussels. The production machinery was soon transferred to the new factory in Bruges. After the war, production was first concentrated in Herstal, then distributed worldwide.

The Pré-Madame factory under construction. The style is decidedly modernist, with a reinforced concrete frame (1928-1929).

First return to the core business

Illustration of the promotional leaflet for the new FN 11 hp car (1930), by the Liège artist and painter Milo Martinet (1904-1988).
Originally a poster artist, he worked under contract for FN from 1929 until 1934, and on an occasional basis thereafter.

As far as the company was concerned, the crisis lasted until 1935. All sections had been hit, some of them so seriously that they never recovered and were closed.

The motorcar won, and lost

It was mainly the motorcar division which was most affected. After FN had become practically restricted to the domestic market after the neighboring markets were closed off, in February 1935 it even lost the protection of import restrictions which had allowed it to face the competition from importers of foreign cars to a certain degree.

In view of this situation, the board decided on 17 September 1935 to abandon the production of motorcars and to direct the factory towards the construction of special vehicles.

This decision was as necessary as it was painful. The quality of the vehicles manufactured by FN - especially the two latest models, the 42 Prince Baudouin (1933) and the 42A Prince Albert (1934) - had in fact remained high and the successes in sporting competitions had never been so brilliant.
The following prize list speaks for itself : in 1930, an 11 hp car (1929 model) wins Liège-Madrid-Liège. In 1931, three 8-cylinder cars (1930 model) win the Alps Cup.

A car of the same type covers the distance between Gibraltar and Brussels in a very short time. In 1932, three 8-cylinder cars win the Royal Cup at the Twenty-four Hours of Francorchamps. In 1933, similar victory for three cars of the 42 Prince Baudouin type (1933 model), and an 8-cylinder car wins Liège-Rome-Liège.

The patrol of the International Exhibition in Liège (1930) with their FN side-car motorcycle. In the background, the "Cycles, arms and motorcycles" exhibition hall, housing a collection of widely acclaimed specialities of the Liège region, in which FN had a large share.

Assembly lines for the M70 motorcycle, car chassis and the car upholstery workshop (1930).

With over 300 victories riding Belgian motorcycles to his name, René Milhoux was one of the greatest motorcycle racers of the twentieth century. He died in 2003.
He is pictured here with the world record winning machine at Bonheiden near Malines (Belgium). A "posed" photograph taken in front of FN's management office building on 17 April 1934.

FN motorcycle M67 D 500 cc (1930).

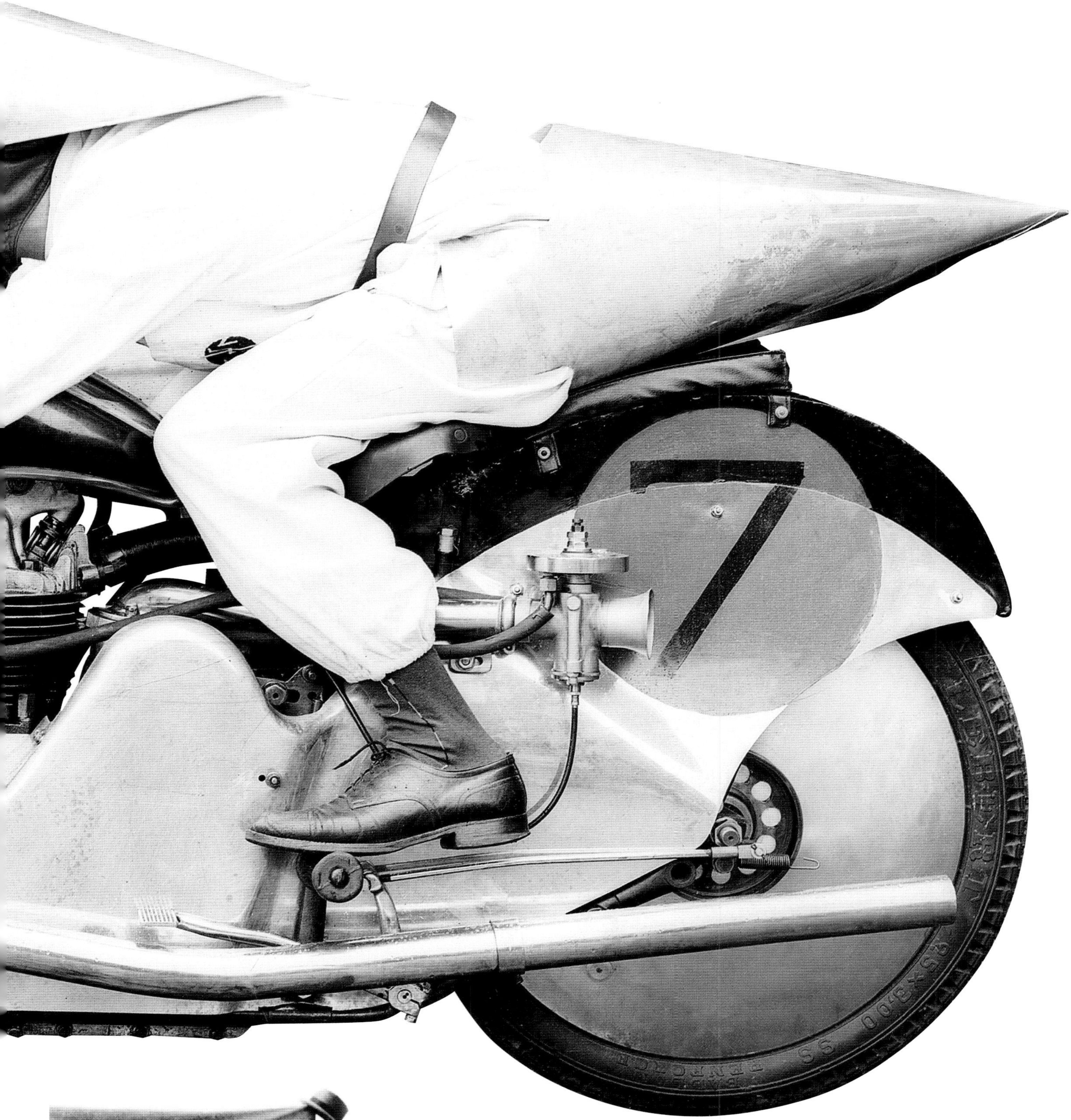

In Bonheiden on 22 April 1934, the champion René Milhoux broke the world speed record on an FN motorbike reaching 244.019 km/hr. Both the motorcycles and motorcars of FN won many brilliant international successes in the thirties.

FN 500 cc 86 Super Sport

(Herstal Group Collection)

Engine : single cylinder engine block, 4-stroke. Bore : 80mm. Stroke : 99mm.
Cylinder capacity : 497cc.
Head valves controlled by protected rocker arms, hand adjustable. Flat bottomed piston made of special aluminium alloy.
Lighting : by Bosch dynamo-breaker.
Electrical equipment : Bosch 50 watts comprising : headlight with three intensity modes, rear light, control light, horn, ignition key, battery.
Odometer : set into the headlight.
Carburettor : automatic, operated by a handle and a twist grip located on the right of the handlebar.
Clutch : multiple disc in oil bath, operated by a lever on the left of the handlebar.
Gearbox : 4 speeds operated by a foot pedal.
Lubrification : completely automatic engine lubrification, performed by double pump bathed in oil.
Transmission : by helicoidal gears from the engine to the gearbox (engine block). By roller chain from the gearbox to the rear wheel with shock absorber at the hub.
Foot rest : rubber blocks.
Fuel tank : at the seat, chromed with knee grips and large gauge, rapid-closing fuel cap. Capacity : 4 gallons.
Tyres : with bead wires 27 x 3.25.
Rims : chromed with hollow base.
Mudguard : enamelled.
The rear mudguard is in 2 sections, very easily removable.
Frame : cradle-type A-frame.

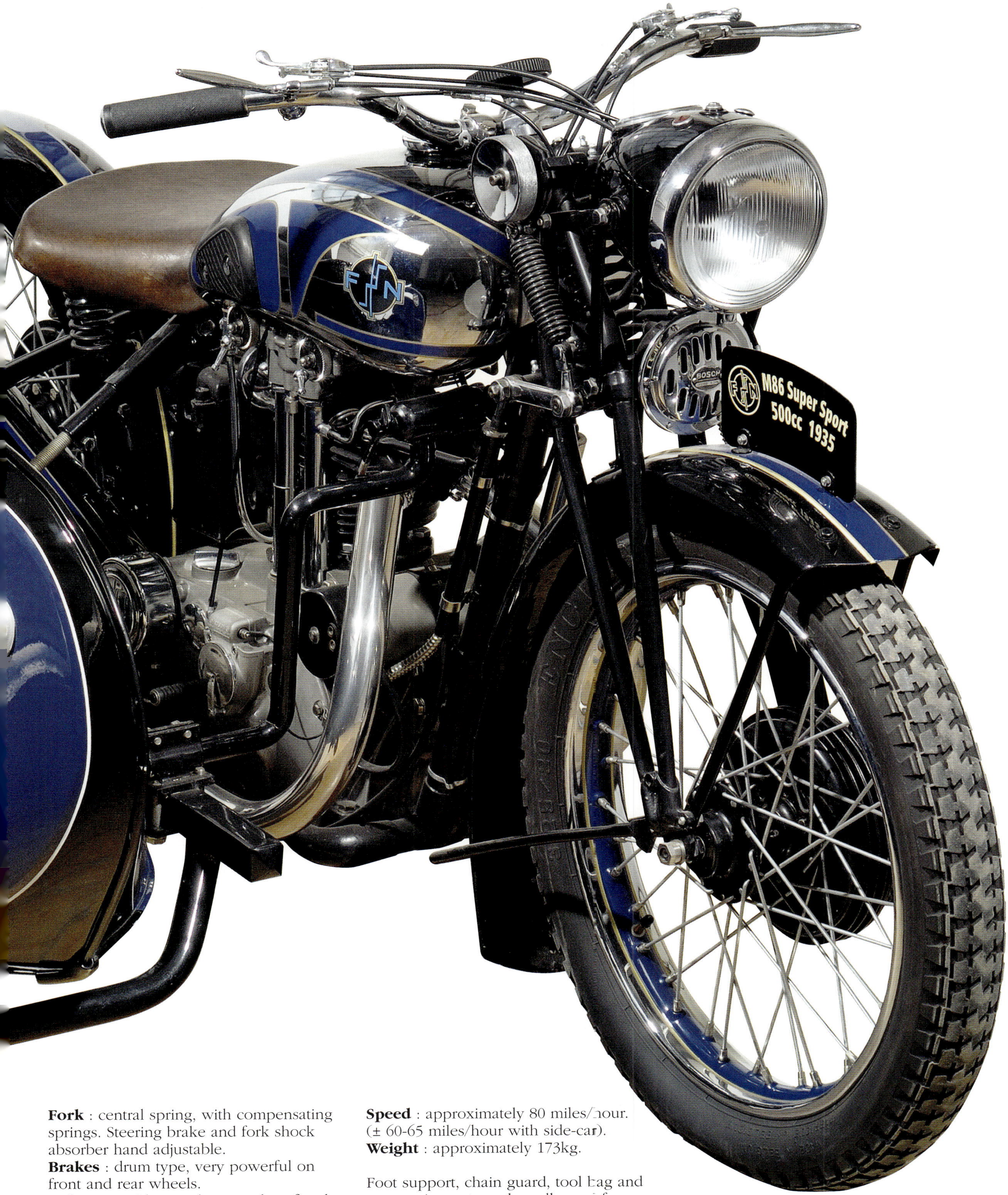

Fork : central spring, with compensating springs. Steering brake and fork shock absorber hand adjustable.
Brakes : drum type, very powerful on front and rear wheels.
Exhaust : with two chrome tubes, fitted with chrome silencers.
Fuel consumption : petrol 1.3 gallon and oil approximately 100 cc per 60 miles.

Speed : approximately 80 miles/hour.
(± 60-65 miles/hour with side-car).
Weight : approximately 173kg.

Foot support, chain guard, tool bag and accessories: extremely well cared for.

(Technical data taken from the 1935 FN Motorcycle catalogue.)

As the Board of Directors had announced at the general meeting of 24 October 1935, the unilateral suspension of import restrictions had been dictated by the "erroneous opinion that henceforth there was only room in Belgium for a motor car assembly industry and no longer for a motor car manufacturing industry".

As we have seen, the management had decided to direct the factory towards the manufacturing of special vehicles. Since 1932, FN manufactured lorries of two and a half tons, one-ton delivery vans, and since 1933 trolleybuses which were used in Liège and Antwerp, some of which were still in service after the Second World War after having clocked more than a million kilometers.

The manufacturing of motorcycles continued, with sales figures which, in comparison with those for motorcars, appeared quite satisfactory. All the models which left the factory at that time – the M90 500 with side-valves from 1931, the M200 with two-stroke engine and the M86 with overhead valves – were of an excellent quality and won the company a great many successes.

The 500 cc type 86 motorcycles carried off an unbroken series of victories, of these the F.I.C.M. European Grand Prix of 1934 in Assen (Holland) and the French Grand Prix of 1935 in Montlhéry are the most famous. By the end of 1935, the FN racing motorcycles had won eighty-one world records.

These successes certainly bore witness to the know-how of the engineers and technicians of FN. This does not detract from the fact that in a Europe which had fallen prey to protectionism, the factory was no longer able to rely on the manufacturing of motor vehicles for its livelihood.

Pol Demeuter, 1934 European Champion, killed in an accident at Chemnitz on 1 July 1934.

F N

L'IMPRESSIONNANTE LISTE DES

81 RECORDS DU MONDE

DETENUS PAR LES MOTOS F. N.

15-10-35

Catégorie	Distance	Vitesse en km. H.	Catégorie	Distance	Vitesse en Km. H.
350 cc.	5 kms	179,60	750 cc.	10 miles	184,93
solo	10 kms	168,77	solo	100 kms	178,86
id.	5 miles	179,35	id.	500 kms	167,21
id.	10 miles	171,98	id.	1000 kms	155,05
id.	50 kms	175,71	id.	500 miles	155,67
id.	100 kms	169,40	id.	2 heures	169,05
id	500 kms	141,92	id.	3 heures	167,21
id.	1000 kms	140,04	id.	4 heures	155,91
id.	50 miles	169,18	id.	5 heures	155,95
id.	500 miles	141,12	id.	6 heures	155,66
id.	1000 miles	138.53	id.	7 heures	147,79
id	4 heures	142,06	id.	8 heures	140,27
id.	5 heures	141,86	id.	9 heures	140,02
id	6 heures	141,06	id.	10 heures	139,76
id.	7 heures	139,97	id	11 heures	138,97
id.	8 heures	140,27	id.	12 heures	138,41
id.	9 heures	140,02			
id.	10 heures	139,76	1000 cc.	500 kms	167,21
id	11 heures	138,97	solo	1000 kms	155,05
id.	12 heures	138,41	id.	500 miles	155,67
			id.	2 heures	169,05
500 cc.	10 kms	181,91	id.	3 heures	167,21
solo	5 miles	192,61	id.	4 heures	155,91
id.	10 miles	184,93	id.	5 heures	155,95
id.	100 kms	178,86	id.	6 heures	155,66
id.	500 kms	167,21	id.	7 heures	147,79
id.	1000 kms	155,05	id.	8 heures	140,27
id.	500 miles	155,67	id.	9 heures	140,02
id.	1000 miles	138,53	id.	10 heures	139,76
id.	2 heures	169,05	id.	11 heures	138,97
id.	3 heures	167,21	id.	12 heures	138,41
id.	4 heures	155,91			
id.	5 heures	155,95	350 cc.	5 kms	150,75
id.	6 heures	155,66	side-car	10 kms	142,52
id.	7 heures	147,79	id.	5 miles	151,03
id.	8 heures	140,27	id.	10 miles	145,50
id.	9 heures	140,02	id.	50 kms	147,78
id.	10 heures	139,76	id.	50 miles	141,86
id.	11 heures	138,97			
id.	12 heures	138,41	600 cc.	5 kms	166,67
			side-car	10 kms	155,04
750 cc.	10 kms	181,91	id.	5 miles	166,10
solo	5 miles	192,61	id.	10 miles	157,87

F. N. HERSTAL

FN motorcycles catalog (1936).

FN motorcycle M70 Sahara 350 cc (1930).

FN motorcycle M86 500 cc (1934).

FN motorcycle M90 500 cc (1930)

FN motorcycles catalog (1936).

1930s' posters for FN lorries, cars and motorcycles. (Herstal Group collection).

Advertisement picture of the FN M70 de Luxe, 1931 model. An idealistic and storybook picture of the motorcycle, in those days still considered as a utility vehicle. FN also had assembly plants in Paris (France) and Aachen (Germany), in addition to its installations in Herstal, which in 1930 were transferred to the new factory in Pré-Madame. The commercial network was equally extensive.

The motorcycle of the Paris-Saigon Race (1931). A young woman of 21, Justine Tibesar from Arlon (Belgium), finished a 22,000 km race alone, on an FN M70 Touring Standard 350 cc motorcycle. She is seen here after returning, welcomed by the managers of the Motorcycle Division, in front of the management offices of FN.

FN 500 cc Bécassine motorcycle - engine block (1931).

FN motorcycle Bécassine 500 cc (1931).

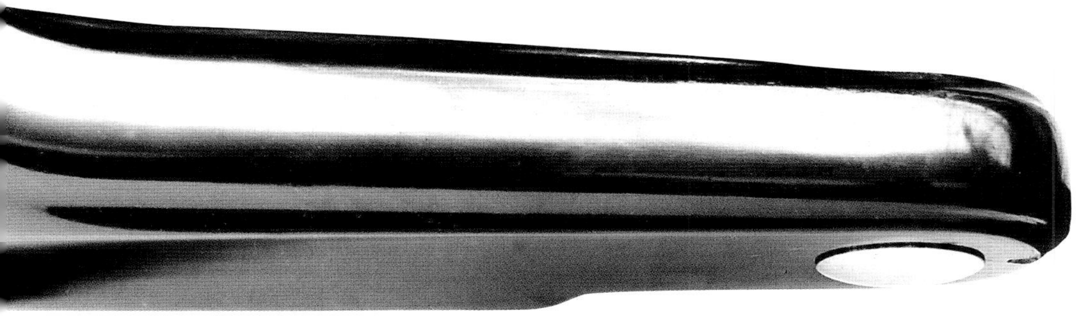

During his brief incumbency at the head of FN's Racing section (1929-1931), the British engineer and competition rider Dougal Marchant designed a 500 cc motorcycle prototype with modernist lines, the 'Bécassine'. This was never produced.

FN 500 cc Bécassine motorcycle. Frontal view (1931).

185216

FN eight cylinder sedan (1931). A photograph retouched by Albert Chavepeyer and laid out by M-lo Martinet in the publicity brochure for the FN eight cylinder.

Preceding double page and below : FN eight cylinder, seven seater sedan, bodywork by Mathieu Walsh, Liège (1931).

Plan for a highly innovative FN car: the Prince Albert 'Surprofilée' (Super-streamlined) (1934). At this time, car design was beginning to take the laws of aerodynamics into account.
Its designer was the engineer Philippe Questienne (1902-1997), then Technical Secretary of the car design office.
A prototype would be manufactured by the body builder Pritchard and Demollin, of Liège. At least one car, of a pre-production series of ten, was then made by another local body builder, Laurent Colson. *See page 162.*

After the 'eight cylinders' in 1932, three FN Baudouin cars won the Royal Cup at the Francorchamps 24 hour race in 1933.

Chassis and 'Airflow' bodywork of the FN 42 automobile, called the 'Prince Albert' (1934).

FN 42 car, called the Prince Albert (1934). Prototype by Pritchard and Demollin of Liège.

Samples of the fabrics used to upholster the seats and interior of the car.

Artist's impression, for publicity purposes, of the FN eight cylinder car (around 1931).

FN 42A car, called the Prince Albert (1934). This was therefore named after the King's youngest son, who later became the current King of Belgium, Albert II. This was the last automobile designed by FN. The limited size of the national market, the world recession and the removal of customs protection forced this sad decision on the Herstal company.

FN 42 standard Roadster car. Bodywork by Reiterman in Liège (1930).

8-cylinder FN motorcar model 1931.

Meeting of the automobile representatives on 25 October 1930.

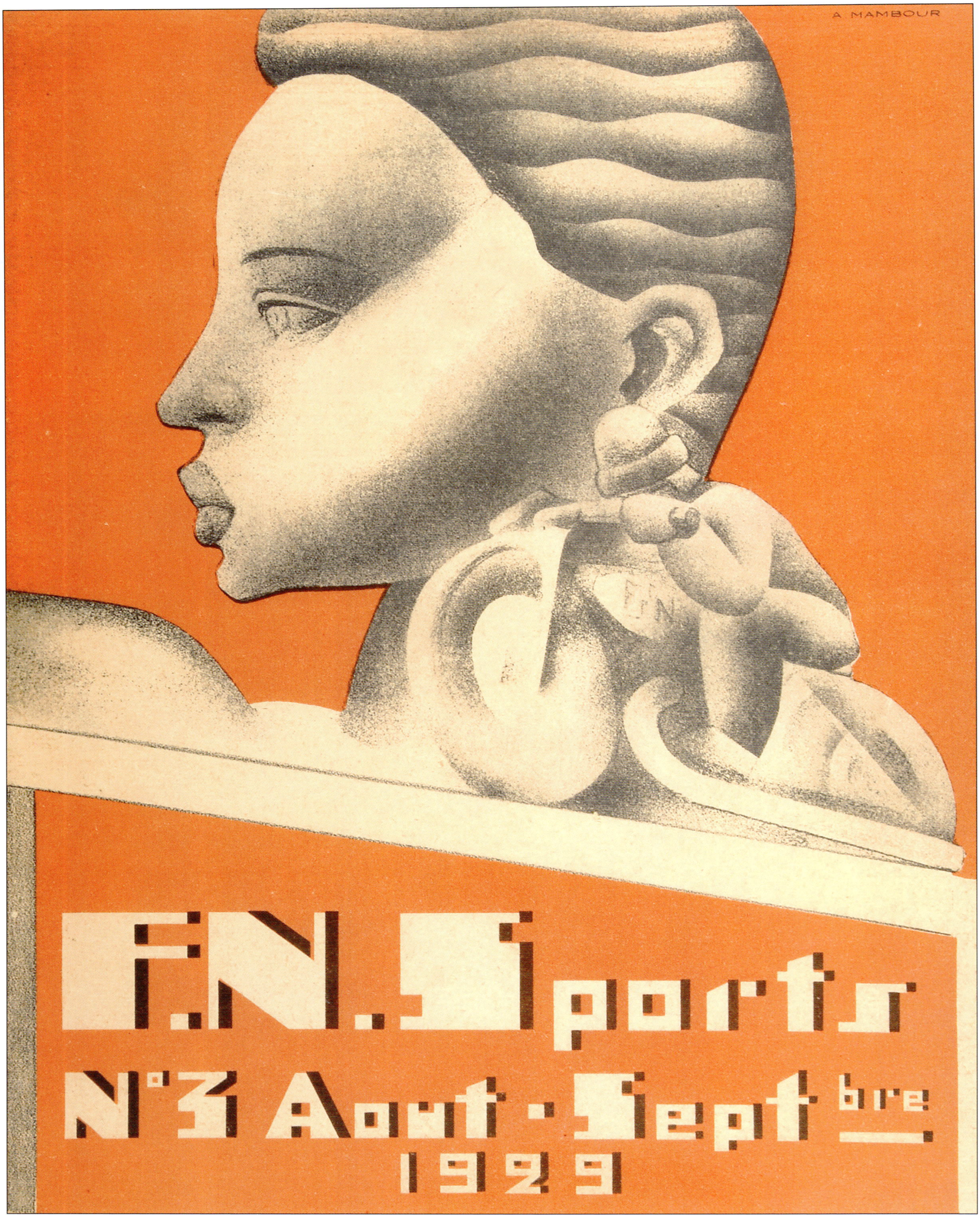
A MAMBOUR
F.N. Sports
N°3 Aout - Sept bre
1929

Covers of the magazine "FN Sports" (1930-1931). During the inter-war years, FN's 'Sports' department (vehicles) published this magazine on motorcars, motorcycles, aviation and motor sports. The presentation was the work of a number of excellent graphic designers. Covers n° 3 and 4 illustrated on this double page, are by Auguste Mambour (Collection A. Vanderliaden, USA).

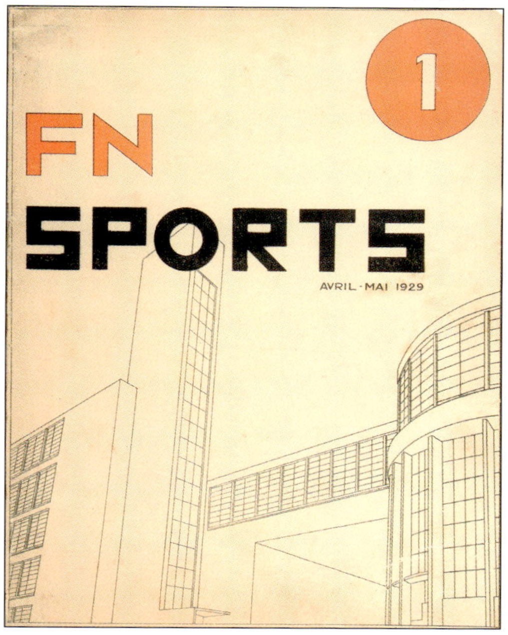

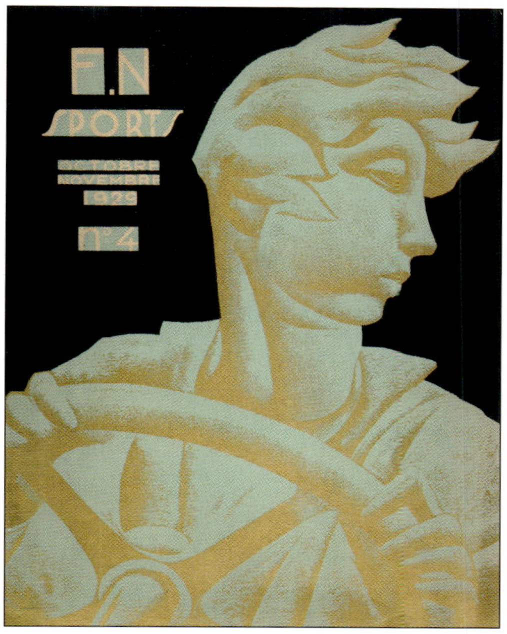

A demand for arms and ammunition

With the energetic support of René Laloux, son of Georges Laloux and Deputy Manager since 1 April 1932, Gustave Joassart, whose acumen nowadays appears to have been prophetic, was to concentrate from that moment all his energies on the development of the arms and ammunition business.

The research department, now headed by Dieudonné Saive, developed new models. It fitted the Browning automatic rifle first with a firing rate reducer (1930) and then with a quick-change barrel (1932). Subsequently Browning's small and medium caliber infantry machine guns were modified to increase the firing rate for fighter planes.

In the domain of pistols, a new and smaller model, the 'Baby', was developed, along with a locked breech High Power pistol (9 mm caliber), fitted with a magazine with two rows of cartridges (1934), the 'HP'.

Around the same period, the range of cartridges was extended with armor piercing bullets, tracer-bullets, incendiary bullets and medium caliber cartridges of 12.7 and 13.2 mm (1933).

These efforts were not in vain. The crisis indeed hardly induced governments to appropriate substantial sums of money to the procurement of armaments, while from the League of Nations there came calls for disarmament. In such an unfavorable climate the only arms factories which could still survive were those whose products had proved their value beyond all question. FN was lucky in that respect.

Conditions proved to be very opportune for the company. Having recently abandoned, for infantry equipment, the cartridge with a round-nosed bullet in favour of the FN developed cartridge with a pointed bullet, the Belgian government in fact had to buy in quick succession large quantities of the new type of ammunition: 9,000,000 in 1930, 28,000,000 in 1931, 22,000,000 in 1932, 23,000,000 in 1933 and 14,000,000 in 1934. Most of this ammunition was manufactured in the Bruges factory.
At the same time, the Defense Minister, the Count de Broqueville, had procured for the army the Browning automatic rifle with firing-rate reducer. Three contracts were signed consecutively on 10 November 1930, 12 June 1931 and 18 August 1931, providing for the delivery of a total of 6,000 of these weapons.

It is also worth mentioning that in 1934 and 1935 the Belgian government bought 630 machine guns for fighter planes and 1,000 High-Power pistols. In 1932, Manchuria, under the instigation of Japan, declared its independence. We know that from that moment, , the two great empires of the Far East were in permanent conflict, without war having been declared. FN sent its rep-

Browning automatic rifle mod. 30. Belgian Army. Standing and firing positions photographed in a studio (February 1931).

resentative Edouard Schroeder to China. Resuming the relations which had been prepared more than forty years earlier, Edouard Schroeder concluded a number of agreements with the government of Field Marshal Tchang-Kai-Chek, the importance of which will become clear with the following figures: during these years, FN supplied China with 24,220 Mauser rifles, 6,465 automatic rifles with quick-change barrel, 40 machine guns and 58,185,000 cartridges. Apart from China and Belgium, other governments too procured arms from FN. From 1930 to 1934, Persia bought about 21,500,000 Mauser 7.9 mm cartridges. On the eve of its conflict with Italy, Ethiopia ordered 25,000 Mauser rifles, 425 automatic rifles and about 16,000,000 cartridges (1933-1935). Venezuela bought 16,500 Mauser rifles and 12,000,000 cartridges (1934-1935), the Netherlands, Lithuania, Estonia and Romania also bought machine guns and small and medium-caliber cartridges (1934-1935).

Water-cooled Browning machine gun.

Air-cooled Browning machine gun.
Designed by J. M. Browning to be cooled by water and then by air, available in .30 and .50 calibers, these machine guns underwent a number of variations and were produced in their millions. The heavy model remains one of the Herstal Group's specialities.

Prince Léopold of Belgium (future King Léopold III) trying out the Belgian Army's new FN Browning automatic rifle, at the camp of Beverlo in Belgium (1931).

On the occasion of an order placed by the Polish government in 1927, FN launched itself successfully into the domain of automatic weapons. It initially manufactured the Browning weapons, then improved them, and invented in turn new models.

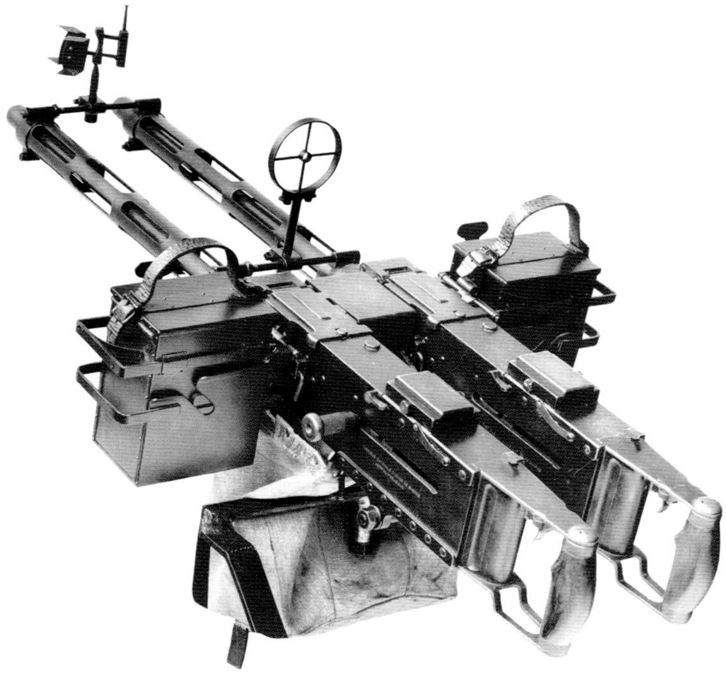

FN Browning machine gun mounted on a Fairy Fox aircraft and operated by the pilot (1933).

A pair of FN Browning machine guns mounted onto the gun turret of a Fairey Fox aircraft (1933). From the thirties, FN undertook the manufacturing of Browning machine guns, with a great number of improvements and adaptations, especially for aircraft. This was the start of a speciality of the Herstal Group, now a leader in the domain of integrated weapons systems.

HP pistol (High Power). Deluxe model produced for Asprey in London, cal. 9 mm Parabellum. Developed from the joint research of John M. Browning and Dieudonné Saive, the HP, which was invented by FN in 1934, has become a great classic in handguns, both for military and civilian use.

Browning pistol, 6.35 mm caliber, de Luxe "Baby" model. Released in 1932, this smaller version of the 6.35 mm pistol was one of the first weapons to be developed in Herstal after the death of J. M. Browning, under the supervision of Dieudonné Saive. Manufactured at FN until 1979, and then until 1983 by its subsidiary, the Manufacture d'Armes de Bayonne, its production was taken over by PSP (Precision Small Parts) of Charlottesville, Virginia, in the following year.

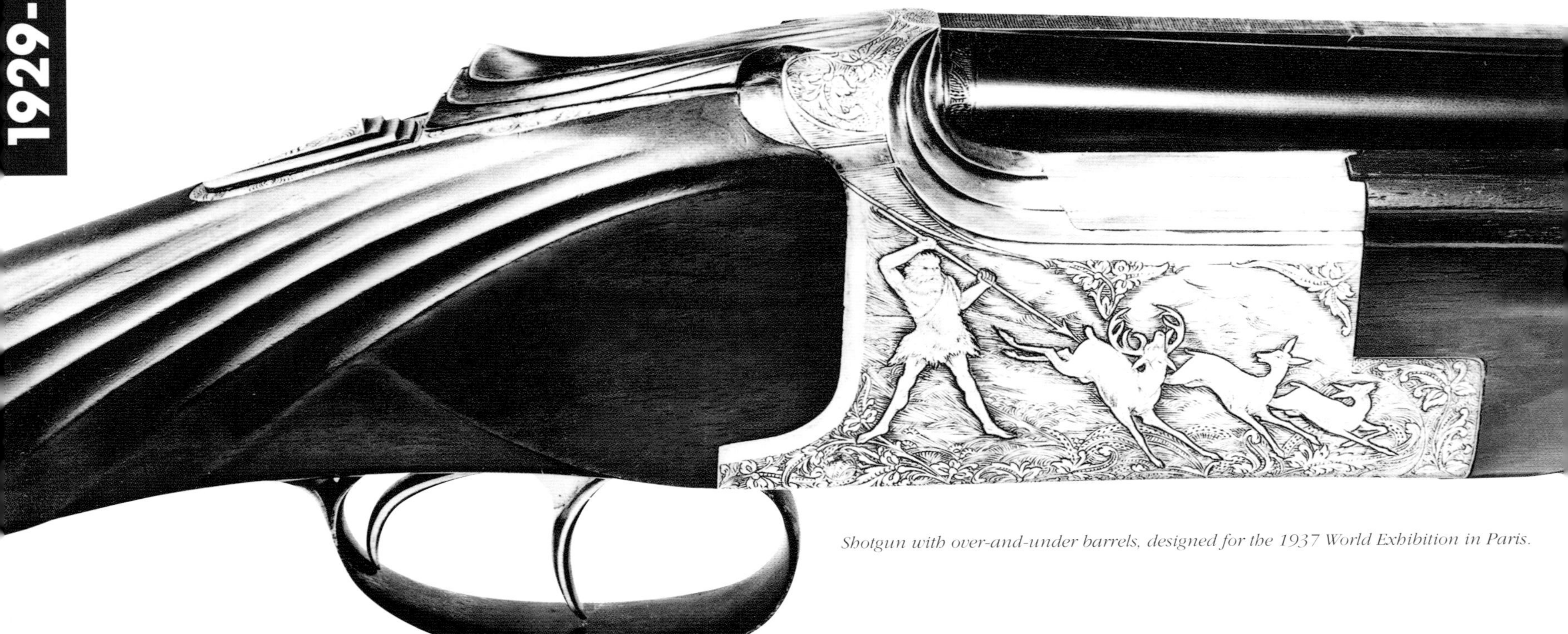

Shotgun with over-and-under barrels, designed for the 1937 World Exhibition in Paris.

The master engraver Félix Funken, in the year of his death (1965). As FN wished to meet the requirements of its expanding production as well as the decline in the cottage industry, the company entrusted Félix Funken in 1926 with the organization of an arms engraver's workshop on the company's premises. Shown above and on next page, some items produced by the master engraver Félix Funken during the thirties, in "art deco" style.

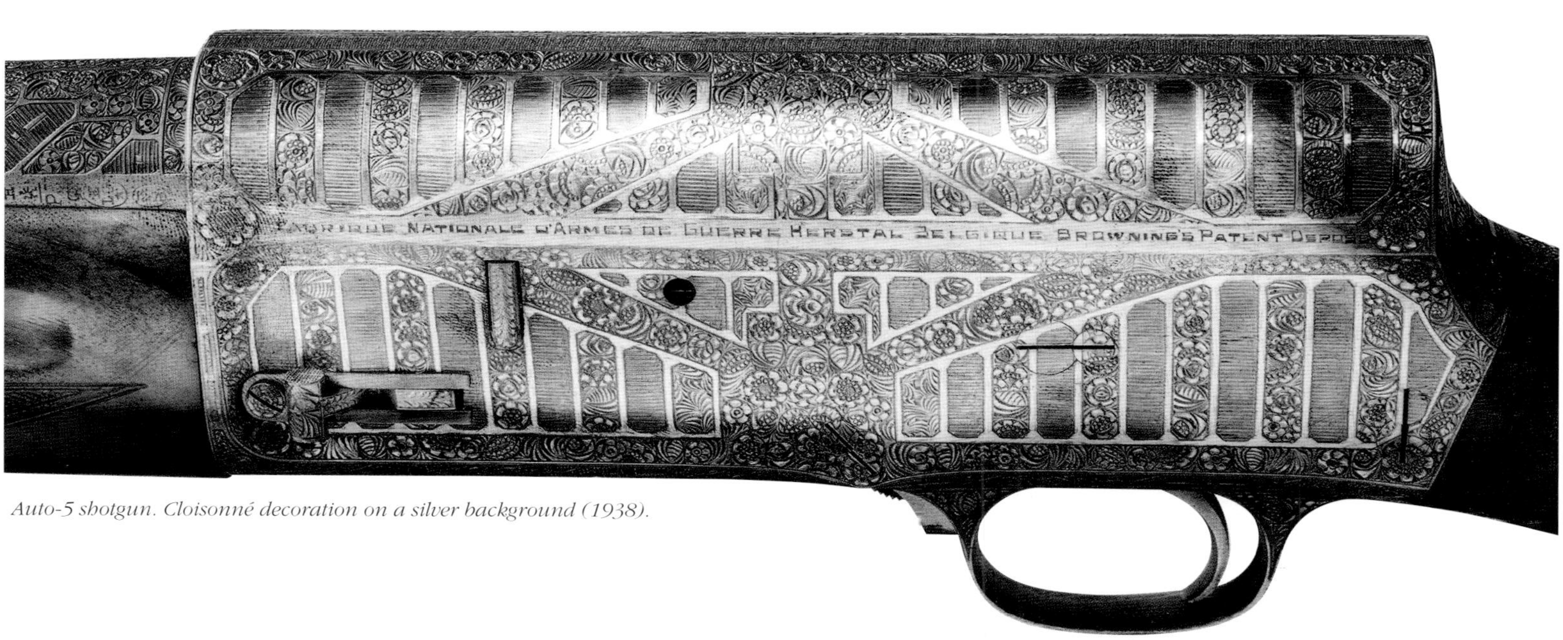

Auto-5 shotgun. Cloisonné decoration on a silver background (1938).

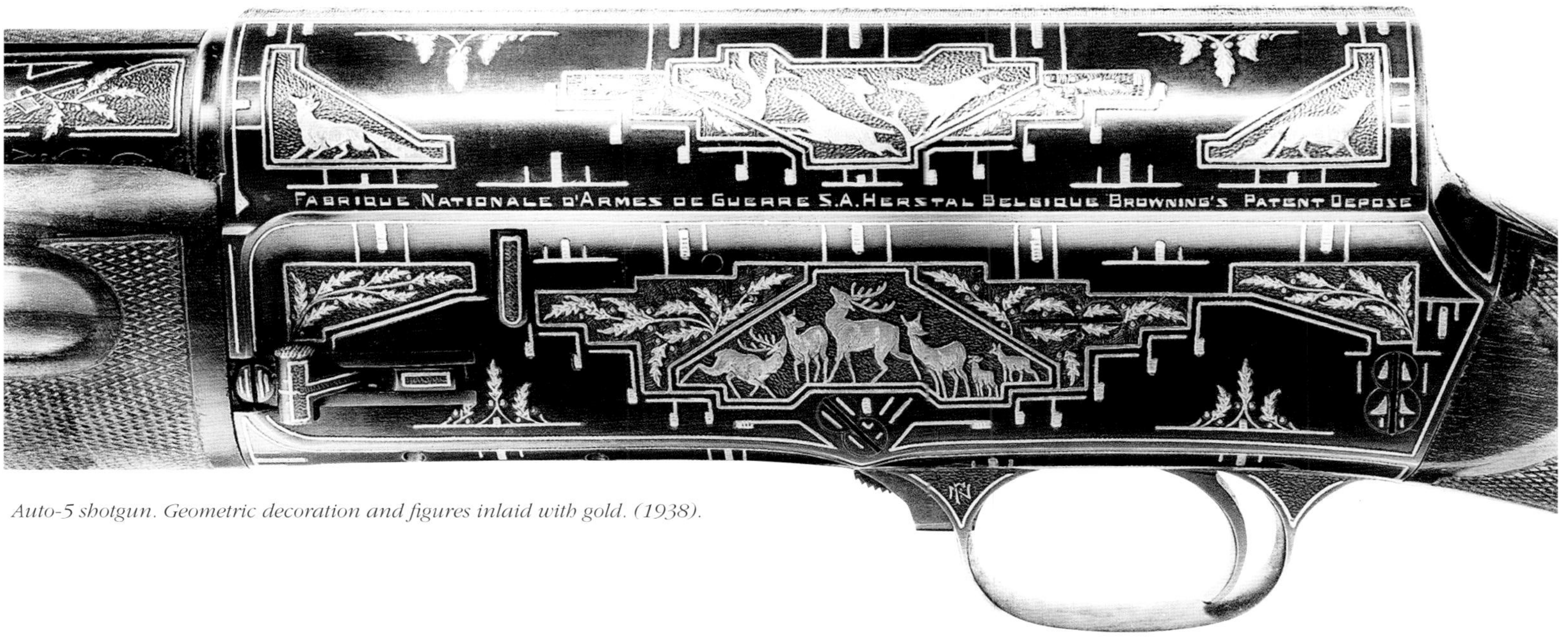

Auto-5 shotgun. Geometric decoration and figures inlaid with gold. (1938).

Auto-5 shotgun. Oak leaf engraving and figures on a silver background. (1936).

Reporting in the workshops in September 1929

Polisher

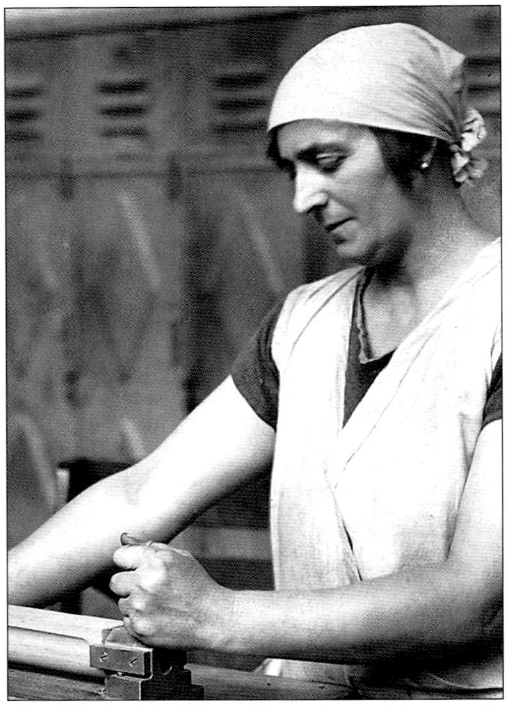
Cutter

Polisher

Polisher

Wood fitter

Wood fitter

Reviser

Barrel trimmer

Driller

Barrel trimmer

Filer

Milling worker

Barrel straitghtener

Milling worker

Milling worker

FN shares at 89 francs on the stock exchange !

A great deal can be learnt from the above dates and events. When recession set in, the motor vehicle division appeared to be on an equal footing with the arms and cartridges division. Five years later, while the manufacturing of motorcars was suspended and that of motorcycles seemed to slow down, the arms and ammunition division was in full swing. In the context of the hard economic reality, the plan to convert FN into a factory for civilian products had failed, and that forty years on, the company had regained its original structure.

It should be added that this failure could have led to the complete downfall of the company. We shall see whether this was indeed the case. In June 1929, the factory in Herstal employed 9,138 people, while in May 1934, this figure had fallen to 2,580, and in addition substantially reduced salaries. The financial burden had become so heavy as to be crushing. On 17 May 1930, a loan of 15,000,000 Swiss francs, at 6% interest , was granted by the Société de Banque Suisse in Basel. This loan, which provided the company with more than 103,000,000 Belgian francs did not suffice to provide the means necessary for its survival : the bank overdraft had increased from 52,000,000 in 1930 to 101,000,000 in 1934. In the meantime, the losses suffered as a result of the decline of the motor car division and the stagnation in the other activities weighed heavily on the overall results.

These were in fact so mediocre – one financial year, that of 1932-1933, even showed a serious loss – that no dividends were distributed after 1929 and the market value of the shares had fallen to 89 francs (30 May 1934).

Moreover, a number of deaths added to the general gloom. In May 1930, Raoul de Terwangne, who had been appointed auditor in 1896 and Director in 1919, died. Two years later, Jean Jadot, who had presided over the board of directors since U.F.I. had bought the shares of D.W.M., also passed away.

Alexandre Galopin succeeded him at the head of the board, which was composed of Georges Laloux, Alfred Andri, Arthur Bemelmans and Jacques Trasenster, replaced after his death by Marcel Nagelmackers (1936).

Industrial Accounting personnel (July 1930).

Management of the cartridge department (July 1930).

Cartridge factory. Technical personnel (July 1930).

Center : arms testing and acceptance board. Bottom : arms and cartridge testing department (July 1930).

Publicity brochure for the Auto 5 shotgun (around 1928).

Cover of the first of Browning USA's publicity brochures, devoted to the 'Over-and-Under' (1931). It shows the American Sporting Shooting Champion, Gus L. Becker, a friend of John Moses Browning.

Art Deco cover of the 1927 FN Browning catalogue. Design by J. E. Goossens in Brussels.

Promotional poster for the FN six cylinder lorry (Herstal Group Collection).

Business recovery

On 30 March 1935, five days after he was appointed Prime Minister, Paul van Zeeland decided to devalue the Belgian franc by 28 % of its value in 1926.
As for Belgium, this measure, which had been necessitated by the new situation in international trade, bore witness to a recognition of the effects of the crisis and marked the beginning of a general recovery of business.

Devaluation and rearmament

As we have seen, FN had shown the first signs of recovery some months earlier, at least in the arms business. Therefore in fact, the devaluation merely helped along a trend which had already set in. However, as the company had debts abroad, this commercial advantage was largely counterbalanced by a substantial rise in long-term debt - the increase amounted to exactly 40,838,760 francs. On the other hand, thanks to the decrees and laws which in the previous year had provided for a partial or complete conversion of the bank overdraft into long-term S.N.C.I. credits, FN was able to reduce this by 67,000,000 francs. Moreover, after the labor upheavals of the summer of 1936, relations between the company and its staff improved noticeably. So all in all, the period we shall be discussing began auspiciously.

There was another event which was beneficial to FN's prosperity. As the years went by, some clear-sighted people were of the opinion that a disarmament by the pacific nations would only stimulate the audacity of the warlike nations, and, as a result, increase the risk of war. It is well known that Hitler, who became Chancellor of Germany in 1933, then Reichsführer in 1934, was preparing his nation for National Socialism and aggression. After the military occupation of the left bank of the Rhine in 1936, the annexation of Austria on 11 March 1938, of Sudetenland in September 1938, and of Bohemia and Moravia in March 1939, most European countries became alarmed and started to rearm themselves.
Many of them were well acquainted with the material produced by FN and now once again they would turn to the company.

Lorries, tractors and army motorcycles

All the divisions of the factory benefited from the rearmament. Even the development of motor vehicles, which had been so badly hit by the crisis, benefited from the reputation of its military equipment which FN had built up with the various governments. In 1935 the factory built a lorry with an 8-cylinder engine which had originally been intended for a car; the Belgian army bought a number of them. But this vehicle, which had been designed for civilian purposes, did not meet the requirements of a modern army.

So FN designed a three-ton lorry with a 6-cylinder engine and four-wheel drive, which was quite uncommon for that time. Several versions of this lorry were manufactured in relatively large quantities, to meet the needs of the Belgian army. Not long before hostilities broke out, the company designed for the Polish army a prototype derived from the same vehicle and which was characterized by the possibility for the driver to lock the differentials. Some lorries, which were to be produced after the war, would be derived from this Polish prototype.

At the same time, FN manufactured, under a license from Citroën-Kégresse, a caterpillar tractor with rubber bands, intended for the Belgian artillery. These vehicles, the construction of which had been commenced by Minerva Ltd., were finally fitted with the 6-cylinder motor which originally had been intended for the above mentioned lorry. Concurrently, FN machined the different parts of the Latil artillery tractor, which was also sold to the Belgian army.

In the meantime, the research department, which had been working on a "cross-country" vehicle called the Colonial since 1932, developed a large number of prototypes, of which we shall mention an FN 47 tractor with two engines and independent four-wheel drive and particularly the 4RM42F lorry. The latter vehicle, which was fitted with a 4-cylinder engine and eight gears, was the model from which the Ardennes lorry (4RM-62F) was derived and subsequently manufactured by FN.

Civilian 'Autocar' version of the FN eight cylinder 3250 cc lorry. The creation of a 'Heavy Vehicles' department at FN, succeeding the automobile department, dates from the 1930s.

FN lorry, the FN 63C six cylinder. Cylinder capacity 3,993 cc. Adapted for a variety of uses, 300 were produced between 1937 and 1940.

FN 4RM army lorries, ready for delivery (1938). To meet the needs of the army, FN developed a three-'on, 6-cylinder lorry with four-wheel drive. It was adopted by the Belgian army not long before the outbreak of the Second World War.

Strength tests of the FN 4RM lorry on the Thier Savary downhill slope in Liège (1938).

FN van on a C10 chassis, with telescope turret, for the Belgian Department of Public Works. During the thirties, FN was branching out into the manufacturing of "heavy vehicles" for civilian and military use. The platform elevators, which were intended for the fitting of and repairs to the overhead wires for trams and trolleybuses, were to be part of the urban landscape until the fifties.

Citroën-Kégresse 63T tractor.
Tractor with rubber tracks produced for the Belgian artillery, fitted with the six cylinder engine originally designed for the FN 63C lorry : 140 were produced. (1936 - 1940).

FN C10 van (10 of "1,000 kg").
Four cylinder engine, 2,269 cc (1934) : 300 of these vehicles were produced.

The motorcycle division benefited even more from the rearmament. Three types merit further discussion. The first was the M86 motorcycle undoubtedly the most remarkable ever manufactured by FN.

Adapted to military requirements, it was mass-produced for Belgium, along with an innovative model with motorized side-car. These vehicles were subsequently armored and supplied to the Venezuelan army. The M86 motorcycle was followed by the M12 with motorized side-car. This vehicle, which was fitted with a 1,000 cc two-cylinder engine and provided with a shaft-drive, was produced in very large quantities. The Belgian army alone had bought several hundreds at the outbreak of the Second World War. Iran also bought a number of them. A little later, as a logical next step after the side-car, FN produced for Belgium in large quantities a cross-country tricycle with the same engine as that of the M12 motorcycle.

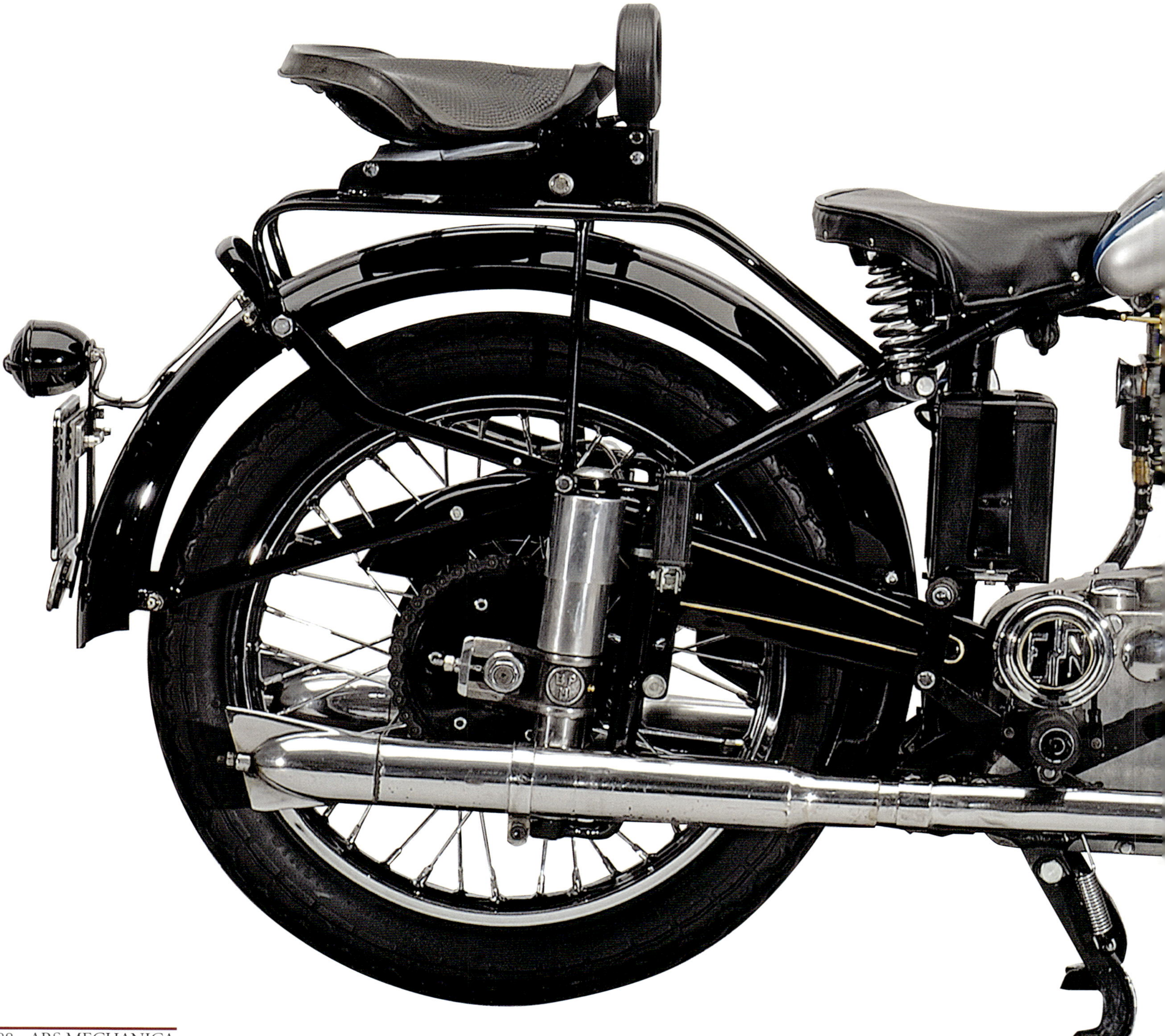

Group of FN M86 motorcycles - 600 cc, military armored type, for the Venezuelan army (1936).

FN M86 motorcycle - 600 cc civilian model, undoubtedly the most remarkable motorcycle ever produced by FN (1935).

119658

FN M86 motorcycle - 600 cc, military model with a drive-wheel sidecar, an innovation at that time. It is equipped with a Browning automatic rifle (1935).

Acceptance at FN by the Belgian army of M12-1,000 cc motorcycles with sidecar.
Between 1937 and 1940, the Belgian government purchased 1,090 units of this remarkable precursor to the Jeep.
Other countries, principally Iran and Greece, also placed orders during the same period.

Prototype of the FN Tricar - 12 T3 1,000 cc 'all terrain' tested by Belgian officials.

FN 12 T3 1,000 cc Tricar able to take passengers, being demonstrated for the East Indies (January 1939).

FN 12 T3 1,000 cc Tricar - Tests in Portugal.

1938 motorcycle catalogue.

Publicity leaflet promoting the FN M71, M11, M86 and M12 1000 cc motorcycles (around 1937). Composition by the Liège artist and caricaturist Ernest Forgeur (1897-1961).

FN M200 motorcycle - 1938 : 1,564 motorcycles of this type were produced (including the M250).

FN M71 motorcycle - 1936 :
2,489 motorcycles of this type were produced.

China, the most important client

Whatever their particular interest may have been, all these various business deals appear nowadays quite unimportant compared to the sale of arms and ammunition. The range of models manufactured by FN had been considerably extended in the course of the previous few years.
The machine guns for fighters were constantly improved to become eventually one of FN's most highly regarded products. Their reliability and their particularly rapid firing rate put them well ahead of all their competitors. In 1936, FN obtained from the company Bofors Aktiebolaget the license for the manufacturing of an automatic 40 mm anti-aircraft gun which that firm had developed. One hundred and fifty of these guns were sold to the Belgian army. Moreover, the factory was now also able to produce ammunition of 20 mm, 13.2 mm and .50 caliber, and the cartridges of the two last calibers could be charged with exploding bullets.

Of all the company's clients, China remained comparatively speaking the most important. It bought 164,500 Mauser rifles, 5,000 HP pistols, 6,470 automatic rifles, 357 machine guns, 87,000,000 infantry cartridges, 13,000,000 pistol cartridges and 1,375,000 13.2 mm and .50 caliber cartridges. Since 1937, China had been at war with Japan, while the government of Field Marshal Tchang-Kai-Chek – whom FN dealt with – still had to fight the communist armies which held the Northern provinces.

All over the world

The procurements by European and certain Latin-American countries continued to increase. To give an extensive list would be an impossible task, therefore we shall restrict ourselves to mentioning the most important. During the period under discussion, not counting the supplies to China, FN produced 2,062 automatic rifles for Belgium; 3,000 small-caliber machine guns for Belgium, the Netherlands, Greece, Yugoslavia, Finland and Portugal; 4,200 Browning aircraft machine guns for France, Lithuania and Romania; 431 heavy machine guns for Romania, Greece and the Netherlands; 56,500 High Power pistols, notably for Belgium, Lithuania, Estonia and Peru; 40,000 1922-model pistols for the Netherlands and Yugoslavia; 143,000 Mauser rifles for France, Greece, Lithuania, Venezuela, Peru, Paraguay and Yemen; 204,000,000 infantry cartridges (7 mm, 7.65 mm, 7.9 mm, 7.92 mm, 11 mm, .30 caliber) for Belgium, Belgian Congo, France, Yugoslavia, Greece, Romania, Lithuania, Persia, Yemen, Paraguay, Peru, Venezuela, Uruguay and Bolivia; 18,000,000 pistol cartridges for Lithuania, Peru and Bolivia; 3,000,000 medium caliber cartridges (12.7 mm, 13.2 mm), some of which were supplied in separate parts, for the Netherlands, Belgium, Romania and Norway and several hundred thousand shells of the following calibers: 40 mm DTCA, 47 mm anti-tank, 75 mm OC, 75 mm DTCA models 1927 and 1936, 105 mm L43, 120 mm, 150 mm L43 and L17, 155 mm model 1924, and 88 mm (56).

Bofors automatic anti-aircraft guns, 40 mm caliber. In 1936, FN obtained from Bofors the license for the manufacturing of antiaircraft guns. One hundred and fifty pieces were procured by the Belgian army. These models were operated by hand. The production of the gun carriages compensated to some extent for the loss of the motor car section.

FN armored motorcycles with motorized side-car, used by the Chinese army (1936). In the thirties, China was one of FN's most important clients. Venezuela also ordered a number of these motorcycles.

Cartridges produced by FN in the late 1930s.

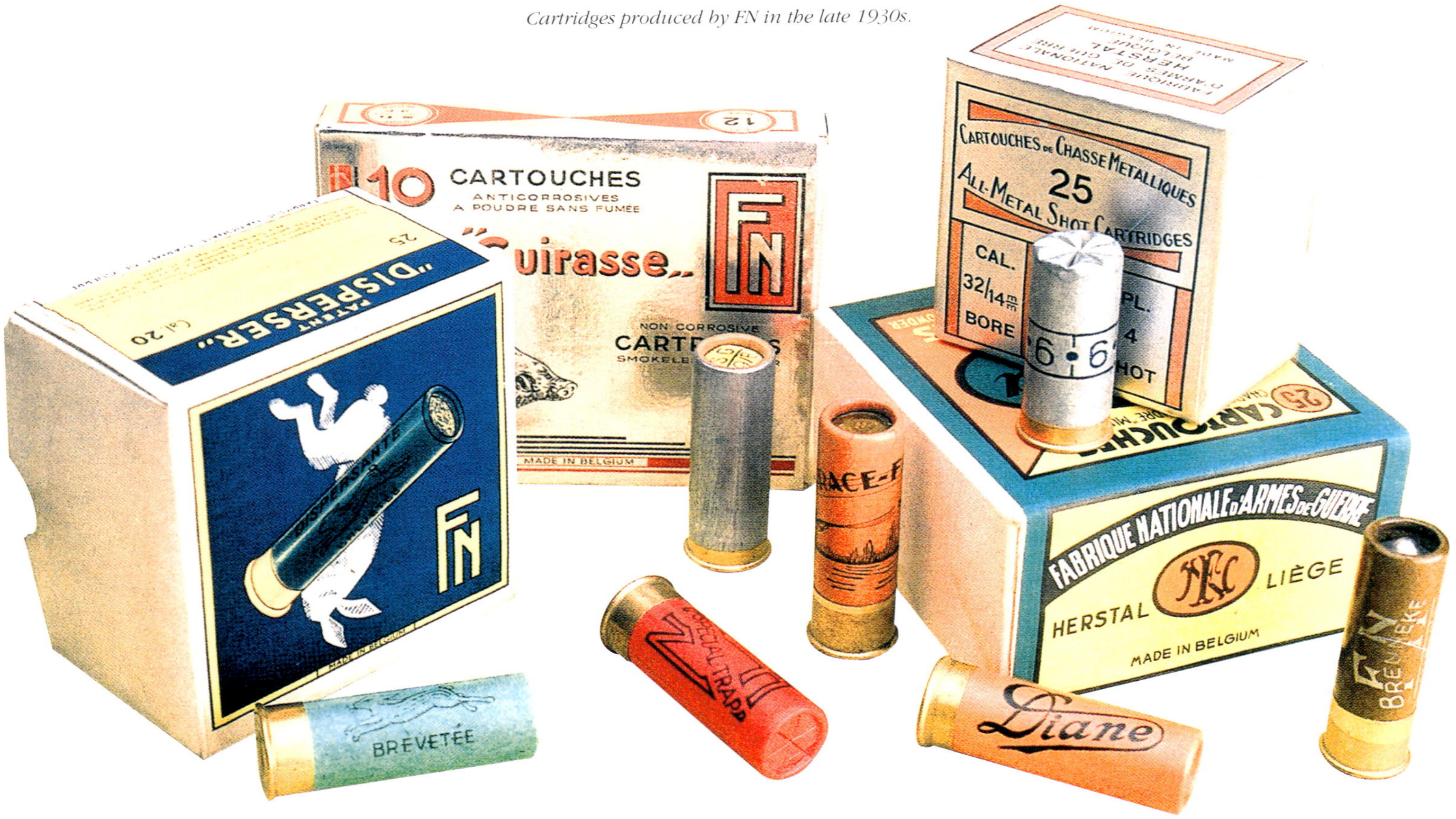

Publicity poster by J. Ochs for the 'Légia' cartridge (1930s).

Publicity leaflet promoting the new Metal Star cartridge (around 1935).

In the meantime, on 1 September 1939, after the invasion of Poland, France and Great Britain declared war on Germany while most of the other European countries - Belgium first of all - took mobilisation measures. Even though the country was legally neutral, FN helped the allies considerably. France and Great Britain were thus able to procure Bofors anti-aircraft guns which were soon to prove very useful.

We shall not recall the first months of the Second World War. We know that after a long period of inactivity on the Western front, Germany invaded Denmark and Norway in the spring of 1940. These events induced the allies, albeit rather late, to step up their war effort.
Thus in May 1940, the British authorities asked FN to build a factory in England for the manufacturing of Browning aircraft machine guns, 13.2 mm caliber. Gustave Joassart, who since January 1940 combined the posts of General Manager and Director-on the board he replaced Alfred Andri, who had just died - and the Deputy Manager René Laloux, , went over to England. They were still there when on 10 May Germany attacked Belgium and the Netherlands, thereby launching the first big offensive against the Allies on the Western front.

Some reflections on a jubilee

At this point we would like to make a brief step back in history. On 3 July 1939, the company celebrated its fiftieth anniversary. This event led the board to reflect on the importance of that half-century. This reflection has been recorded in a speech which Gustave Joassart made to his colleagues. Summarizing the events which we have just been discussing, Joassart pointed out that throughout its existence, FN had been tempted by two fields of activity : "the field of commercial products and that of military or war equipment".
He declared:

"Born to execute the Belgian order and baptized the Fabrique Nationale d'Armes de Guerre, our company was set directly on the road to which it had been destined by its name...
In 1914, the commercial products brought it most benefit.
In 1939, on the other hand, it was the military equipment which brought it most success...
And indeed, we have left this child (from 1889) to return to the activity to which it had been destined by its name ...
We have left this child to return to it's first love. It's a law of nature."

Indeed, when war broke out for the second time in a quarter of a century, FN had become once again almost exclusively an arsenal. This turnabout, which had started during the great depression and had accelerated as a result of political events, had allowed the company to recover financially. Profits increased both in absolute terms and in relation to the sales. Moreover, for the first time since 1929, a dividend could be paid out for the financial years 1937-1938 (30 francs) and 1938-1939 (50 francs) while the market value of the shares varied between 400 and 900 francs. As in 1914, the company was experiencing full prosperity when war broke out.

Gustave Joassart, General Manager of FN (1923-1950)
Bust by Louis Dupont (1953) (Herstal Group Collection).

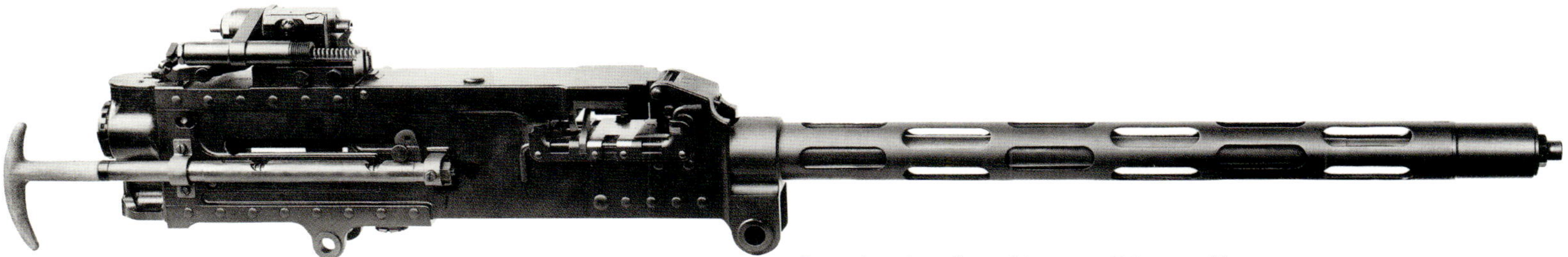

Browning aircraft machine guns, 7.5 mm caliber, with a firing rate of 1,500 rounds per minute. French model (1939).
Above : type of engine cowl with traction synchronizer.
Below : type of wing with solenoid.

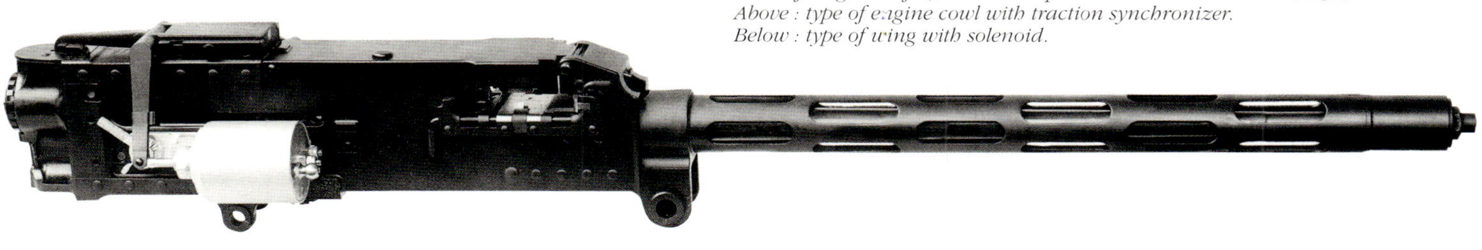

FN M12 1 000 cc sidecar motorcycle with a mount for an FN Browning automatic rifle, of the type produced for the Belgian army (1940).

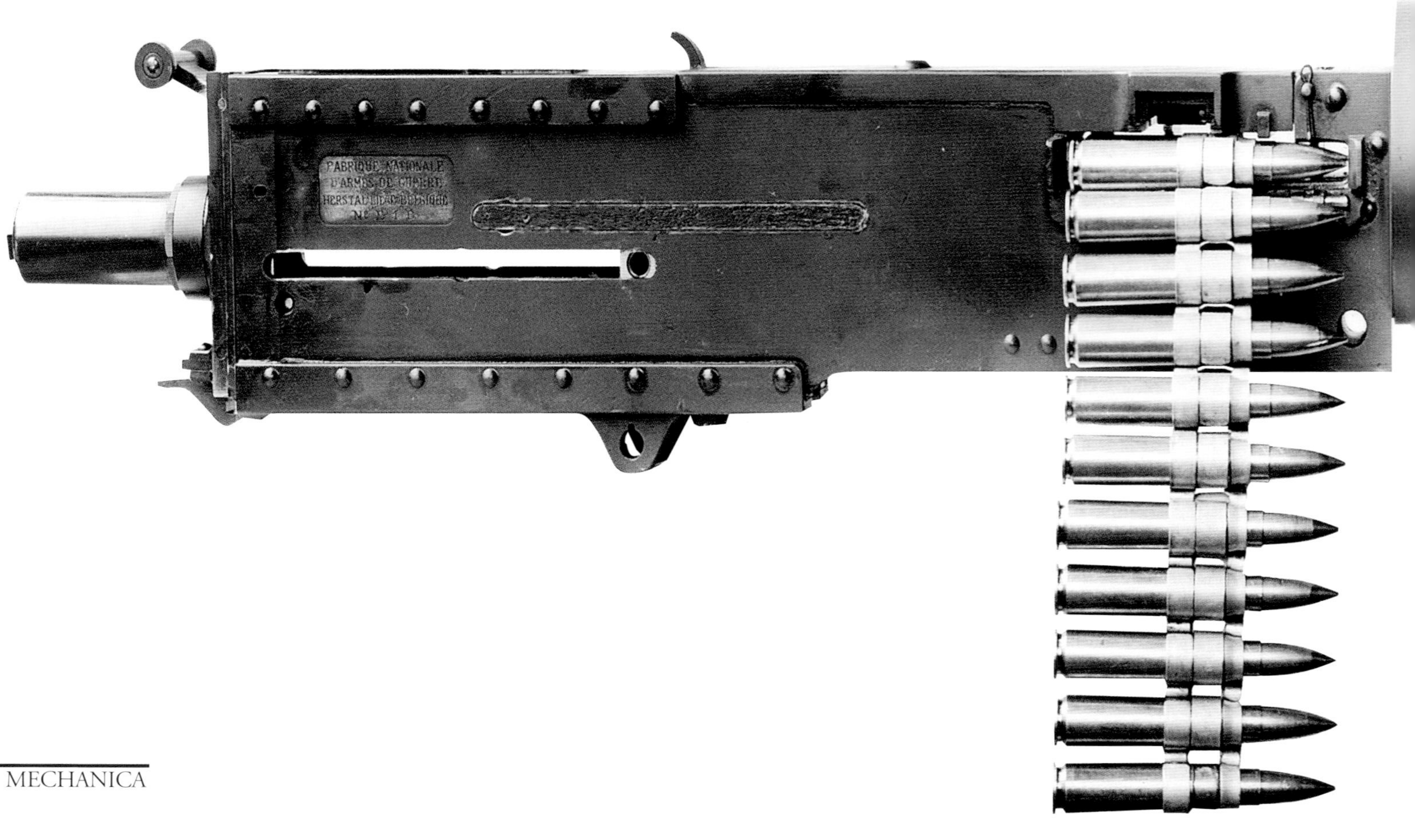

FN Browning automatic rifle, Model D with the new case regulator (1939).

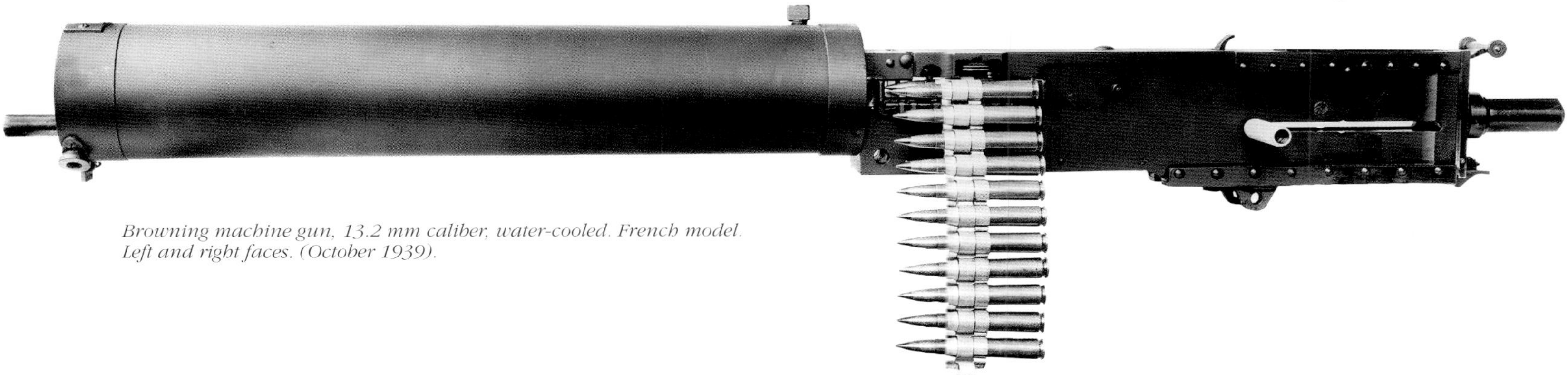

Browning machine gun, 13.2 mm caliber, water-cooled. French model. Left and right faces. (October 1939).

FN under sequestration

Tags worn by the staff of FN during the occupation. These tags bore the German abbreviation of each division. The intention was to strictly monitor the comings and goings of each staff member of the company, as a preventive measure against sabotage.

The war had not caught FN unawares. Since 1939, in consultation with the Ministry of Defense, the board had adopted certain measures. Realizing that the region of Liège could not possibly be defended, it had decided that from the beginning of hostilities the equipment should be transferred to Bruges and to the workshops which had been especially constructed for that purpose in Anzeghem, near Courtrai in Belgium, and at Mécanouty, in the Paris region. Other measures had been taken for the protection of the factories in Herstal and for the evacuation of the equipment.

We are well acquainted with the events which followed the invasion and which rendered all these precautions pointless. On 12 May, the Wehrmacht invaded Liège. On 28 May the Belgian army capitulated, and on 18 June France sued for an armistice. The removal of the equipment from Herstal, which had only just started, was suspended. On 12 May, the speed of the German advance had forced General Joseph Quintin to evacuate to France the entire Bruges cartridge factory , whose production capacity had attained a million cartridges per day. The machinery and equipment were loaded onto four trains and transported to a French national arsenal in Toulouse, where at the same time the staff arrived by car. The cease-fire on the French front happened the day after production could have started up! The equipment was left in Toulouse, but the Germans could not take possession of it until two years later, when they occupied what after the armistice had been called the free zone (November 1942). As for the workshops in Anzeghem, the contents of which had been transferred to Thouars and to Mécanouty, their fate was not such a fortunate one: almost at once, the events had held up production, and soon the equipment fell into the hands of the enemy.

Refusal to collaborate

With the Grand Duchy of Luxembourg and the northern provinces of France, Belgium was soon subjected to the regime of the occupational forces. FN was soon to feel the severity of this situation. Moreover, from 17 May, the managers who had stayed in Liège, namely Henri Pommerenke and René Lecocq, had been requested by the German Ortskommandantur to start up the factory in Herstal. One after the other, both men refused point-blank. The Chairman of the Board, Alexandre Galopin, and Vice-Chairman Georges. Laloux were informed of this refusal and endorsed this position wholeheartedly.

Under the supervision of the deutsche Waffen- und Munitionsfabriken (D.W.M.)

Not in the least discouraged by their initial failure, the German authorities insisted more strongly, while on their requisition the arms and ammunition which were still in the factory were removed. Finally, on 19 June, after a final attempt to reach a compromise, Lieutenant-General Stubb, who was in charge of the industrial equipment division of the army (Industrialle Rustung), informed Henri Pommerenke and René Lecocq of his decision to place the FN factories under the supervision of the Deutsche Waffen- und Munitionsfabriken in Berlin, with an officer appointed by the Oberkommando of the Wehrmacht to represent the company's interests. The next day, an engineer of the D.W.M. came to Herstal and started to recruit the workforce the following day.

FABRIQUE NATIONALE D'ARMES DE GUERRE S.A.
HERSTAL — LÜTTICH
KOMMISSARISCHE VERWALTUNG

KOMMISSARISCHER VERWALTER	BANKVERBINDUNG	POSTCHECKKONTO	FERNSPRECHER	POSTFACH
Dr. F. SCHARPINET	REICHSKREDITKASSE LÜTTICH	92516	VERMITTLUNG LÜTTICH 408.09 und 179.13 EINKAUF : 408.00	HERSTAL 62

FN letter heading, when the company was under sequestration during the occupation. The sequestrator Franz Scharpinet, who was invested with full powers by the German authorities, ran FN from 1940 till 1942. He was replaced by Dr. Holl, who remained at the head until the Belgian management took over again on 1 September 1944.

The regime thus installed was soon modified. On 5 July 1940, in execution of an ordinance decreed on 20 May on the supervision and sequestration of industrial enterprises, the military commander for Belgium and Northern France appointed Dr. Franz Scharpinet, Manager of the D.W.M., to the function of sequestrator (Kommissarischer Verwalter) of FN. According to the terms of this ordinance, the sequestrator was invested with all the powers previously held by the board of directors, the general assembly and the management of the factory. It may be said that on that day the clock had been put back twenty-one years.

In the meantime, Gustave Joassart, who was in Great Britain at the time when war broke out and who had been prevented from returning to Herstal by the swift German advance, had travelled to France. After that country had sued for an armistice, he returned to Belgium (13 July 1940), where he learnt of the death of his son Nicholas, a young and brilliant departmental manager at FN, who had fallen in Calais.

On 17 July, Gustave Joassart met Dr. Scharpinet, and after difficult negotiations obtained from the latter that FN could keep in Herstal a group of about twenty employees of all levels, whose task consisted of defending the interests of the company with regard to D.W.M. FN was also authorized to set up in Liège – in the Rue Louvrex and on the Place Coronmeuse – two offices which would be independent of the sequestrator and which were designed first of all to prepare the manufacturing of hunting equipment and motor vehicles, in anticipation of the time when peace returned, and secondly to wind up administratively the current business.

These agreements came into effect a few months after Gustave Joassart and some of his colleagues – René Laloux, Jean Vogels, Edouard Dufrasne and Dieudonné. Saive – had left Belgium clandestinely (in the summer of 1941) and had travelled through France, and via Spain and Portugal, eventually crossing over to Great Britain (September 1941).

Not long after his arrival, on 19 February to be exact, Joassart was appointed Under-secretary of State (Deputy Minister) at the Department of Refugees and at the Department of Employment and Social Security for the Belgian government in exile in London. He would hold that post until 26 September 1944, while his colleagues put their professional experience at the service of the British Ministry of Defense.

Meanwhile, under the supervision of the sequestrator, the D.W.M. gradually started up the factory in Herstal, a task which proved long and laborious after the practically unanimous refusal of the executive staff and the workforce to collaborate. Also, DWM, who were planning one day to transform the se

Tablet commemorating the death of resistance fighter Henri Mesch, a foreman at FN, killed in 1944. FN, which was sequestrated from 1940 until 1944 by the Germans, was started up again under the supervision of the Deutsche Waffen-und Munitionsfabriken in Berlin. The Belgian management refused to collaborate, and some of its members fled to England where they contributed to the war effort of the Allies. Various acts of sabotage and resistance carried out by the workforce were severely punished.

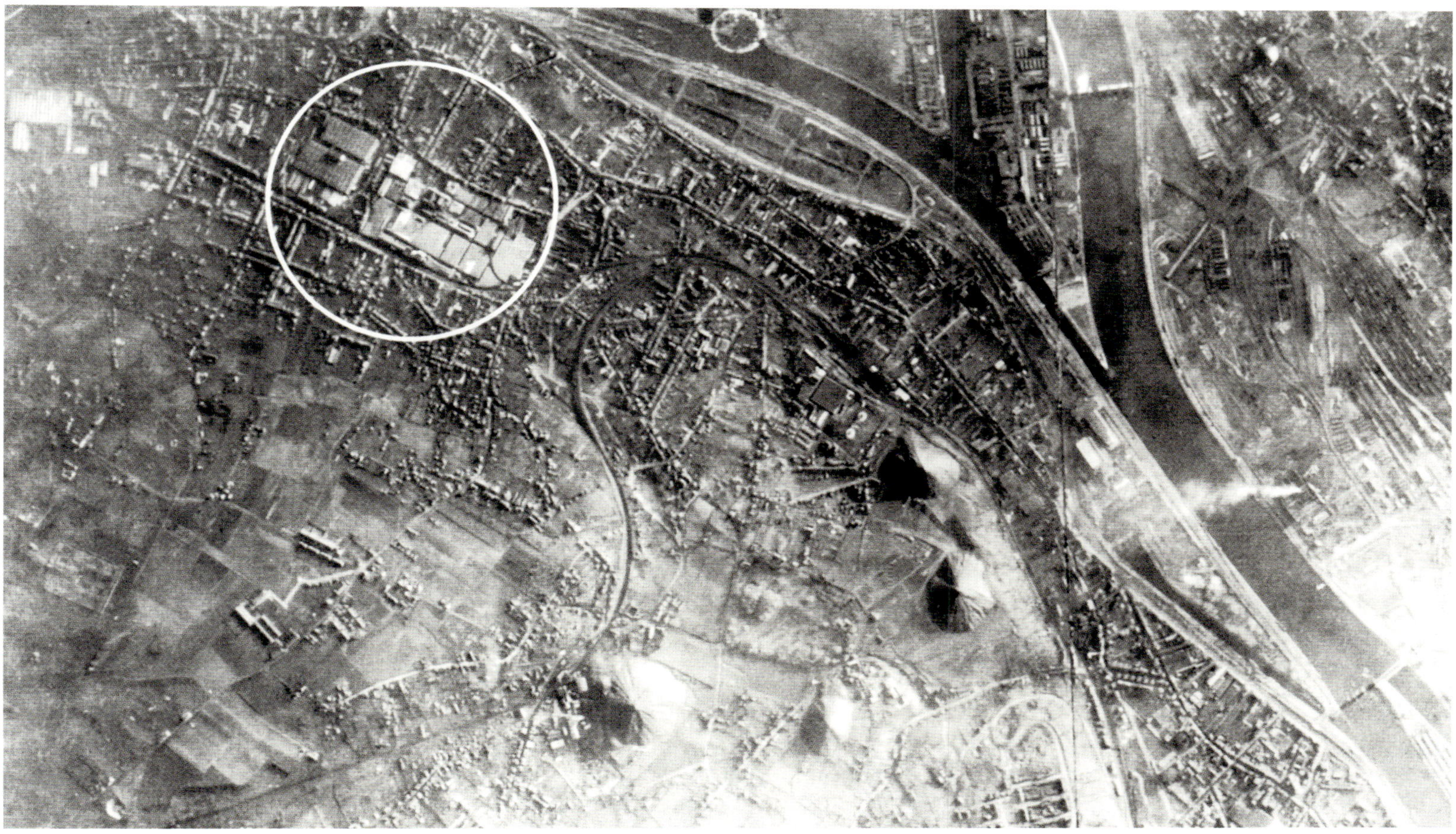

Picture of FN (encircled), the Meuse and the Albert Canal, taken by the RAF in 1941. Even though the enemy had undertaken various military manufactures at FN, the Allies did not bomb the factory in order to save the civilian population living in the neighbourhood, as well as those forced by the Germans to work in the workshops.

questration into a participation in the capital, had a general valuation made of FN's assets by a trustee company, the Schwabische Treuhand. The result was that the share value was set at a ridiculously low figure.

Both this plan and the increasing threat posed to Germany by the British air force induced DWM to link their factory in Herstal more strongly to their own company. The factory of Herstal was matched with their factory in Karlsruhe and to the arsenal in Poznan (Posen), the management of which they had taken over in September 1939. The three companies, which henceforth were entrusted with the manufacturing of complementary products, were placed under the sole supervision of a manager of DWM.
This amalgamation, which threatened to further aggravate the damage inflicted on FN, prompted the German authorities to replace Dr. Scharpinet with Prof. Dr. Holl (in the summer of 1942), who was not associated with DWM. The new trustee concluded with the latter an agreement for a trading concession. The distinction between the sequestrator and the concessionaire thus being, at least legally, more clearly defined, Dr. Holl thought he could renege on certain agreements which had been concluded between his predecessor and Gustave Joassart.
On 14 July 1942, he had the Feld gendarmerie close down the offices in the Rue Louvrex and on the Place Coronmeuse, summoning part of the workforce there back to Herstal and placing René Lecocq and Henri Pommerenke under temporary house arrest.

The two last years of the occupation were also the harshest. Since the spring of 1942, the ordinances decreed by the military government on forced labor had provided DWM with a large, if not competent and devoted workforce. The stocks and financial reserves of FN were still substantial. By incessantly increasing the number of workers and by getting the most out of the machinery, DWM were able to increase production figures. Nevertheless, acts of sabotage equally increased in number. All this threatened to drain the factory's resources down to a dangerously low level.

Restricted production

Placed under German sequestration, FN was integrated into the Nazi military-industrial complex. This was based on sharing out the production of war equipment amongst the various factories in the Reich and annexed territories, assigning only partial tasks to each, with obvious strategic and security objectives. This general concept dictated the assignment of the manufacturing of a quantity of diverse parts for land vehicles and aircraft to Herstal. But mostly, between 1940 and 1944 the following in particular were produced there : 319,000 9 mm HP pistols, known as the Pistole 640(b) (b stood for 'Belgium'), 363,000 10/22 pistols, both 7.65 mm and 9 mm short models (the Pistole 626(b) and the Pistole 641(b) respectively).
These weapons bore a special marking, that of the "Waffenamt" (Office of Military Armaments) with the Nazi eagle.
n addition, the FN workshops also manufactured parts for the Mauser 98K carbine, specifically 1,647,000 barrels and 459,000 bolts. An unknown quantity of frames and slides for the Walther P38 pistol were also manufactured.

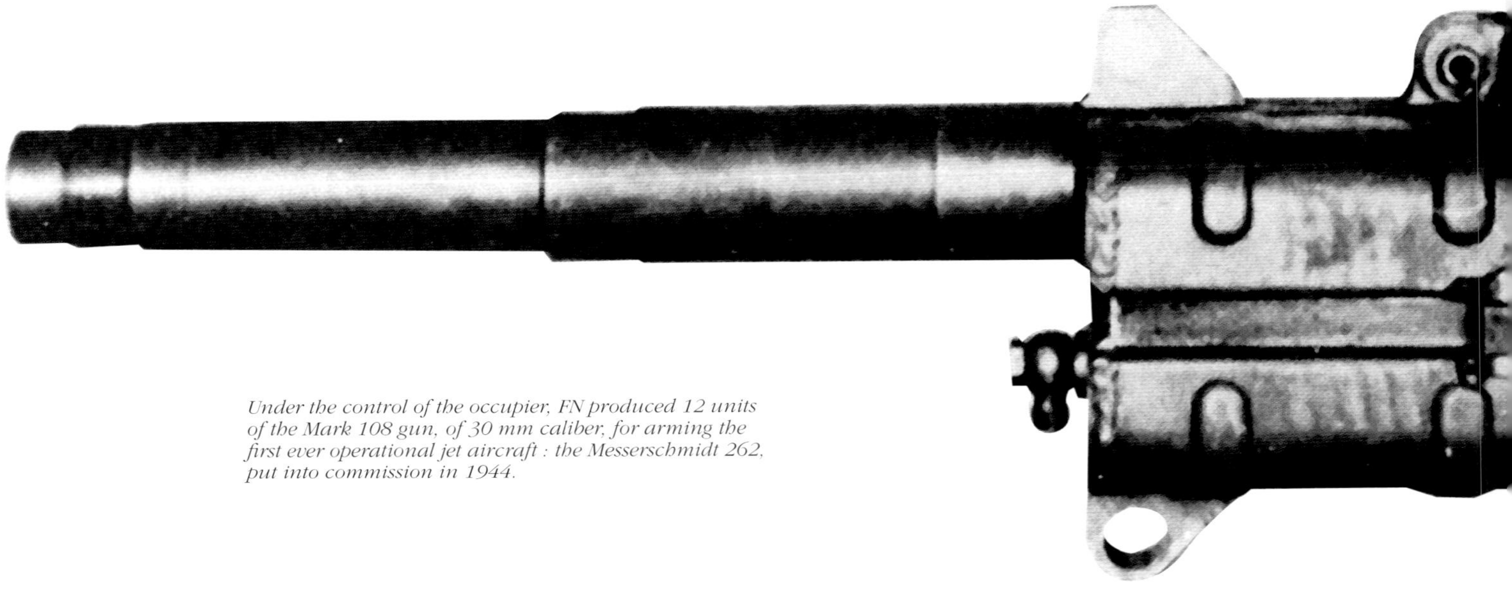

Under the control of the occupier, FN produced 12 units of the Mark 108 gun, of 30 mm caliber, for arming the first ever operational jet aircraft : the Messerschmidt 262, put into commission in 1944.

More specific was the production of twelve parts of the 80 mm Mark 108 cannon used to arm the first operational jet-powered aircraft in history : the Messerschmidt 262, deployed in 1944.

Finally the cartridge factory first produced 115,000,000 6.5 mm caliber rifle cartridges for Sweden, which involved a barter transaction in order to obtain iron ore. Production was then concentrated on the German 7.92 mm caliber cartridge, whose production reached a maximum of 680,000 per day (including steel-cased ammunition) and ammunition for 7.65 and 9 mm caliber pistols.

Apart from a small nucleus of Belgian managers, either collaborators or secretly working for the Resistance, the management of the requisitioned FN was made up of German nationals. Only a small proportion – about 10% – of the manual workers continued to work for the occupier after 1940. The remainder of the workforce, which reached some 12,000 in number, was made up of labor conscripts.

In these circumstances, not only were production levels very low, but a number of acts of passive resistance, even of sabotage, were carried out, sometimes detected and punished by the Nazi authorities. The systematic production of faulty parts – rejects in the order of 25% have been suggested – or rendered brittle by excessive tempering severely affected the reliability of weapons containing such substandard parts. This also applied to stocks of parts abandoned by the occupier at the Liberation, which were initially recovered to fill immediate orders.
As for the Mark 108 guns, the systematic sabotage at FN, and of its slides in particular, contributed to delaying the aircraft itself being deployed until October 1944, thus minimizing the damage it could inflict on the allied forces, deprived as it was of such a valuable weapon.
As regards civilian production, it should be noted that in 1940 FN had obtained authorization to complete an order for twenty-eight TB IV trolleybuses, which had been placed by the Société des Tramways Unifiés de Liège et Extensions (T.U.L.E.) in 1938. The work on the eighteen outstanding vehicles was carried out in difficult conditions at the Liège depot in the Rue Richard Heintz, under the direction of engineer Ernest Demaret (+1991).
When this work was completed in 1942, the small team created for the purpose was authorized to repair and maintain the fleet of Liège trolleybuses before rejoining FN after the Liberation two years later.

Trolleybus, type TBIV (or TB38) : only ten cars were built between 1938 and 1940. In difficult circumstances, the order's remaining twenty-eight vehicles were completed between 1940 and 1942, under the direction of Ernest Demaret.

The G.I.s in Liège

Meanwhile, the occupation was approaching its final stages. On 6 June 1944, the British and Americans landed in Normandy. After their front had been breached (30 July 1944), the Germans started to transport all the most precious equipment and machinery to their own country.

The speed with which the Anglo-American forces moved into Belgium prevented the Germans from completing this task. The first units of the American 1st Army entered Belgium on 1 September 1944.
On that day, Dr. Holl restored to Henri Pommerenke and René Lecocq the powers which had been conferred on him by virtue of the ordinance of sequestration.

On 5 September, at 20.00 hours, and accompanied by the last of his collaborators, he left Herstal. On 7 September, after the occupational forces had themselves left the factory, the Belgian management retook possession of the company. Shortly afterwards, American troops liberated Liège and the surrounding area.

The liberation of Liège on 7 September 1944 by units of the American 3rd Armored Division, cheered by the crowd in the Rue de l'Université. (Print from the Media Library of the Province of Liège. Fonds Desarcy-Robyns).

Note the presence of Browning machine guns on the armored.

The arrival of the Americans at the Place du Marché in Liège. (Print from the Media Library of the Province of Liège. Fonds Desarcy-Robyns).

Distribution of autographs by the GIs, who were welcomed as stars by the population of Liège. (Print from the Media Library of the Province of Liège. Fonds Desarcy-Robyns).

General George S. Patton (1885 - 1945), a legendary figure of the Second World War. His troops relieved the American soldiers who had been surrounded in Bastogne during the Battle of the Bulge. Liège had been liberated on 7 September by the US Army 3rd Armored Division (General Rose of General Courtney H. Hodges' 1st Army).

FN dresses its wounds

By September 1944, after the liberation, the company was able to assess the damage which it had suffered. First of all, there was the loss of manpower, particularly Alexandre Galopin, Chairman of the Board since 1932, who was assassinated in February 1944 by Belgian Nazi collaborators.
There was also the loss of equipment as a result of the military requisitions, the management by the sequestrator and the removal of equipment in August 1944, all of which had severely affected production capacity: of the 8,113 machines, 1,976 had been either destroyed or had disappeared, while those that were still left were in such a state of dilapidation that extensive repairs were needed.

A dispersed or imprisoned workforce, the dismantled workshops in Bruges and Zutendaal, the Herstal factory robbed of a substantial part of its production apparatus and soon after damaged by buzz bombs (November 1944 - February 1945), such was the sad legacy of the war.

Opposite and following pages : destruction of part of the Herstal factory by a buzz-bomb. After the retreat of the Nazi forces, the area of Liège was subjected from November 1944 to January 1945 to heavy retaliatory bombardments, mainly by V1 buzz-bombs.
FN suffered severe damage at that time.

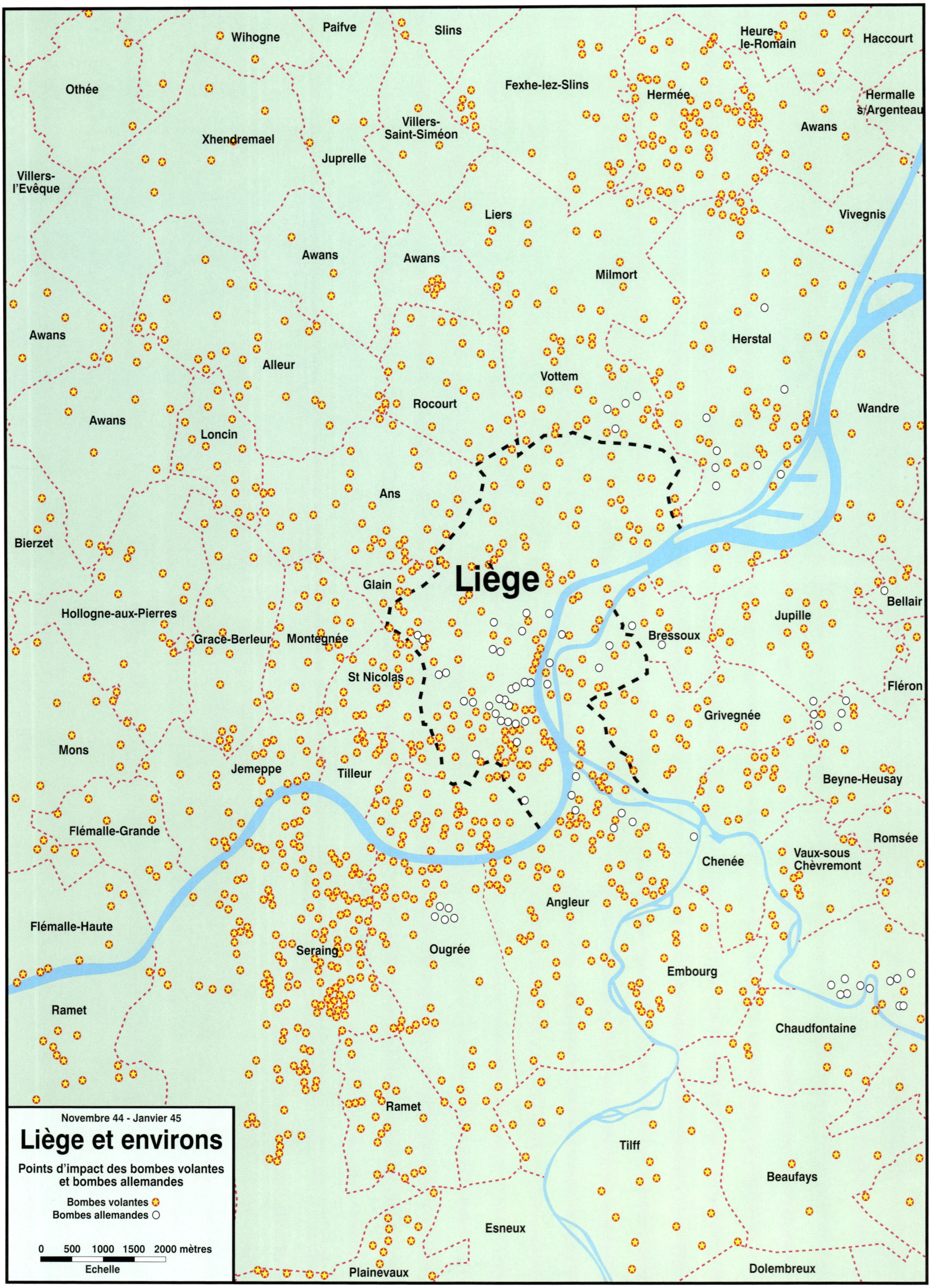

Points where V1 and V2 bombs were dropped over the Liège region. The white circles indicate the impact points of bombs dropped by German aircraft.

V1 flying bomb.

Between June 1944 and March 1945, 22,384 V1s (see photograph above) were dropped, 10,492 over Great Britain (London in particular) and 11,892 over Belgium, especially Antwerp, Liège and Brussels. Much more insidious, as they were supersonic and therefore inaudible before impact, 3,170 V2s (photograph below) were fired, 1,403 of them on London and 1,664 on Antwerp and Liège.

Launch of a V2 flying bomb. (Photograph Tracy Dungan).

The bombardment of 25 December 1944.

Restoration immediately commenced under the vigorous impetus of Gustave Joassart with the support of Georges Laloux, who had succeeded Alexandre Galopin as Chairman of the Board. As was the case after the First World War, the company was to benefit to a substantial degree from government assistance.

This became clear after the capitulation of the Third Reich. As we have already seen, during the last days of the occupation the Germans had taken away all the most up-to-date equipment. With the help of the Department of Economic Recovery of the Belgian Military Mission, 1,228 machines could be recovered between the months of May 1945 and June 1946, 331 from Germany and 897 from Belgium. Unfortunately, DWM had installed part of the equipment taken from FN in their factories in Poland and East Germany, and the Russian authorities, which occupied that zone, refused to return the equipment to FN. As a result, some 115 machines were lost.

Capital was required above all

The reconstitution of the company's assets on the other hand, constituted a problem of such magnitude that the application for a bank credit would barely alleviate the damage. In 1946, the company obtained from the Société Nationale de Crédit à l'Industrie a long-term loan of 100,000,000 francs.
This amount was subsequently raised a few times and eventually reached 320,000,000 in 1948. That same year, the SNCI agreed to share its general lien on FN's assets with the Caisse Générale d'Epargne et de Retraite and with the Société de Banque Suisse, who on their part granted each to FN an advance of 50,000,000 francs.
On 12 November 1948, the capital was tripled by incorporation of the so called revaluation reserves, the constitution of which had been approved by virtue of the Regency Decree of 15 October 1947. On 9th January 1950, by virtue of the Act of 1 October 1947, FN received from the government compensation of 350,000,000 francs for damages sustained during the war.
This compensation allowed the company to repay the Société de Banque Suisse and the Caisse Générale d'Epargne et de Retraite, while the company's debt to the SNCI was reduced to 220,000,000 with a reduced interest rate of 2.75 %.

At the same time, FN had unsuccessfully tried to recuperate a number of sizeable credits. DWM had abandoned some equipment and various items at Herstal. This property, which could quite legally have been given to the company in compensation for the material which had been removed, was instead sequestrated by the Belgian government, while FN was permitted to lay a claim to this property. In spite of repeated efforts, the situation thus created remained unresolved for many years...
The company was slightly more successful in its actions against government clients who after the war had still failed to pay for previous deliveries.
We wish to point out that Greece did honor its signature in 1956, while the People's Federal Republic of Yugoslavia made only a very small payment for transaction (1959-1961).

Replica of the FN hunting shotgun, Anson 1928 system, presented to Field Marshal Montgomery of Alamein in 1946.
The engraving in 'Liège Louis XVI' style, is by Sylvain Dorval and Félix Funken. Victor of the decisive Battle of El Alamein in North Africa, Montgomery was also one of Belgium's liberators in 1944. The arms engravers of Liège were very popular with the Anglo-Saxon liberation troops, especially those who visited FN.

On the other hand, the People's Republic of Romania has always refused to meet the commitments made by the previous regime.
In the autumn of 1944, the management of the company had started to bring the factory back into operation. At the request of the American army, which had a supply center in Liège, more than 400,000 mud grips were shaped and welded into caterpillar tracks for tanks, so that they could get a better grip on muddy soil. As it was an urgent order, people had to work twenty-four hours a day. One night, during the Ardennes offensive, the welders were subjected to a German air raid: six workers were killed and seventeen were injured.
Meanwhile, work was gradually resuming in the arms section with the finishing and assembly of the automatic pistol parts which DWM had left behind in Herstal. Several thousand weapons of this type were sold to American soldiers, while 30,000 of the model 10 were supplied to the French police. In May 1945, the supply services of the American army entrusted FN with the inspection, cleaning, repair and packing of all small arms owned by American troops on the Continent. The factory, whose workforce had been partially reformed but where material resources still remained limited, was able to complete this task within a year (6 June 1945 to 23 June 1946); more than 2,100,000 arms were inspected in this way.

Once peace had returned, the workshops could once again be equipped for their task. The cartridge factory, which had been badly affected, was also restored but according to methods which were quite different from those which had been necessary before 1940.

The factory in Bruges was closed and save for dangerous operations - fusing, preparation of explosives - which were from then on carried out in Zutendaal in the workshops adjoining the shooting range, everything was concentrated in Herstal.

A. Regout, R. De Ridder and Captain Prevot examine a mud grip developed for the tracks of American tanks.

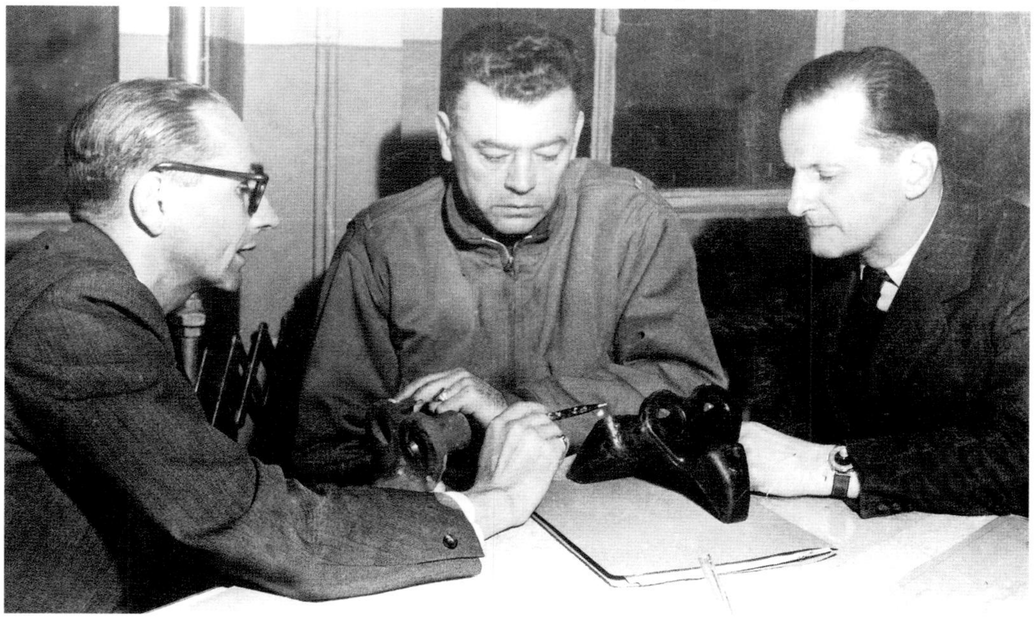

Tanks of the American liberation forces equipped with 'mud grips' (encircled). FN produced more than 400,000 steel grips, which were intended to improve the mobility of the American tanks on muddy terrain.

Manufacturing of 5,000 gasoline cans for the American army (1945).

In 1945, the American army placed an order with FN for the cleaning, repair and packaging of all its hand weapons in service in Europe. This program, named "redeployment", involved over 2,100,000 weapons. Above : the American weapons depots at Herstal (October 1945).

Repair of stocks on the M1 carbine - Group 41 (August 1945).

Fitting partial assemblies for the M1 and M2 carbines at Group 47 (August 1945).

Cleaning, repair and reconditioning of Browning .50 machine guns (August 1945).

Inspection of M1 Garand rifle parts before repair (August 1945).

Assembly of M3 machine pistols - Group 48 (August 1945).

Gustave Joassart, Managing Director, signs a document relating to the repair of the American weapons. Behind him, René Laloux, Director.

Photograph taken on the occasion of the reconditioning of the two millionth weapon for the USA.

Renewing the ties with Browning and others

In the meantime, the company was renewing its relations with its traditional clients. First of all with the Browning Arms Company. From 1946, after a visit by Val A. Browning to Herstal, this company started once again to buy large quantities of hunting shotguns from FN and, in the following years, business improved to a level beyond that of before the war.

As for the sale of military arms and ammunition, this picked up somewhat more slowly.

Peace had been acquired at such a cost that it was considered to be definitive, while just about everywhere the economic recovery required immense resources.

In fact, the first orders which FN received from Belgium, France, the Netherlands, the German government and Greece were mainly destined for their police forces and were only for pistols and ammunition.

So the prospects did not appear particularly encouraging. Gustave Joassart, who combined political wisdom with business acumen to an exceptional degree, nevertheless continued to pursue the development of new types of arms.

In this way, Dieudonné Saive developed a self loading infantry rifle which he had designed at the beginning of the last war, the plans for which he had taken to Great Britain in 1941, where he continued to work on them. Meanwhile, events outside Europe confirmed the expedience of such a decision. Already the empire which the Netherlands had created for themselves in Indonesia was becoming plunged into turmoil.

With a great deal of courage, Holland was preparing to fight, and for this purpose bought from FN large numbers of Mauser rifles, automatic rifles and infantry cartridges. In China too, after a brief lull, the communist forces resumed their campaign against the government of Field Marshal Tchang-Kai-Chek, who placed substantial orders with FN. Finally, from 1948, Egypt, which had severed the political links which had united it with Great Britain, agreed to purchase its arms from Belgium: FN supplied it with automatic rifles, self-loading infantry rifles of the latest model which we have just discussed, as well as with cartridges.

During that same period, FN had resumed its relations with the South American governments, particularly with Venezuela, which bought massive quantities of Mauser rifles, automatic rifles and HP pistols as well as the first self-loading rifles to be manufactured.

Paraguay, Argentina, Columbia, Ecuador and Peru also bought various types of arms and ammunition.

To return to the Old World, we will add that Saudi-Arabia ordered a substantial batch of Mauser rifles.

FN SAFN self-loading infantry rifle (Herstal Group collection).
Between 1948 and 1961, Herstal produced 176,212 rifles of this type. The principal purchasers were Belgium, Egypt and Indonesia.
This weapon was used by the Belgian battalion during the Korean War from 1951 to 1953.

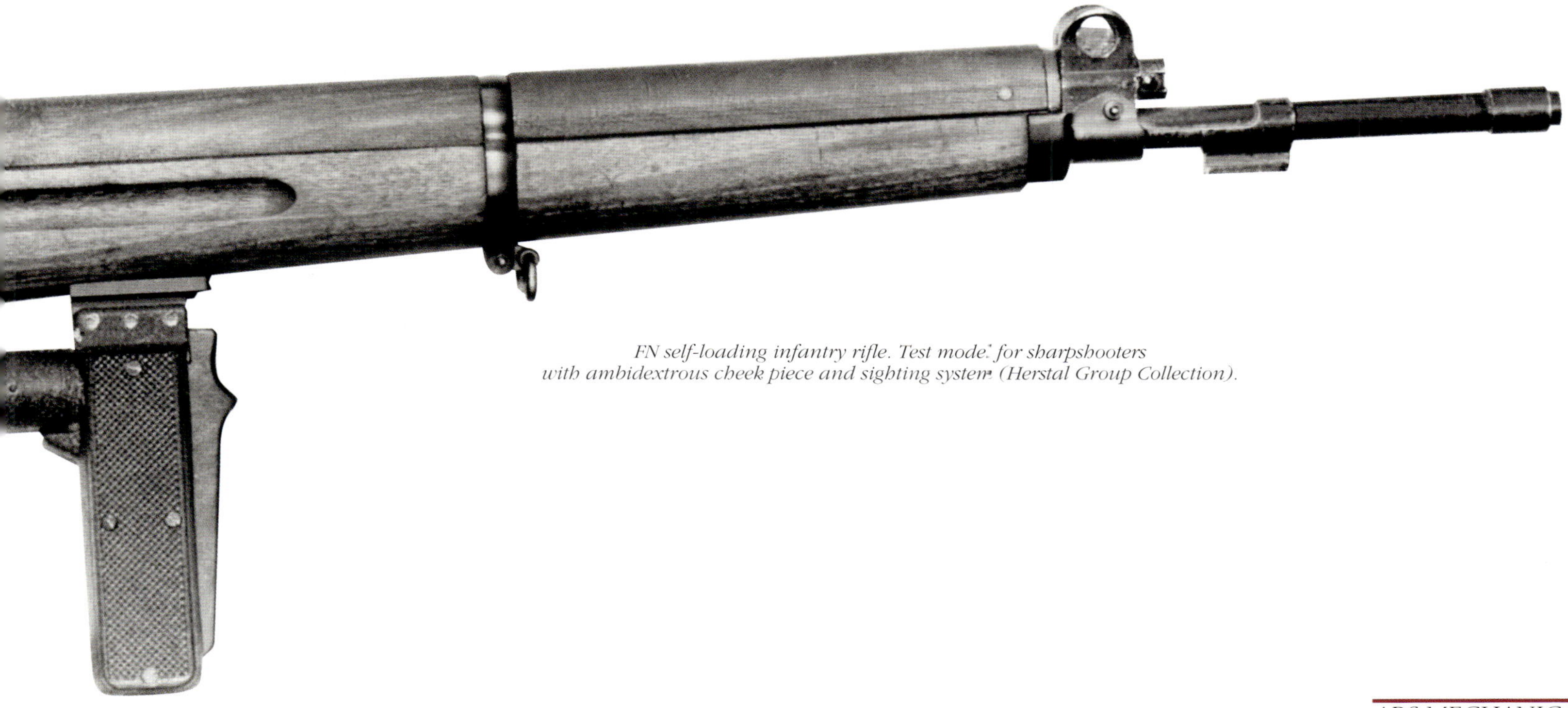

FN self-loading infantry rifle. Test model for sharpshooters
with ambidextrous cheek piece and sighting system (Herstal Group Collection).

FN at the motor races, on the farm, in the air

FN had also restarted the motor vehicle section, particularly that of motorcycles. The company manufactured vehicles of medium cubic capacity and took part in various motocross competitions where it achieved many brilliant victories. At the same time (1946-1947) a section manufacturing agricultural equipment – first milking machines, then aluminium milk churns – was set up in the workshops of Pré-Madame.

We have said that after the First World War, Alexandre Galopin had dreamt of converting FN into a car and motorcycle factory and how circumstances had forced the plan to be abandoned. In the period now under discussion an event occurred which, as the years went by, appears to have brought about a considerable revolution in the traditional structure of the company.
In 1948, the Belgian government ordered from FN quite a large number of jet engines for the Meteor fighter planes of the Belgian and Dutch air forces.

For the construction of these engines – Derwent turbo jet engines with centrifugal compressor, developed by Rolls-Royce Ltd. – FN had to train specialized staff and buy state-of-the-art and costly production equipment. The sequel to this story will show just how valuable this effort was.

Financial recovery, reconstitution of the former sections, creation of new and important divisions, all tasks which were carried out at the same time and with astonishing speed. On 26 October 1950, the board of directors was able to announce to the general meeting of shareholders that "with respect to the industrial and commercial aspects, no traces remain in this company of the events of 1940-1945". It was also able to announce for the first time since 1938 the distribution of a dividend fixed at 175 francs per share, but deferred for one year.

On that same day, the shareholders learned that one of the chief architects of this remarkable recovery, Gustave Joassart, had just turned the general management of the company over to René Laloux, who had been his deputy since 1948, although he was to remain Managing Director.

Tri-car FN T8 (1947).

FN motorcycle MXIII - 250 cc (1948).

The former "Daughters of the Cross" boarding school in Wandre (Liège). Accommodation center for FN's female Italian workers (June 1947). In view of the recovery in the national economy, the Belgian government resorted extensively to Italian 'immigrant labor, particularly in the Liège region, where the coal, steel and engineering sectors benefited from this influx of labor.

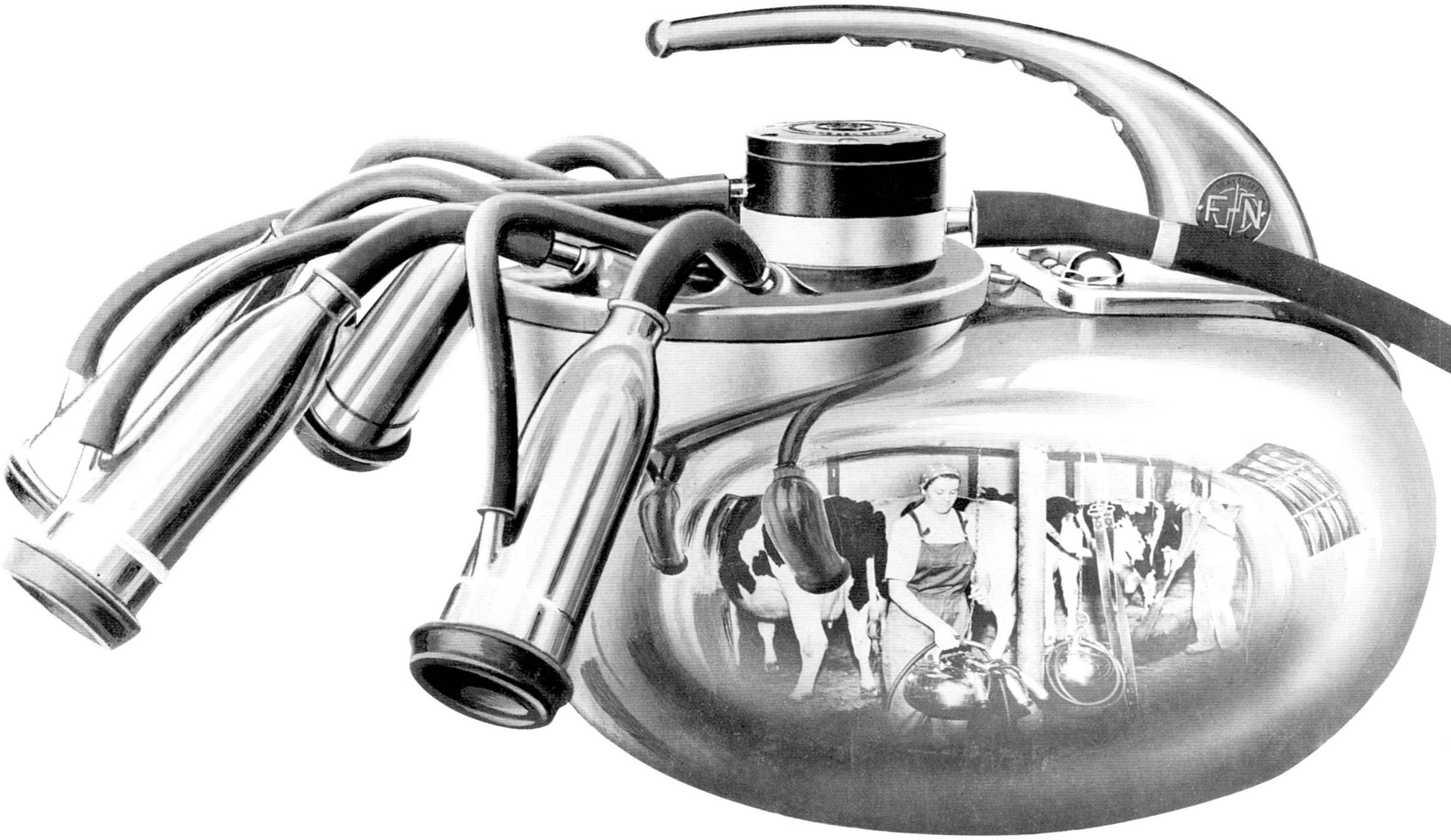

FN milking churn, suspended model.
In 1946, FN, which was seeking to diversify its production, set up a department for agricultural machinery, one of the most successful of which was this suspendable milking churn, which consisted of two stainless-steel shells welded together, with the seam neatly polished away.
In 1957 this creation obtained the 'Signe d'Or' prize for industrial design.

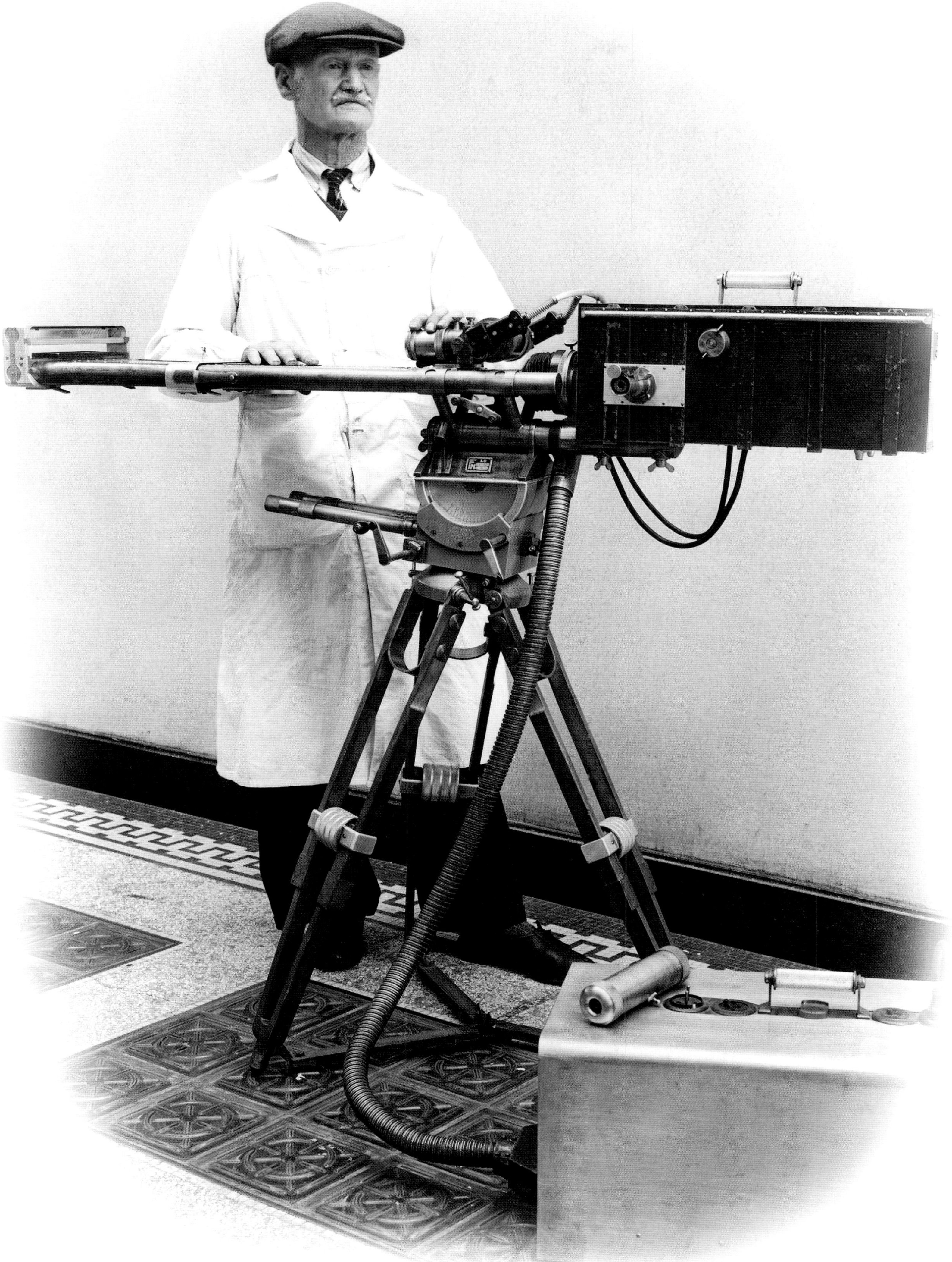

A device for recording the movement of an arm, and its user Monsieur Pirotte (1950).

Between 9 December 1950 and 15 February 1951, seven Belgian servicemen completed a journey of approximately 12,000 kilometers on FN 450 cc motorcycles to establish a connection between Brussels and the new base of Kamina in the south of the former Belgian Congo.

Passing through Tadmaït (Sahara) during the long-distance rally between Brussels and Kamina. Engraving by Fernand Vetcour. Illustration created for an FN New Year greetings card.

A reproduction of the map of the Brussels-Kamina Race by seven Belgian parachute officers, on FN motorcycles (1951).

The Gloster Meteor: the Belgian Air Force's first jet-propelled aircraft (1948).
The government ordered the plane's Derwent engines from FN, thereby allowing the Herstal company to launch into a new and promising domain.

Assembly line for Derwent engines at FN. These engines, which were manufactured under license from Rolls Royce, represented at that time a technological advance which FN made successfully, relying on its fifty years experience in the field of land vehicles.

Publicity photograph (1948) of a female worker at FN lifting a light alloy (Hiduminium) tube.
As a particular result of the contacts forged by Herstal engineers who had taken refuge in Great Britain during the war, new techniques practiced by their Anglo-Saxon allies were brought back to Belgium. Light alloys, carbide and tungsten cutting tools, aviation engines etc.

Harmony at FN in 1950. The characteristic paternalism of the old companies, which persisted at Herstal until the 1970s, created amongst the personnel a certain camaraderie and a feeling of belonging, although this did not prevent the appearance of serious social conflicts on occasion.

Tradition and prosperity

For the historian of FN, 1950 marks not only the completion of the reconstruction and a change of management. Various events, both within and external to the company, have given the era which we shall be describing a particular significance.

As a military arms manufacturer selling its products all over the world, FN has always been particularly sensitive to changes in the political situation. Moreover, it is well known that no sooner was the Second World War over than omens of new conflicts began to emerge.

The aspirations for world dominance as displayed by Soviet Russia had compelled the leading nations of Europe to conclude in 1949 a military alliance with the United States of America, the so called North Atlantic Treaty, which provided for the creation of a permanent organization, called NATO. This organization immediately proceeded with the consolidation of the armed forces of the member states.

The war which broke out in Korea and which highlighted the lack of preparation displayed by the free nations, induced them to step up their efforts. But it soon became clear that a military alliance did not suffice to face the threats which preyed on Europe.
This consideration led the industrialized nations of our continent to co-ordinate their actions on the economic level. After the constitution of the European Coal and Steel Community (18 April 1951), there came also a treaty constituting a European Economic Community and one for a European Atomic Energy Community (27 March 1957).

These treaties, which incorporated the markets of Belgium, the Netherlands, Luxembourg, Germany, France and Italy into a single homogeneous economic common market, tended to recreate, at least as far as the economic relations between these countries were concerned, the pre-1914 situation.

The concerted disarmament of the free nations and the unification of Europe could however not prevent the fragmentation of the colonial empires which was to mark profoundly the last years of the period we are currently discussing.

*René Baeten, winner of the 1958 world championship motorcycle race, with an FN motorcycle.
In spite of decreased post-war production (less than 1,000 motorcycles of high cubic capacity per year), FN retained and even reinforced its international reputation in this area thanks to the success of its engines in sporting contests.
It secured the collaboration of leading motorcycle-racing champions, but at the end of 1958 it withdrew from competition because of the fall in sales of Belgian motorcycles. Production of the two other Herstal manufacturers, Gillet and Saroléa, ceased at around the same time.*

Employers and workers

So while the economic and political context in which the company was developing its external activities was changing, a slower and less perceptible movement was affecting its internal development.
We are referring here to the changes affecting the relations between employers and workers. In the post-war years, circumstances had forced the two parties to co-operate. Indeed, discounting the strike which in July 1950 coincided with the return of King Léopold III to Belgium , we can safely say that between 1944 and 1950 industrial disputes were few in number and of very limited gravity. But now that prosperity had returned creating issues about sharing the benefits of this prosperity, questions began to arise as to how relations between the company and the workforce should evolve.

We shall now see how this development unfolded for FN. It is well known that for half a century the region of Liège, where the vast majority of the workforce came from, had had a very low birth rate, while the working population was growing progressively older. The consequent scarcity of human resources compelled the company to take on immigrant workers.
All these phenomena directly affected FN, especially as its products required a large number of specialized workmen.
Also the labor cost constituted a considerable proportion of the prime cost of the weapons.
Moreover, the effect of abundant orders as well as the general prosperity of the company certainly had a part to play in driving up wages.

Also the company's concern to secure the services of diligent collaborators had considerable effects. With this in mind Gustave Joassart decided to allocate to the workers an annual premium as a reward for their loyalty to the company. As time went by, this premium was increased and reached 7.5% of the overall annual wage bill in 1964. For 4,000 workers, that is, approximately a third of the total blue color workforce, this premium represented more than 9 % of the contractual wages.

The importance of this premium must not, however, overshadow other more recent company institutions. It may also be added that the older institutions have developed considerably. By way of example, in 1952, the welfare center was completed by a dental surgery where the staff could have all but free treatment, and the total figure of building subsidies paid out by the company during that period topped four million francs per year, etc.

On 14 September 1953, the board of directors set up a fund for scholarships, called the 'Joassart Fund', with capital fixed at two million francs and subsequently increased to eight million. From 1 June 1956, the staff could join a pension scheme for their retirement pension and death-in-service benefit.
On 22 October 1959, a resolution by the general meeting of shareholders constituted a redundancy fund for workers laid off or unemployed.

Joint enterprise

The increase in the remunerations, be they wages, salaries, bonuses or other benefits, could not have been possible financially without a corresponding increase in output, and specifically in the relation between the value of the goods produced and the corresponding labor cost, which is called productivity. In a company like FN, productivity depends first and foremost upon the quality of the equipment and the organization of the production process. But it is clear that even the most sophisticated equipment and the most rational organization require the constant collaboration of the staff.

This fact, dictated by common sense, points to the connection which exists between the general prosperity of a company, the productivity of the work and the size of the remuneration. It also serves to underline the idea that although the short-term interests of the various economic entities may be at odds from time to time, their long-term interests always converge.

The painful experiences of the great depression and of the Second World War had in no small measure underscored belief in this already well-established idea, among those who were considering the future of modern companies.
More and more philosophers, sociologists and economists were proclaiming the belief that the inevitability of social conflict must be substituted by a practical philosophy based on solidarity and co-operation.

Cartridge group. The retirement of a member of the personnel (1955).

One event, which occurred at around the beginning of the period being discussed, represented the hope that this philosophy would from then on bring about a change in the industrial relations between FN and its workforce. On 6 April 1950, a works council was set up within the company which, as Gustave Joassart strongly underlined in his inaugural speech, "had for its principal objective the completion of the economic and social framework of our institutions, by creating the necessary organisms for the establishment of a fruitful collaboration between the social groups who are working together in the economic process."

From this point of view, it is true to say that the results of the works council's first fourteen years have not been as favorable as might have been hoped. After a period of industrial peace which lasted until June 1957, a large proportion of the workforce became involved in several strikes.

(It is worth noting that some of these strikes were the result of outside influences: the most serious strike, the one during the winter of 1960-1961, had been motivated by the passing of the so-called 'Loi Unique, i.e. General Act). We know from Montesquieu that ideas work in a subtle manner and also that these conflicts should not obscure the fact that slowly and almost imperceptibly the relations between employers and workers were becoming characterized by a new style.

While events were causing an upheaval in the framework in which the company was moving, the two personalities who had marked most profoundly the character of the company had passed away within a few years of one another.

On 4 June 1953, Gustave Joassart suddenly died in his study at his home in Sarolay, near Herstal.

On 30 January, Georges Laloux died in Liège after a brief illness. A most likeable man of the world, with a discerning and graceful spirit, Georges Laloux had upheld in FN the intellectual and moral traditions which had inspired the company's founders.

A severe but nevertheless freer and more audacious genius and invested with a rare shrewdness and an even rarer diligence, Gustave Joassart had led the company masterfully through the 30 most difficult years of its history.

In spite of the changes which events had brought about in the company's make-up, the style which both men had given to FN would remain visible for a long time to come.

Celebrating Saint Hubert's Day in the Cartridge factory (1955).

As the most important employer in the Liège area after the steel and coal industry, FN's chief concern has always been the recruitment, training, management and remuneration of the workforce. In 1965, the company employed a record 13,000 people in Belgium, mainly at Herstal. Below: a group of pipe fitters (October 1955). Sitted in the center : the group's social worker.

Cartridges for NATO

FN had always remained faithful to its traditional fields of activity. The manufacturing of arms and infantry ammunition in particular had enjoyed almost uninterrupted growth. Around 1950, most free nations had felt the need to modernise and, if possible, to standardize the equipment of its armed forces.

Great Britain, where since the Second World War the top men of FN had maintained some solid friendships, invited the company to take part in the development of a new type of infantry cartridge for the NATO armed forces, thus replacing the American Army's .30-06 ammunition, the British Army's .303 , the French Army's 7.5 mm, the Belgian Army's 7.65 mm and many others.

There was a great deal of heated discussion among the experts of the different countries as to the specifications of the new cartridge. Some, like the Americans, wanted at all costs that the stopping power of the existing ammunition types be matched by the new the standard.
Others, however, were of the opinion that a less powerful and therefore less bulky cartridge would be better adapted to the conditions of modern warfare.

The range of cartridges used by NATO countries before the adoption of the 7.62 mm ammunition :
1) 9 mm Parabellum
2) .30 Carbine USA
3) 7.9 mm short German
4) .280 or 7 mm short
5) 7.5 mm France
6) 7.65 mm Mauser
7) .303 U.K.
8) 7 mm Mauser
9) 7.9 mm Mauser
10) .30/M2 U.S.A.

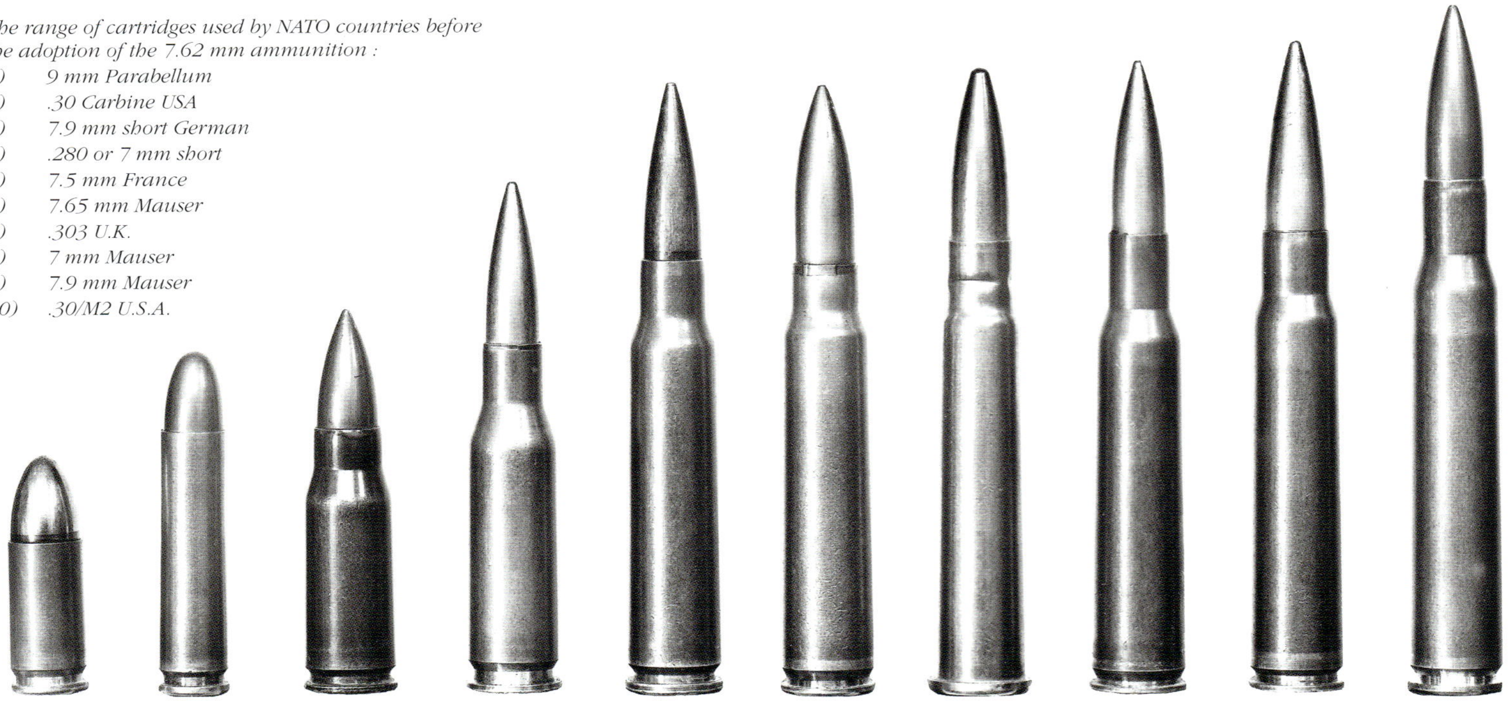

Various types of FN 7.62 x 51 mm cartridge. Developed in the 1950s by FN, the 7.62 x 51 mm cartridge was approved by NATO in 1957, and was intended to replace the different types of infantry ammunition which had previously been used by the various countries of the Atlantic Alliance.

Modified, diversified and improved over the years, the LAR was a resounding success in the countries of what was then called the free world, as well as in some non-aligned countries.

The extraordinary adventure of the LAR was the result of a favorable combination of factors which is, as is nearly always the case, neither totally reproducible nor transferable: the creativity and capacity for adaptation of FN's technicians, starting with the inventor himself, the farsightedness and abilities of its manager René Laloux, the experience and capital of sympathy developed in Great Britain during the war, the redefinition and renewal in armaments under the sponsorship of those very Anglo-Saxons whose dominance was to be affirmed in the context of the Cold War, and finally – a related factor – the absence of geopolitical pretensions of a small country like Belgium.

In the light of all the above, the LAR was undoubtedly one of the most significant weapons of the twentieth century.

The first production series LAR produced in Herstal's workshops was presented to the King of Belgium. The model presented above bears the number two and this was presented to the Governor of the Société Générale de Belgique, the majority shareholder at the time. Today it forms part of the Herstal Group's heritage collection.

Dieudonné Saive, arms designer with FN (1888-1970).
A former collaborator of J. M. Browning,
he headed the research department in Herstal.
In 1930 he improved the Browning Automatic Rifle, he developed the HP and "Baby" pistols as well as the self-loading rifle shortly after the war. But to him we owe most of all the invention of the light automatic rifle (the LAR), one of the most significant small arms of its generation.

A Canadian soldier operating the LAR (1960).

Instructors of the Australian Infantry Training College showing target positions with the LAR (1960).

Manoeuvres of the Peruvian Navy Infantry, equipped with LARs (1960).

Training the 1st Coldstream Guards in the use of the LAR, at Chelsea Barracks, 28 August 1959. The elite regiment whose origin dates back to the Restoration of the British Monarchy, in 1660, mounted guard for the first time with this rifle at Buckingham Palace, on 1 September that year.

Prince Philip, Duke of Edinburgh, inspecting the new LAR rifle in the hands of the Welsh Guards, Caterham, Surrey (1959).

A LAR presented by FN to Sir Winston Churchill in 1955, shortly after the adoption of the "Belgian rifle" by the British Army the previous year. This decision was decisive for the subsequent success of the weapon, in the Commonwealth countries as well as in other nations. The inscription engraved on the magazine receiver is a quotation by the great British statesman with regard to the LAR. London, Churchill Museum.

Destruction testing the American T48 light automatic rifle at Fort Benning, Georgia (April 1956).

Procession of a military detachment, armed with the LAR, past His Majesty King Baudouin 1st of Belgium (1960).

A military review in Canada (1960). Canada was the first country to adopt the LAR rifle, in 1953.

Jean Van Rutten (1925-1982), one of the best known demonstrators of FN arms throughout the world, here presenting a LAR to His Imperial Majesty the Shah of Iran on 9 November 1958, outside Tehran.

Using the LAR in Germany (1960).

Demonstration of the LAR in the Federal Republic of Germany in the presence of Chancellor Konrad Adenauer (1876-1967), on 22 February 1960.

Chancellor Adenauer reviews troops armed with the LAR (1960).

1950-1964

The MAG, the general purpose machine gun

The MAG is a by-product of the Second World War. In fact that conflict had highlighted the need for a new type of machine gun. It appeared that troops now needed a machine gun which would be both sufficiently lightweight to serve a squad support role and sufficiently heavy to deliver sustained deterrent fire.

Several countries, including Germany, had shown the way in this regard, but after 1945 there was still no weapon which could muster sufficient support.
It fell to Ernest Vervier (1908-1986), Dieudonné Saive's successor at FN, to find the most effective synthesis between the two approaches – offensive and defensive – in automatic weaponry.

During the 1950s, he developed a gas-operated machine gun, called the "Mitrailleuse à gaz" or the "Mitrailleuse d'appui général" (General Purpose Machine Gun), shortened to the MAG, which met the criteria for multiple use.

Ernest Vervier (1908-1986), inventor of the MAG. He succeeded Dieudonné Saive at the head of the research department at Herstal,

It combined an action derived from the BAR automatic rifle (then manufactured by FN) but with a different locking system, and a feed principle which was similar to that of the German MG42 machine gun.

Like the LAR, the machine gun had undergone thorough tests, particularly in the snow of Kiruna, in the north of Sweden, and in Great Britain.
These two countries adopted the weapon forthwith and obtained a manufacturing license from FN.

The first country to acquire this weapon was Sweden, although in a special caliber, the 6.5 x 55 mm, which was changed to the NATO 7.62 caliber in 1962.
All in all, the MAG was to be adopted by over 70 countries, many of which manu facture it themselves. Interest in this weapon was renewed when the USA adopted it in 1976, under the name M240.

Since 1978, FN has made mounting these machine guns on land, air and sea vehicles a speciality, in increasingly sophisticated configurations.

To complete our list of small arms developed by FN, we will add that since 1958 the company had been producing a sub-machine gun called the UZI, designed by Captain Uziel, an officer attached to Israeli Military Industries. This sub-machine gun had already been adopted by the Federal Republic of Germany.

To the four corners of the world

In the opinion of the specialists, the LAR and MAG – like the HP pistol – had become what Liddell Hart had said once of the Mauser repeating rifle and the Bren gun, that is to say, the "classics of infantry armament".
It should therefore not surprise us that these products had brought FN the most remarkable commercial successes in its history. Without violating the rule of discretion which this kind of business imposes, we list below the nations to whom the company supplied considerable quantities of arms from 1950 to 1964, discounting a number of smaller countries who had not bought such large quantities.

Europe : Austria, Belgium, Denmark, Federal Republic of Germany, France, Great Britain, Grand Duchy of Greece, Ireland, Italy, Luxembourg, the Netherlands, Norway, Portugal, Sweden.

North and Central America : Canada, Cuba (before 1960), United States.

South America : Argentina, Brazil, Chile, Columbia, Ecuador, Paraguay, Peru, Venezuela.

Asia : Cambodia, Iraq, India, Iran, Israel, Lebanon, Malaysia, Pakistan, Saudi Arabia, Syria, Thailand, Turkey.

Africa : Burundi, Democratic Republic of Congo, Egypt, Federation of Rhodesia-Nyassaland, Ghana, Liberia, Libya, Morocco, Mozambique, Rwanda, Union of South Africa, Zambia.

The length of this list becomes significant if we remember that most countries do not procure weapons before they have been submitted to extensive technical and operational trials.

Moreover, these orders allowed FN to build up close relations with the highest military and civilian authorities of the interested countries.

where he invented the MAG machine gun (1957) and conducted early research for 5.56 mm caliber infantry rifles.

MAG 7.62 mm NATO machine gun. (January 1958).

MAG machine gun in action. Since it was invented, this weapon has been considered to be one of the best of its kind in the world.

The 1st Battalion of the Royal Ulster rifles on patrol in Malaysia. These men are armed with LARs and a MAG machine gun (1965).

One thousand three hundred million infantry cartridges

Before 1965, more than 650,000 arms of this type had been manufactured at Herstal and about 500,000 others in Great Britain, Canada, Australia, Argentina, South Africa and Austria. As soon as the LAR was adopted, these countries applied for a license to manufacture it for their own use.

Considering that the MAG is a more complex and larger weapon, the production figures before the mid sixties – 22,000 at Herstal alone – are not to be sneezed at either. By then, twenty-two countries had procured this machine gun, two of which, Great Britain and Sweden, produced the gun under license.
It is also worth noting that in the same period 200,000 Mauser rifles and 250,000 H.P. pistols were produced at FN and supplied to a large number of countries.

The procurement of arms was usually accompanied by the purchase of ammunition, with the difference of course that the latter were bought in far greater numbers. It is indeed in terms of millions and even hundreds of millions that FN manufactured infantry cartridges.

From 1950 to 1964, one thousand three hundred million of these cartridges were produced. This figure includes all types : with standard, armor piercing, tracer, incendiary or propulsive bullets, blank cartridges or exercise cartridges.
Half of them are at least 7.62 mm caliber, that is to say the model standardized by NATO, sales of which exceeded the hopes even of its advocates.
At the same time, more than 430,000,000 pistol cartridges, 60,000,000 medium-caliber cartridges (12.7 mm, 20 mm Hispano, 20 mm Oerlikon, 20 mm Vulcan, 30 mm Aden) and almost 2,500,000 parts for the 40 mm (complete) artillery shell were produced in the factories of Herstal and Zutendaal.

A considerable proportion of this ammunition, especially the heavy ammunition, had been manufactured during and after the Korean War for the United States Administration (so-called Offshore contracts).
Infantry cartridges and artillery shells are FN's traditional products. Meanwhile, advances in applied ballistics were also contributing to a revolution in ammunition technology. From 1957, FN manufactured various models of grenades shot from rifle-barrels (so-called rifle-grenades).
These models had been developed by the Société Technique de Recherches Industrielles et Mécaniques (S.T.R.I.M.) in Paris. Subsequently FN designed various types of rockets. Also, a testing stand was set up within the precincts of the shooting range in Zutendaal.

It is worth remembering that before the Second World War FN had built parts for the Bofors Aktiebolaget-designed anti-aircraft guns. After the war, this company invented a more powerful gun – a 57 mm – which was fitted with a remote-control device. It was a very complicated weapon, and Bofors granted FN the license for the manufacturing of a major component of the gun for the Belgian Army (1953-1956).
FN was similarly entrusted by a Dutch company with the construction of the platforms for the Bofors 40 mm L70 guns (1955-1960).

Bofors anti-aircraft gun, 57 mm caliber, built by FN from 1953. More powerful version than the 40 mm, which was manufactured before the war, and fitted with a remote-control device.

Successes of the hunting and sporting arms

The manufacture of hunting weapons certainly developed in a less spectacular way than that of armaments, but nevertheless with a regularity which was economically no less satisfactory. In this respect, two characteristics have typified these 14 years. First of all, there was the fact that the United States remained a very loyal customer. Secondly, there was the success of FN's arms on the European markets after the Treaty of Rome started to break down all commercial and administrative barriers.

The United States remained FN's most important client. In August 1963, the millionth Auto-5 shotgun since the end of the Second World War was delivered to the Browning Arms Company.
The collaboration between FN and John Browning's descendants was, however, not limited to commercial exchanges.

Successively, the inventor's son and grandson had brought into the company their own arms designs. In 1950, there was a new type of automatic shotgun (the 'double automatic') by Val Allen Browning, and in 1962 an automatic .22 caliber pistol by his son, Bruce W Browning. By the end of 1962, furthermore, FN had taken a holding of 35,000 shares (± 3%) in the capital of the Browning Arms Company.

However, the importance of the American market should not obscure the fact that Europe, or at least the six signatories to the Treaty of Rome, had become a significant outlet for hunting weapons. Here there was a tendency which could be of prime importance to the commercial orientation of the company.

Even more so than the arms race, the progressive realization of a common market in Europe could be felt in the sale of hunting ammunition. Where before this had been limited to Belgium, FN would now export increasing quantities of hunting ammunition. In this respect, the development of a powerful cartridge – the Légia Star High Speed – provided with a plastic cup wad invented by an FN engineer, André Laloux had given the company a very good reputation among shooters and hunters, which contributed in no small degree to the above-mentioned commercial successes.

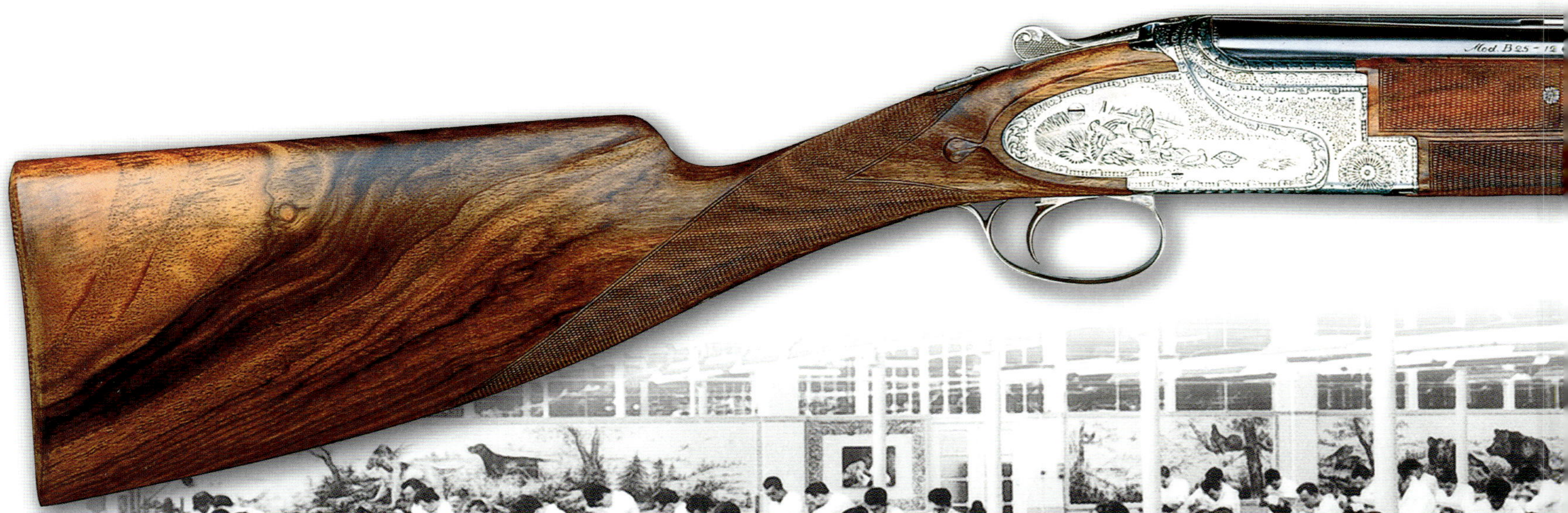

The arms engraving workshop at FN as it looked in the sixties. The FN arms engraving workshop was by far the most important in the world :

Automatic .22 caliber pistol, "Match" model.
A modern version, with engravings and sculpture in Renaissance style, of the sporting pistol invented by Bruce Browning in 1962.

Browning B25 "over-and-under", Special Game 13 model cal. 12. M2 side-plate engraving. 3 piece fore-end. Renamed B25 to distinguish it from its competitors and derived models, the Browning "superposed" met with an increasing success as soon as the sporting arms department was reactivated after World War II.

more than 180 craftsmen once worked there. In 1987, it was transformed into an independent co-operative society.

Browning Over-and-under shotgun. Gold inlays with chimeras. Composition by Louis Vrancken (1974).

Browning Over-and-under shotgun. Renaissance engravings and figures by Louis Vrancken (1975).

Browning Over-and-under shotgun. Chased, engraved and inlaid with gold by José Baerten.
Diana, hunting goddess, plant motifs and stylized animals. Composition by Louis Vrancken (1975).

Commemorative model of the two millionth Browning Auto-5 manufactured at FN since 1945.
Gold inlays by André Watrin, José Baerten and Gaston Vandersmissen (1970), inspired by the model developed for the International Exhibition in Liège in 1930, by Félix Funken and C. Oukelinse.

Publicity brochure from 1952. Artwork by Jacques Ochs.

Cover of the publicity brochure for FN hunting cartridges, by Maurice Mathy (1965). This highly talented Liège graphic artist was one of the FN Publicity department's main collaborators during the 1960s and 1970s.

Publicity erased from history !

From its invention in the mid nineteenth century until the 1930s, the hunting cartridge, with its brass cup and primer, consisted of a cardboard tube holding the lead shot.
The disadvantage of this system was that the tube was permeable to humidity. Long storage in unfavorable atmospheric conditions could result in swelling of the ammunition and deterioration of the powder.
To overcome this problem, especially acute in rainy and tropical regions, cartridges with light metal tubes – in theory impermeable and unwarpable – were designed in the 1930s. Examples were the 'cuirasse' (armored) and chiefly the 'métallique' (all metal) types.
These were manufactured along with the cardboard types until they were both replaced, over the course of the 1960s, by ammunition with plastic tubes.

To praise the modernity of the 'Métallique' cartridge, as compared to the traditional Légia type with a cardboard tube, FN commissioned the Liège artist Jacques Ochs (1883-1971) to create a publicity leaflet intended to promote this innovation.

A former military airman and Director of the Liège Académie de Beaux Arts, Ochs was also a talented caricaturist, known for his satirical illustrations on the covers of "Pourquoi Pas" magazine. Happy to return to allegory, he chose to represent the old cartridge as an old monarch making way for the new product, personified by a young sovereign. The height of poor timing. The leaflet came out in 1951, a year when Belgium was rocked by a political crisis known precisely as the "royal question" ! In actual fact, on 16 July that year, King Leopold III was forced to abdicate in favour of his son Baudouin I.

Afraid that this advertising material might be construed as taking a position on the controversy of the moment, FN's management decided to destroy the leaflets and to prohibit their distribution.

Taken from the company archives, this leaflet – which has never existed officially – is shown here for the first time.

Sky riders

We have pointed out already that at the end of the 40s, FN had undertaken the manufacturing of aircraft engines of the Rolls-Royce Derwent type. More than 1,000 engines of this type were produced, a large number of which were subsequently repaired and overhauled at Herstal. In 1954, when these works were almost completed, the company obtained directly from Rolls-Royce Ltd. a license for the production of turbo-jet engines with axial compressor, the Avon model (versions 113 and 203).
These engines, which were constructed for the Belgian and Dutch air forces, were placed in their Hawker-Siddeley Hunter fighters. In addition, FN was entrusted with the repair of the engines purchased by the Belgian government as well as some of those procured by the Netherlands.
A little later, the company undertook to repair and overhaul other jet engines, such as the Canadian Orenda (on the Canuck CF 100) and the French Marboré (on the Fouga Magister).

The improvement and associated rise in price of the fighter planes, as well as the tightening of relations between the European nations had induced them to review their rearmament policies. In this respect, West Germany (1958), Belgium and the Netherlands (1959) and Italy (1960) adopted the fighter-bomber F104 G (Starfighter) in their air forces and agreed on various protocols, the main one of which was signed in Wahn, near Cologne, on 5 March 1960, to allocate the manufacturing of this aircraft to the factories of each country, so as to reduce as much as possible the cost price of the whole plane.

FN, whose reputation in aeronautics was already firmly established, was also entrusted with the manufacturing of a large section of the engine (44%, later to become 46%) – the turbojet engine with axial compressor General Electric J79 – while the rest was allocated to the Bayerische Motorwerke (BMW) in Munich and to FIAT in Turin, so that apart from the assembly and testing, which were assigned jointly to the three companies, there was no double use of equipment.

It is worth pointing out that in this respect it concerned only principles which were jointly agreed by the various states, but which strictly speaking did not commit them as far as their future suppliers were concerned. FN did not hesitate to make the necessary preparations for this work.

Within a period of a few months, several license and technical assistance contracts were signed, first of all with the General Electric Company which had designed the engine, the workshops were equipped with the necessary machinery and installations, while raw materials, certain finished parts and accessories were conditionally purchased from the United States, and co-operation agreements were concluded with FIAT and BMW. Finally, on the strength of a letter in which the Belgian government officially informed FN of its intention to place an order for a certain number of engines, the company decided to commence production (2 June 1960).

For ten months, production was carried out under the sole responsibility of FN and was solely financed from the company's own funds. It was only on 11 August 1961 that the interested governments regularized the existing situation by signing a temporary contract which authorized the first payments.

Finally, on 24 October 1962, a definitive agreement was concluded. The first engine had been delivered six months earlier, and sales had already reached 20 % of the transaction volume. The boldness shown by FN did not go unnoticed by its clients.
The subsequent execution of the agreement went smoothly, while contracts for engine repair and for the delivery of spare parts complemented the initial order.
We can safely say that the transaction was a great success for FN.

The same applied to the sixty or so European companies who took part in the production of the F104 G plane under the same conditions.
This served to demonstrate to certain pessimists that it was perfectly possible for serious and dedicated companies to collaborate in a straightforward fashion in the successful execution of the most difficult and important manufacturing processes. The lesson has not been forgotten.

In terms of sales, the engine division had become, at least temporarily, the most important in the entire factory. However, this fact should not obscure the development of the other sections at Pré-Madame. These sections had experienced markedly different fortunes.

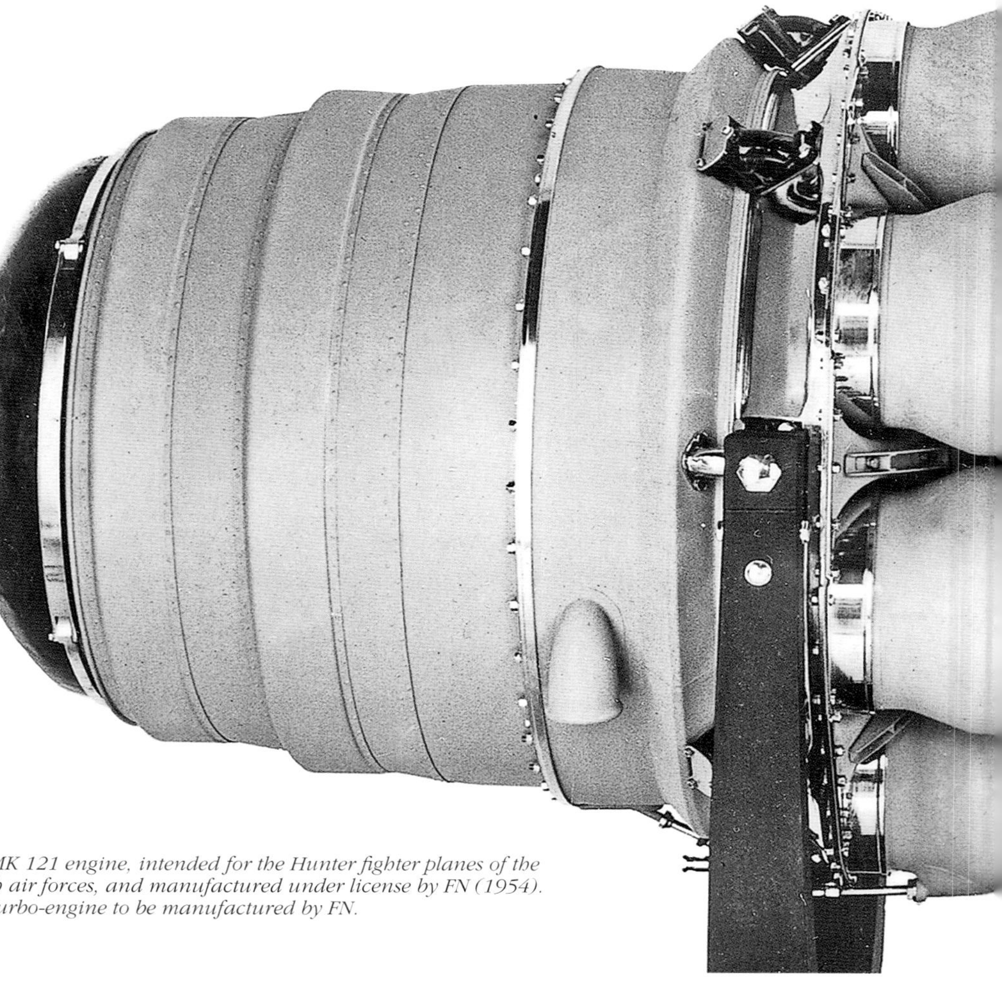

Rolls-Royce Avon MK 121 engine, intended for the Hunter fighter planes of the Belgian and Dutch air forces, and manufactured under license by FN (1954). It was the second turbo-engine to be manufactured by FN.

Built for the Belgian Air Force and for the Nederlandse Luchtmacht, Rolls Royce Avon jet engines with an axial-flow compressor equipped both countries' Hawker-Siddeley Hunter fighters. Moreover, FN was tasked with the repair of the engines purchased by the Belgian State and of some of those acquired by the Netherlands. (Photograph from 1956).

Fighter bomber F104 G Starfighter. From 1959, FN has been involved in the production of its turbojet engine with axial compressor General Electric J79.

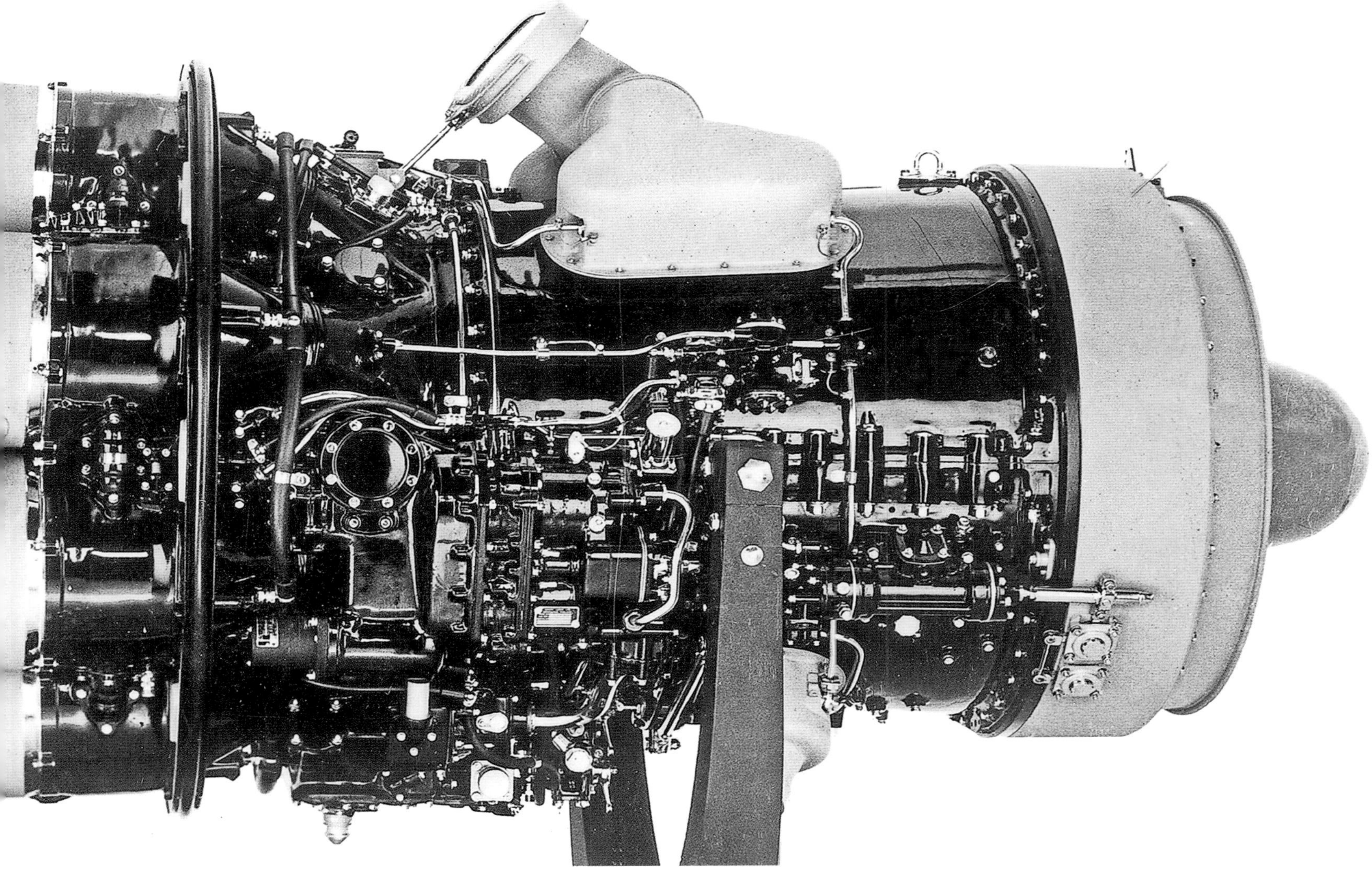

Partial assembly - turbine assemblies (1955).

Machining gas distributors (1955).

Final assembly of the AVON engine (1955).

AVON MK113 jet engine (1956).

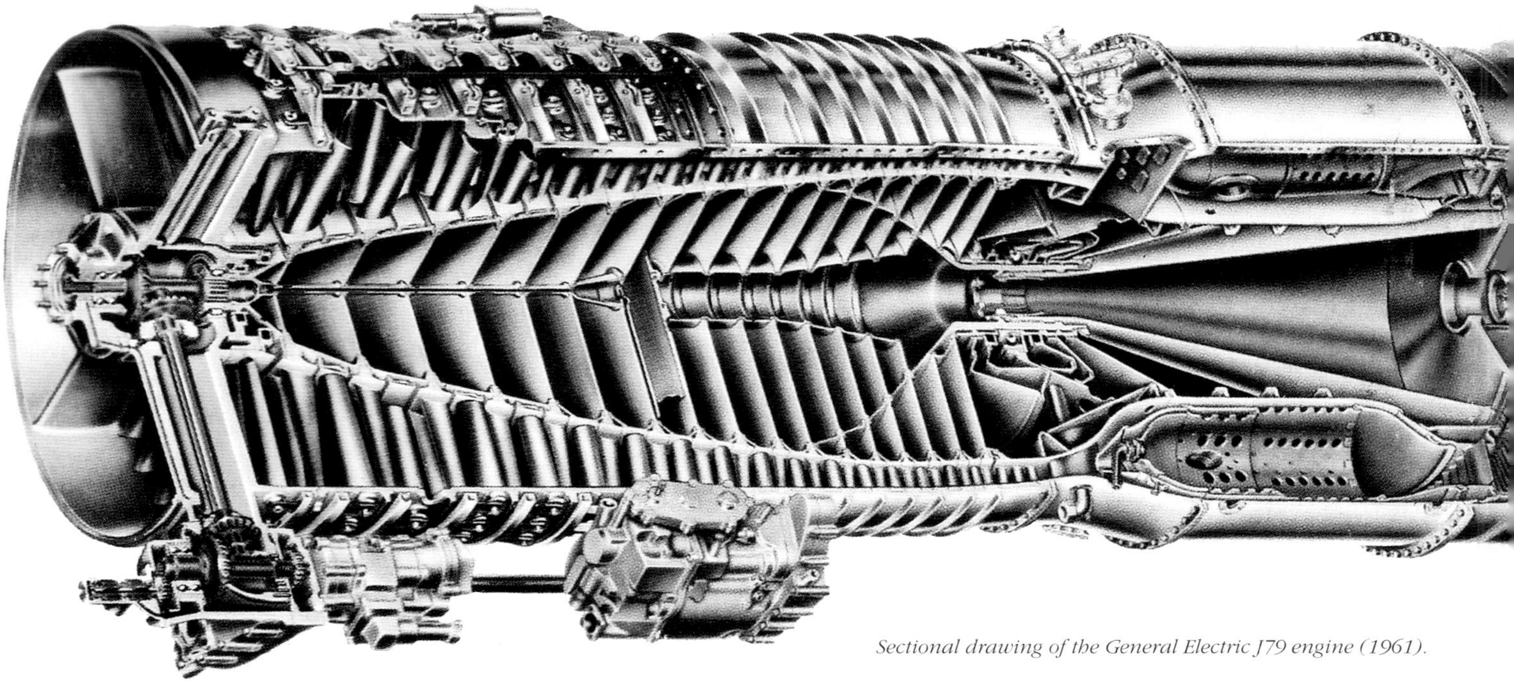

Sectional drawing of the General Electric J79 engine (1961).

Assembly of the General Electric J79 11A engine for the F104 G fighter-bomber. The manufacturing of this engine in 1960 was FN's first experience of participating in an integrated production project of international scope.

AVON jet engine (1958).

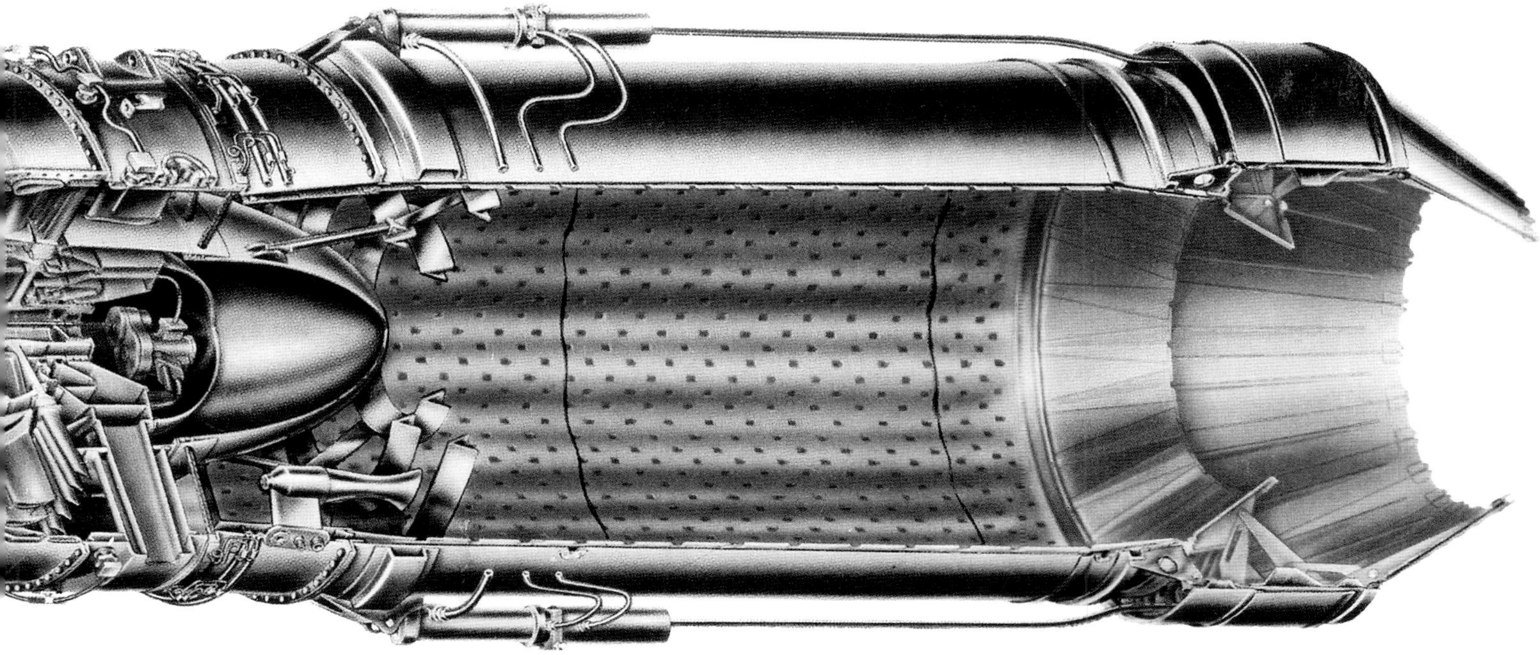

The operation was a resounding success and it prepared FN for subsequent involvement in large-scale ent?rprises (1961).

The end of the motorcycles

One of the oldest sections at Pré-Madame, the motorcycle section, had gone through a very low period due to the decline in popularity of this type of vehicle.

Nevertheless, FN had made a considerable effort to adapt to the tastes and requirements of its clientele. Apart from the heavy motorcycles, the company successively released medium-sized motorbikes, such as the 175 cc M22 (1953) and the mopeds (1959).

The success of these vehicles in the face of competition once again bears witness to the quality of these products. However, the motorcycle was losing ground to the motorcar as a utility vehicle and at Herstal the culture of enterprise mitigated against designing and manufacturing a leisure consumer product in place of a durable item, which was considered more worthy.

Foreign competition and market contraction on the one hand, and inflexibility in adapting to new "two-wheel" trends on the other, combined with the costs of outmoded production methods, inevitably led to FN's decision to cease production at the end of 1964.

Publicity poster by a distributor based in Tervueren near Brussels, for the 'full suspension' FN motorcycle.

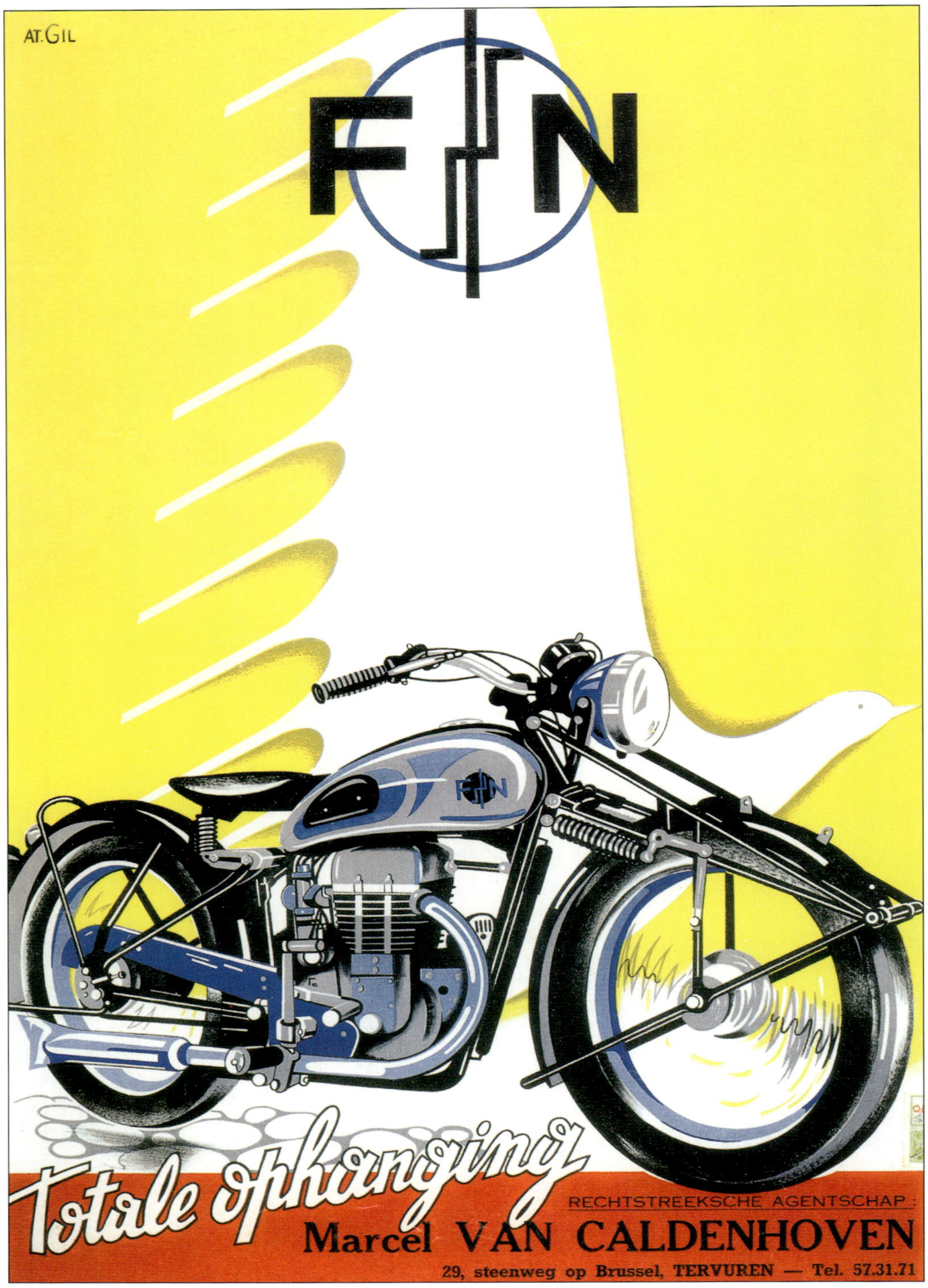

Shop window of the Ledent company, an FN motorcycle distributor (1954).

Publicity poster from 1956, by the painter and illustrator Pol-François Mathieu (1895-1979).

FN motorcycle "Oiseau Bleu" – 125 cc (1954)

FN motorcycle MXIII - 450 cc OHV (1955)

FN motorcycle MX 15 - 500 cc (1953).

FN motorcycle M22 (1957).

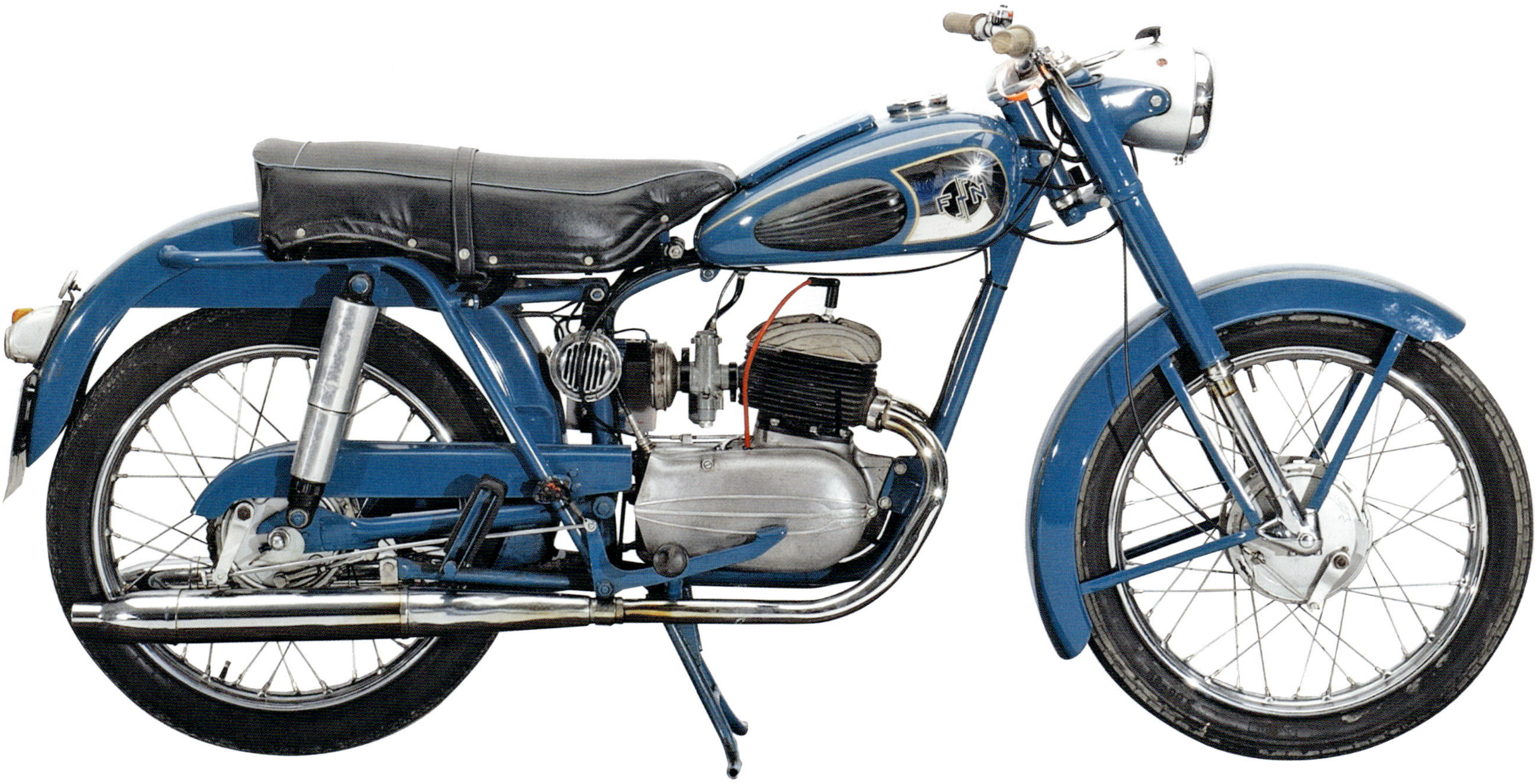

FN motorcycle M22 (1957).

FN motorcycle, model "Utilitaire" 50 cc (1964).

Publicity poster by Albert Fromenteau, in praise of the racer Victor Leloup, 1952 European Champion, riding an FN motorcycle. Having joined FN as a test rider in 1949, Victor Leloup (1924-2006) won more than 150 motorcycle competitions in the course of his career.

Auguste Minguels, European Motocross Champion on an FN motorcycle in 1953 and 1954. Between 1947 and 1961, FN and its Racing group gained a reputation equipping well-known Belgian riders, who between them won hundreds of national and international victories in this very popular sport.

FN PROTOTYPE
Capacity : 500 cc – Year of making : 1955
This 2 cylinder prototype was made to compete with the British production. It was shown at the Brussels motorshow in 1955, but was finally never produced, FN ending its motorcycle production and concentrating on mopeds. Herstal Group collection.

LES SP
12

Above and preceding page: spectacular images of FN motorcycles taking part in races during the 1950s.

FN 50 cc Princess moped (1961).

FN 75 cc Scooterette moped (1962).

Above: the FN Princess Scooter with Miss Belgium (1958). In response to the general craze for this type of practical leisure vehicle, FN began producing mopeds in 1959. In 1965 it permanently ceased the production and sale of motorcycles.

An "Ardennes" 4 x 4

On the other hand, the construction of military vehicles continued more auspiciously. Between 1951 and 1953, FN had supplied 4,050 three-ton lorries with four-wheel drive to the Belgian army. These lorries, which were manufactured in collaboration with other firms, like Brossel, Miesse, Jonckheere and La Brugeoise, were assembled in the workshops of the old cartridge factory in Bruges, which were from then on fitted out especially for this type of production.
Not long afterwards, FN resumed the production of trolleybuses for the Société des Tramways Unifiés de Liège et Extensions (1953-1956). From then on, numerous studies were brought to a successful conclusion and new products were being manufactured.

Among all the various models of four-wheel-drive lorries, we will mention the exceptionally sturdy Ardennes model (1958-1960).
FN also developed an A.S.24 fold-up tricycle, fitted with a 250 cc motor and intended for airborne troops (1960). This little vehicle had an original design and special tires, and could easily be cast off with a parachute and could transport four men across all kinds of terrain. It could also be used as a special vehicle for airport fire brigades and for fighting forest fires.
The agricultural equipment division enjoyed steady growth. The range of products was extended to include, apart from milking machines, refrigerators, milking stalls and deep freezers.
The realization of the Common Market created some interesting outlets for this department. In December 1959, FN took over from the "Société d'Optique et d'Instruments de Précision" (O.I.P.) the manufacturing of Cotton hosiery looms, as well as Kinrex horizontal looms and Quartex vertical looms. After a difficult start, this department, which was set up in the suburbs of Ghent, developed satisfactorily.

FN 4RM-62C 'Ardennes' lorries for the Belgian army (1954). The famous order by the Belgian Ministry of Defense, the so-called '4,000 lorries' (1951-1954), entrusted FN with the construction of these army vehicles, with the assistance of the Brossel and Miesse compan˜es and in co-operation with numerous subcontractors. The main part of the work was carried out at the factory in Bruges. The manufacturing of FN lorries was ceased in 1965.

A "composite" logo for the three companies : FN, Brossel and Miesse. The lorry shown here is of a special model, as the basket, set in action by various hydraulic jacks, can be emptied on three sides : back, left and right. (Herstal Group Collection).

Assembly of knitting machines at the FN division in Ghent. In 1959, FN had set up a knitting machine division in Ghent after it had taken over the textile machines division of the Société Belge d'Optique et d'Instruments de Précision. These activities were then concentrated in Bruges in 1974. Finally they were abandoned four years later for competition reasons.

FN automatic pipeline milking machine. After the "Matagri" division (Agricultural equipment) had existed for more than thirty years – at the beginning of the sixties a section of electric household appliances ("Matelem") was added – FN ceased its activities in that field in 1975 for reasons of profitability.

The FN TBVI trolleybus carrousel in the Place Saint Lambert in Liège on 20 January 1959.

FN TBVI trolleybus in front of the provincial palace in Liège (1954). This was the last trolleybus model to be developed by FN and adopted by the public transport authorities. It measured more than 12 meters in length and could carry 100 passengers. FN started manufacturing trolleybuses in 1932.

AS24 airborne trike. This vehicle, which could be folded up and parachuted, was developed by FN in 1958, based on a patent registered by Nicolas Straussler. (In the abbreviation AS24, 'A' signifies 'airborne', 'S' 'Straussler' and '24' refers to the engine of the M24 motorcycle). Adapted to a variety of military and civilian uses, it was acquired chiefly by the Belgian and Peruvian armies.

AS24 airborne trike used by the Belgian army in operation in the Congo (1964). The soldiers are armed with the LAR, the MAG and the Vigneron machine pistol.

Nuclear energy and turbines

Just as it had after the First World War, the company set out to extend its range of activities by participating in the creation of new companies. The most important one was the limited liability company Métallurgique et Mécanique Nucléaires (M.M.N.) which FN set up on 26 August 1958 together with the Société Générale Métallurgique in Hoboken.

The factory, which was equipped with all the installations and security systems peculiar to the nuclear industry, had been built in Dessel, close to the Mol research center, under the supervision of the former head of the FN laboratory , the engineer Roger Mossoux (d. 1963), succeeded by Ernest. Demaret, former head of department with FN.

In addition, on 6 April 1964 a company called FN-Boeing Turbines was set up in co-operation with the Boeing Company of Seattle. Established at Herstal, the company's object was the construction, sale and servicing of small gas turbines.

Contrary to what had happened after the First World War, the post Second World War reconstruction and even expansion of the company occurred in a particularly favorable economic climate without industrial relations being severely tested, at least during the first few years.

These two facts contributed in no small degree to the prosperity which FN was to enjoy from the moment its restoration was complete.

General view of the industrial facilities of the M.M.N. company in Dessel-Mol.

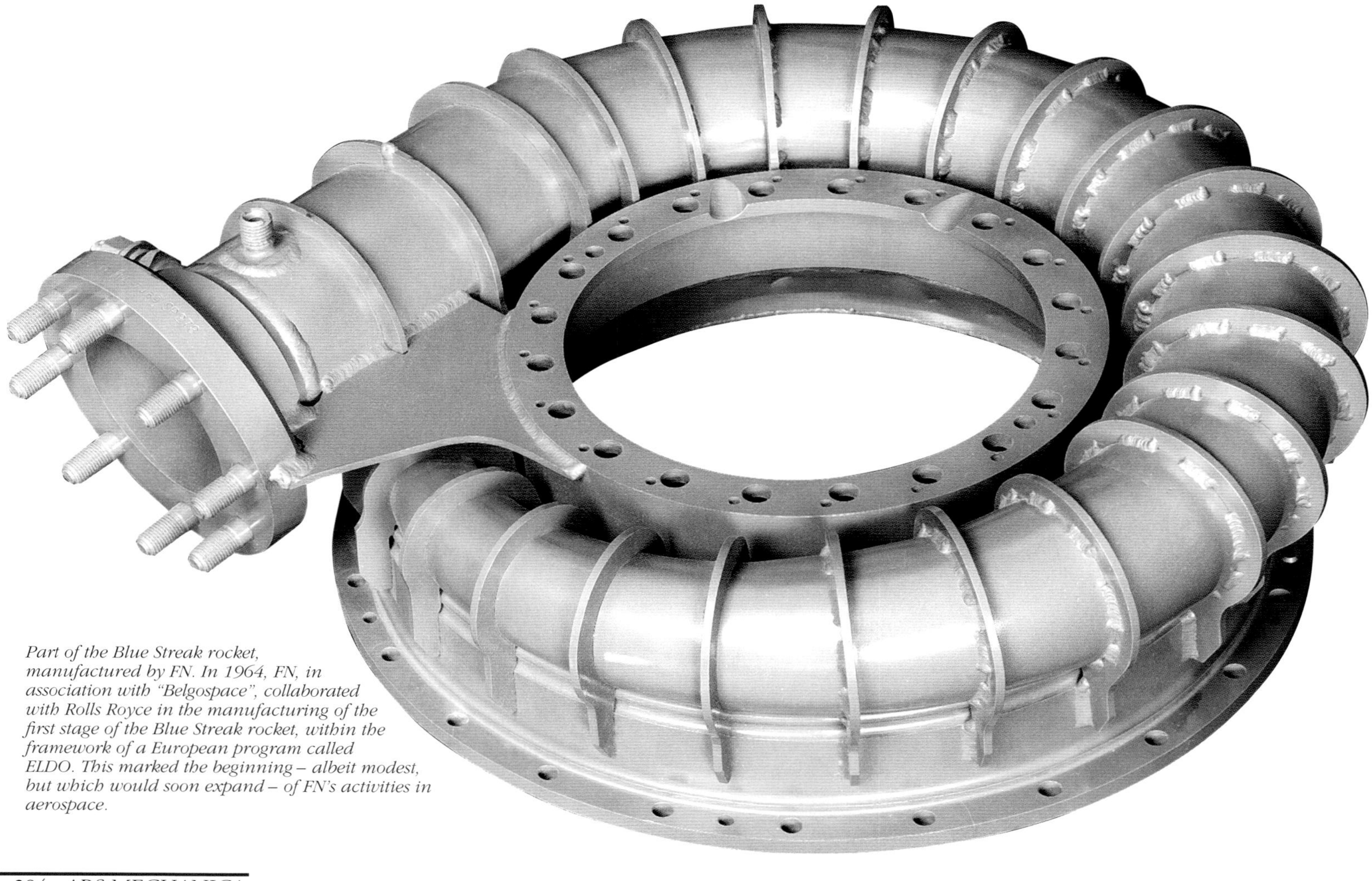

Part of the Blue Streak rocket, manufactured by FN. In 1964, FN, in association with "Belgospace", collaborated with Rolls Royce in the manufacturing of the first stage of the Blue Streak rocket, within the framework of a European program called ELDO. This marked the beginning – albeit modest, but which would soon expand – of FN's activities in aerospace.

Visit by H.R.H. the Prince of Liège to M.M.N., on 11 March 1969, led by the manager, Ernest Demaret.
The S.A. Métallurgie et Mécanique Nucléaires, which was founded in 1958 with the participation of FN, was established in Dessel, near Mol (Belgium), in 1962.
For some fifteen years, it allowed the company in Herstal to carry out activities in high-tech areas, such as mechanical productions for nuclear power stations.

FN Boeing turbo-starter for starting up the engines of the F104 G. In 1964, together with the Boeing Company in Seattle, FN had set up the company FN-Boeing Turbines for the manufacturing, sale and servicing of low-powered industrial gas turbines.
These activities were taken over by the Caterpillar company.

Dividends

The prosperity of a company is reflected in its financial situation. In this respect, FN had developed considerably since the fifties. On 17 June 1957, the capital was increased by incorporation of revaluations, from 315,000,000 francs to 630,000,000 francs, while the number of shares was doubled and fixed at 420,000, whereby the new shares were allocated free to the old shareholders. Since 1951, the company's treasury position had been healthy, thanks to the advance payments agreed by its government clients.

Finally, the loan taken out with the SNCI and the last pre-war debenture loan were paid back in March 1959 and July 1962 respectively. The shareholders were to benefit from this favorable situation.

Dividends were distributed each year from 1951 and if we compare the 1951 figure – 200 francs per share – with that of 1964 – 220 francs per share – we can see clearly that the dividend has always been high.

Even though the internal organization of the company during the period under discussion had become more complex, it nevertheless had hardly changed in its main features.

FN was essentially composed of three parallel divisions under one and the same management : arms - ammunition - engines, as well as a certain number of common departments : the laboratories, the foundry, the forge, the department of thermal treatments and the construction and servicing departments.
Meanwhile, the number of engineers and other university graduates had increased. On the other hand, at the highest level in the hierarchy, the management and board of directors had remained unchanged, discounting the replacements which the passing of the years had necessitated.

René Laloux, who had become Chairman-General Manager on 20 October 1962 was appointed Chairman-Managing Director on 24 October 1963, when Othon Drechsel succeeded him as General Manager .

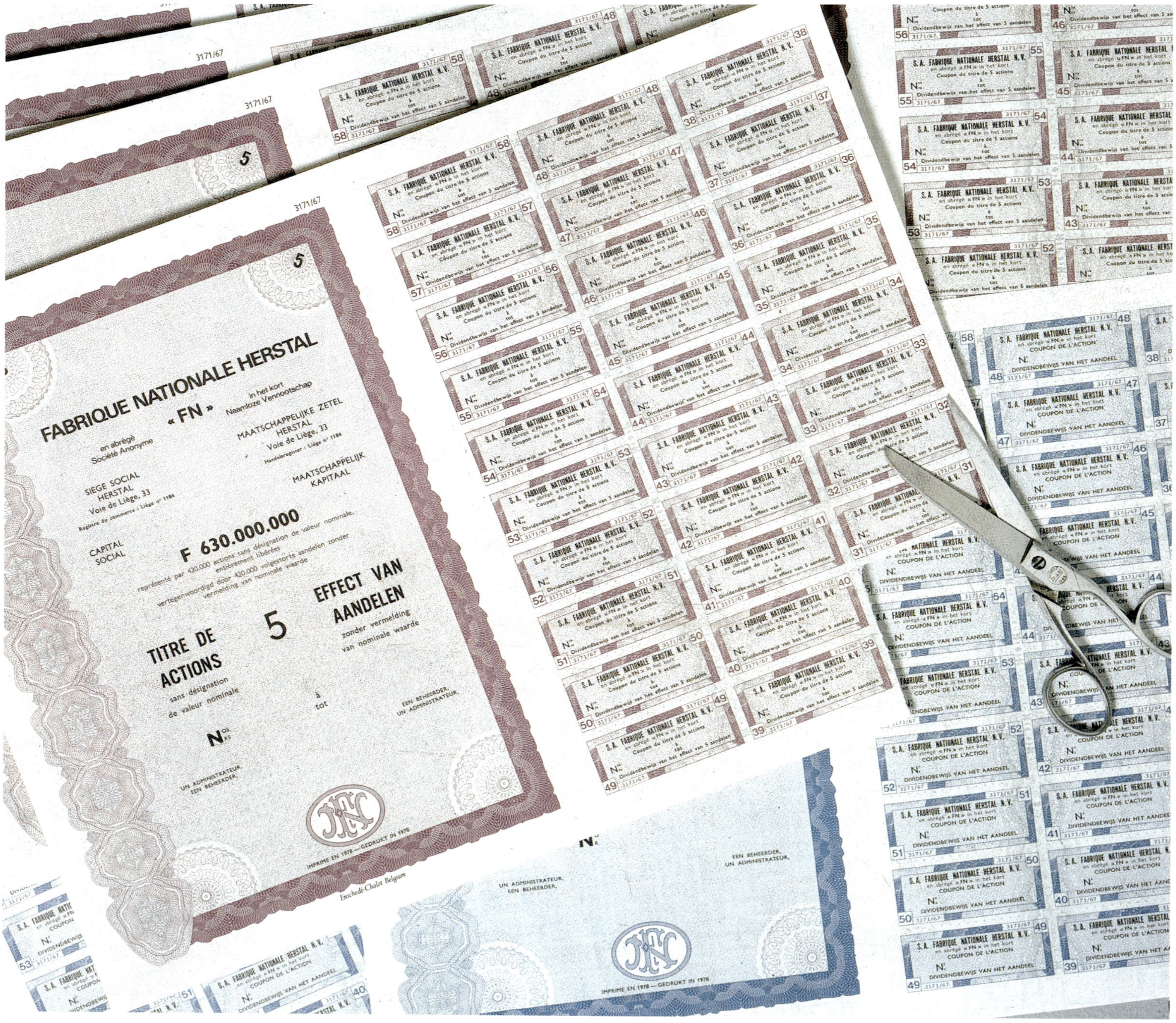

Becoming an international group

The realities of an industrial society are manifold, complex and involved. There are various ways of seeing them. And the most reliable way is one which takes account of the five following essential parameters: the economic progress of a company, that is to say, its ability to make profits as well as the way these profits are distributed, the types of products or services which it has put at the disposal of the public and the success they have met with, the personal role of the managers, the way it has satisfied the human aspirations of its collaborators and the community spirit it has kindled and, last, the image, whether true or distorted, it has created for itself in the eyes of the public and the media.

Without trying to recapture the whole truth, but with a reasonable measure of optimism about the ability of historians to restore the essential facts and driving forces of the past, an analysis according to these facts may be considered relevant. The following pages are inspired by these considerations.

The twenty-five years which separate the celebration of FN's seventy-fifth anniversary in 1964 and the commemoration of the centenary of its foundation, have for the purposes of the account been divided into four periods : from 1965 to 1973, a period characterized by rapid changes in the markets and the economic climate, the modernisation of the production apparatus, as well as by stagnation and even a decline in the profitability of the company; from 1974 to 1980, which was characterized by a spectacular boom and a kind of transfiguration; finally, from 1981 to 1988, an era of new environmental and structural challenges.

However, and this remark goes further than a simple methodological statement, there is no break between the different stages of a development which did not just start in 1964. The past always serves to explain the present, even if this differs constantly from what has gone before. In this way, success contains the seeds of disaster, which in turn may engender new triumphs. Life represents continuity as well as change.

Of course, this still needs to be understood...

1964 : an anniversary

Celebrating the seventy-fifth anniversary of its foundation in full splendor in 1964, FN could pride itself on a clear and unequivocal tradition which made it the showpiece of the Belgian metallurgical industry. Firmly anchored as the company was in the domain of arms and ammunition, both for military and for sporting purposes, the oldest members of the management still considered this to be FN's 'manifest destiny', which they only strayed from to follow a variations in the international political scene or the fluctuations of a seasonal market.
This standpoint, which indeed over time was frequently supported by a superior sales figure in the arms division, would gradually be qualified in favour of a new direction, that of aircraft engines, which was to be vindicated definitively by the success of the J79 contract.

It contrasts with other activities which were either in decline or were characterized by unreliable profitability, and which after the war had been trying to earn a place next to those sacred giants which were the arms and ammunition divisions. From that moment, the Pré-Madame division, which had been the traditional base for the manufacturing of two-, three- and four-wheeled vehicles, was to become in the course of the sixties the home for aviation production.

A temple to technology

Like most of the large Belgian enterprises of that time, FN was a company of engineers. Its development, which resulted from the stupendous rise of mechanization towards the end of the nineteenth century, was largely due to those generations of recruits who graduated from the colleges: the civil engineers, who after World War II were joined by industrial engineers (then called 'technical engineers'). All this trained personnel was in charge of the design and adaptation of the products, the development and application of manufacturing methods, as well as the management of the financial and human resources, and whenever necessary, commercial prospecting.
This preponderance of engineers made FN a temple to technological know-how. Visitors and trainees from all over the world came to its central laboratory, while the size of its machinery stock (more than nine thousand machine tools) never failed to amaze the initiated or lay public.
It seemed as if there was nothing the Herstal Company could not do successfully. FN which was rightly proud to be a recognized nursery for specialists, did not however always fully understand the economic aspect of its ventures, impregnated as it was with the mentality of a manufacturer, which paid attention to the perfection of the products. The opinion prevailed that it sufficed to do a proper job for the product to sell.

Another consequence of this ascendancy of technology was the comparative self-sufficiency of FN, where it was thought not only that everything, or almost everything, could be produced, but that it was better, according to Taylorist logic, to have everything done by the company, rather than subcontract certain, less profitable work to outsiders. On the other hand, this concentration of talent and competence allowed the company to have at its disposal, at a regional level, a store of human resources and means which would provide both FN and its social environment with an additional asset and a favorable international image. However, the rapid development of the economic situation within a period of a decade would soon highlight the dangers of this "technological mono culture".

Carbon content analysis at the Central Laboratory. The Central Laboratory, superbly equipped for testing and control operations, was also the place where young engineers were trained and where they completed their practical training before being taken on in the company. All these different tasks have now been decentralized, with a view to developing a 'quality management' system which is more closely involved with the manufacturing process.

An important employer

In 1964, FN employed more than 13,000 people, the highest number in its history. This figure included some 1,700 salaried staff : management executives (of whom more than 150 had the benefit of higher education) and foremen, designers and other "white-collar" workers. The 'blue-collar' workers included specialized workers who were restricted to very particular areas. The skilled workforce included approximately 600 workmen : regulators, toolmakers and specialists in the mechanical workshops – who incidentally were the best in the country, if not in the whole world – as well as the well-known gunsmiths, who were concentrated in the factory but who, like their predecessors, worked with the bare and skilled hands of a craftsman. In those supposedly sexist days, there were very few women in executive positions, especially in the world of engineering, which was traditionally closed to female candidates. There were about three or four female laboratory engineers and a few social workers, who were devoted representatives of the management's paternalist attitude, but who at the same time were the skilful, semi-official mouthpieces for the individual or collective desires of the rank and file. On the other hand, thirty per cent of the manual workers were women, mainly machine operators, confined to repetitive tasks. The high level of mechanization had allowed FN to engage female workers, who came to earn an extra income. Due to their low level of qualification, they could not hope to get into skilled jobs, which were traditionally reserved for men.

The Liège area, the scene of prodigious industrial development, had for a long time also known a shortage of local labor, and this caused a massive influx of workers from other parts of the country and from abroad. FN too followed this tendency : there were more than a thousand workers from Belgian Limburg and more than 20% immigrants, for the most part Italians.

For most of FN's arms products, and particularly its hunting and sporting weapons, labour accounted for 80% of added value compared with only 20% for materials. This is one of the reasons why FN was such a large employment provider. Its workers and gunsmith crafstmen worked to time-honoured methods which had amply proved their worth. Today, they are used in the production of fine weapons.

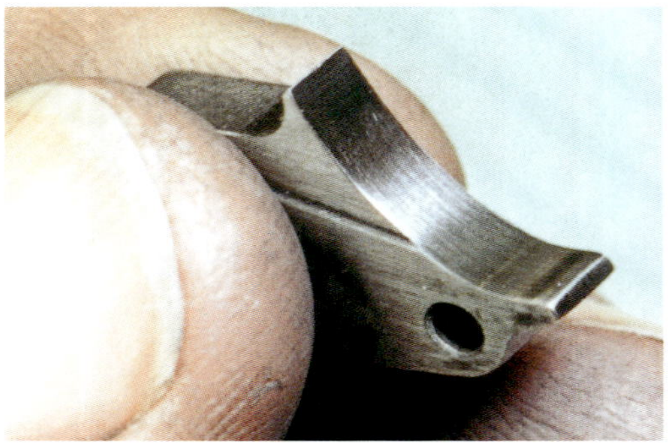

Each part is timed

Barrel fitting

Final assembly

Woodwork

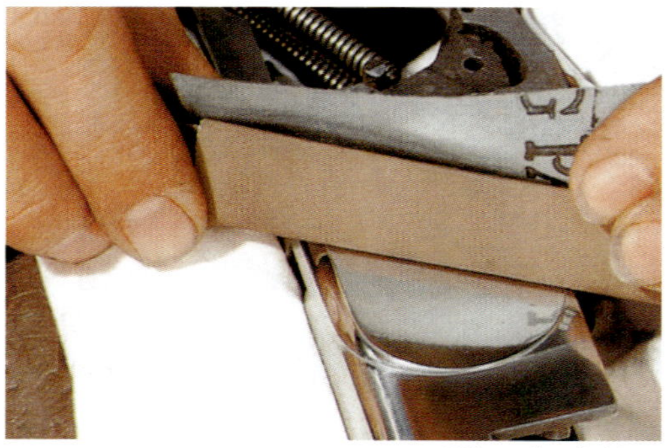

"Mirror" polishing

Checkering

Customized stock.

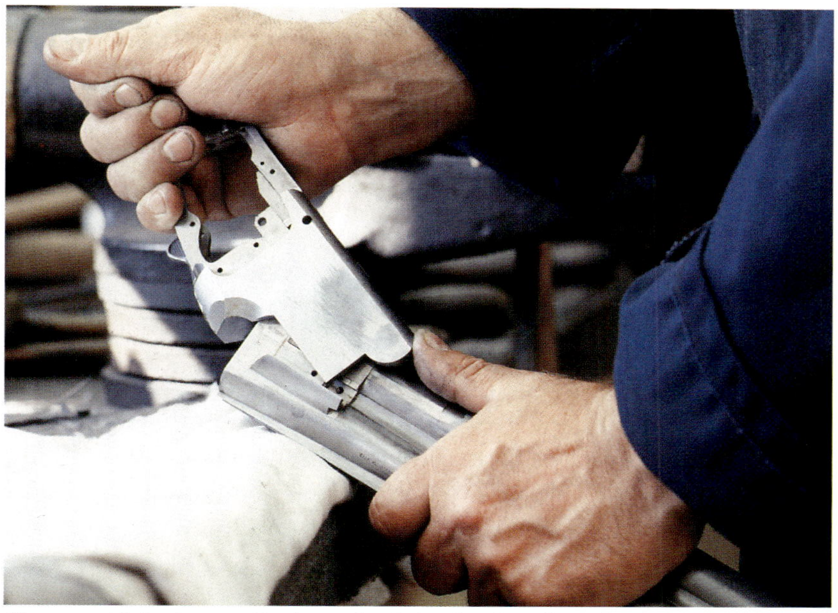

The gunsmith fits the parts firmly together before breaking them cleanly apart.

If there has been any friction between the parts the lampblack will have disappeared.

The gunsmith can thus check the precise fitting and timing of the parts and make any necessary adjustments ... by hand. This operation is repeated until the fitting is perfect.

Example of the craftsmanship still followed at Herstal : the technique of lamp blacking

The gunsmith who assembles the three principal elements of the gun, that is, the action frame, the barrels and the fore-end iron, uses the technique of lamp blacking. An ingenious method which is both simple and efficient which is the only way to "see" the fitting and timing of the parts inside the closed gun.

The gunsmith covers the parts in a fine layer of lampblack produced by the flame of an oil lamp (thickness: one thousandth mm).
He then fits them firmly together before breaking them cleanly apart.
If there has been any friction between the parts the lampblack will have disappeared. The gunsmith can thus check the precise fitting and timing of the parts and make any necessary adjustments ... by hand.
This operation is repeated until the fitting meets the very strict standards imposed at Herstal.

The aim of this extremely fine and precise assembly of the mechanical components is to make each gun "self-locking". When a cartridge is fired the increase in chamber pressure causes a very slight increase in size. Since the barrels are "set down" into the action frame with such extreme precision this swelling means that the barrels are clamped against the side faces of the action frame. One could say that firing a cartridge in a Browning locks the gun automatically by this "self-locking" feature.
The life of a Browning is thus increased since it markedly reduces the stress on the bolt and the barrel lumps.

The fitting and timing norms were traditionally (and still are !) described as follows : "The opening and closing of the gun should be smooth and firm". It is by using this lampblack technique that a fitting and timing is obtained which can guarantee just such a smooth and sound action.
This is the secret of the characteristic sound that we can hear every time we open or close a Browning.

The gunsmith covers the parts with a fine layer of lampblack produced by the flame of an oil lamp (thickness : one thousandth mm).

The old paternalism

This diversity in origin as well as in statute, professions, salaries and, consequently, in mentality, had always characterized the staff of FN. It only added to the extreme complexity of the company's management, especially as the collective mentality generally varied from one department to another, and indeed from one division to another, and dynamism abounded, as well as opposition and cliques.

The whole of this gigantic structure was highly hierarchical and stratified. The limited company FN took the shape of a monolith, implanted in the Herstal soil (Bruges and Zutendaal were its only distant satellites). In principle, everything started from and ended at the top. The Chairman and Managing Director , René Laloux, was a powerful figure in the Liège arms business, and he held a very tight grip on this community. The power of decision rested with him. He was assisted by the General Manager, the Company Secretary and the heads of department or division. The shareholders, who were very dispersed indeed, rarely had a say in the management of the company, while the power of the principal shareholder, the Société Générale de Belgique, remained discreet and restrained. The style was authoritarian: everybody knew his place and his job, which nevertheless did not prevent ambitious projects from being realized - as the facts have proved. However, the rigidity of the system gave rise to a certain routine and failed to exploit potential energy.

Both the orders from above and the reactions of the rank and file had difficulty in filtering through the hierarchy, but the higher ranks, and even the general management, made up for these shortcomings by spending some time visiting the workshops to explain, encourage or reprimand. For the rest, news from both inside and outside the company was out of date and out of place. The adage "To live happily, live in splendid isolation" was the official policy adopted towards the media, which at that time did not expect any further openness anyway. After all, did not FN, powerful as it was, demand the greatest respect?

In spite of its enormous size, FN succeeded in preserving that old paternalist spirit, which was a subtle blend of authoritarianism and solicitude, yielding a particular quality of life and good-natured human relationships which everyone, sometimes secretly, longs for. In spite of the working conditions which in certain departments for unqualified work were soon to prove obsolete, at the beginning of the sixties FN had the reputation of a good employer and good payer, which provided its staff with non-statutory company benefits and which, in addition, proved stable. With such a reputation, the company could be considered rightly to belong to the top brass of Belgian industry.

The burden of industrial relations

The employees' and workers' unions, both Socialist and Christian, but mainly the Socialist, already played a major role in the company, as has always been the case traditionally in the Liège area, where the large majority of workers is affiliated to one or other of these organizations. In spite of the reconciliatory effect of the works councils, the trade unions of 1964 remained above all institutions of protest and even confrontation. Supported in their actions by the emancipation of the workers, which had been secured gradually through hard struggle since the previous century, they still kept their old attitudes of suspicion and watchful scrutiny towards the employers, solidly conscious as they were of their political power in a Wallonia where they had secured a substantial majority. For them, FN constituted a privileged territory, where thousands of workers were concentrated in approximately fifty acres, in a region which was strongly marked by the burden of industrial disputes. Matching the paternalist attitude of the management, was a method of organization on the part of the unions which was no less so. As soon as disputes broke out, generally in the form of walkouts, either localized or throughout the whole factory, of either short or long duration, tough negotiations were conducted in order to try and smooth things over. Blow after blow was dealt. Distrust was felt on both sides of the negotiating table, but nevertheless it should be pointed out that relations between management and union representatives showed an abundance of good will, even of cordiality, outside the places and moments of friction. The common tasks, the personal relations and sympathies tended to channel the aggression, but they did not remove it. This state of affairs was accepted as a fate, and little was done to avert disputes.

These disputes almost invariably pertained to wage claims. In the West, the sixties were in fact characterized by a high level of affluence, the deliberate introduction of a consumer society, which was marked by a high standard of living, the pursuit of high

Female machine operators working in the 'Mauser' hall (1965). The female employees at FN traditionally account for 25 % to 30% of the workforce. In those days they were poorly qualified and worked under abhorrent conditions, but since the end of the sixties efforts have been made to improve the training and position of these workers.

Below, these examples of raw forged parts (on the left) being finished after machining (on the right), illustrate the importance of the work carried out by generations of workers, men and women – the world's finest in their field – gathered together in a factory yet still working as their forebears did, with vices, and putting their heart and soul into their work.

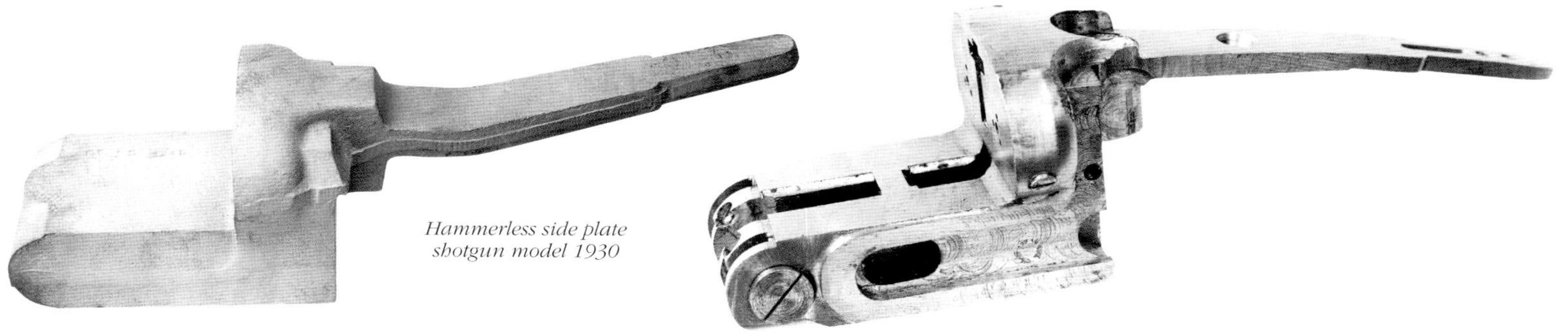

Hammerless side plate shotgun model 1930

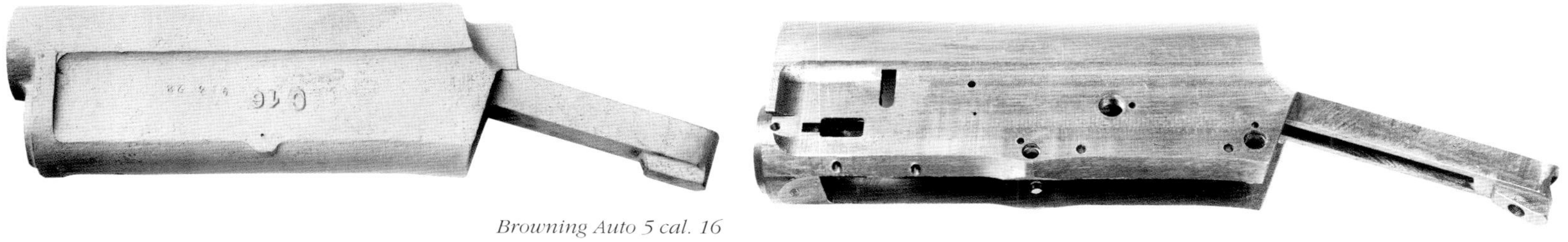

Browning Auto 5 cal. 16

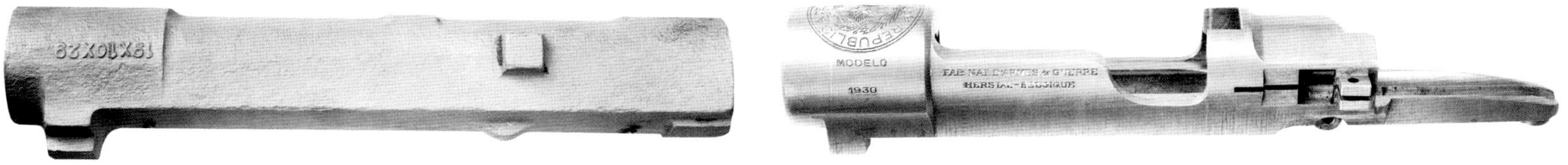

FN Mauser rifle

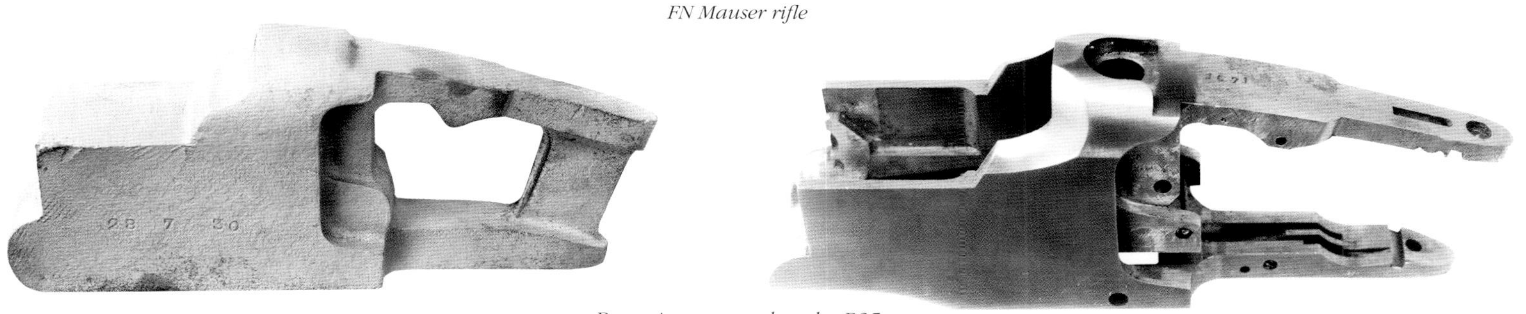

Browning over-and-under B25

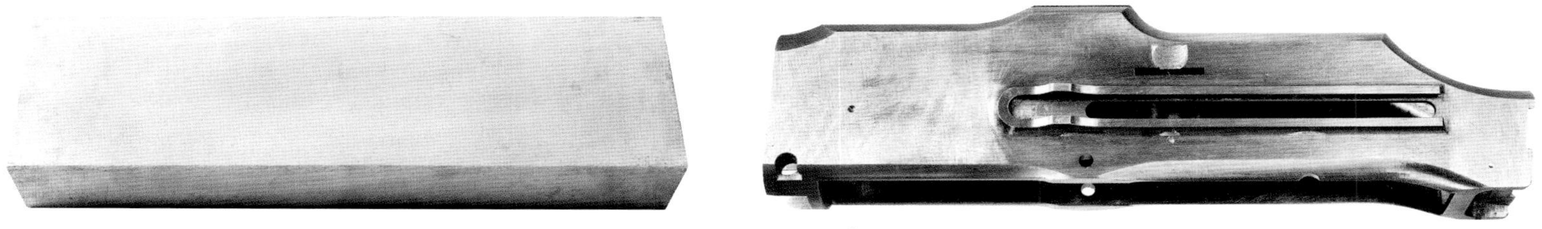

Browning automatic rifle

purchasing power and by the abundance, if not over-abundance, of products. So what, in such a situation, could be more normal for the unions to do than to claim for their members an ever-increasing share of the profits of the company? But in this unrelenting pursuit of the surplus value of labor, a company like FN found itself in an unfavorable balance of power. Engaged as it was in a manufacturing industry with high salaries - especially at a time when handicraft still prevailed in certain divisions - an increase in salaries would seriously prejudice its market position.

On the other hand, it could not afford to oppose this trend without risking a major paralysis of its manufacturing activity. Considering this dilemma, or should we say vicious circle, FN, which was a target and even a testing ground for the Walloon and national union movement, would often yield to social demands, while, paradoxically, the company was considered in employers' circles, especially those of the small and medium-sized enterprises, sufficiently powerful to resist this type of pressure and to defend their common interests. Hence their disappointment at what they considered to be a kind of 'class treason', which they were not prepared to forgive easily.

A company in good health

Judging by the financial results of the mid sixties, which constitute our point of departure, the situation of FN appears healthy and stable. And this is how the business world as well as public opinion saw it. It is true that sales figures evolved favorably after the war, with one peak at the beginning of the fifties and another ten years later. In 1965, sales approximated 3,500 million francs. It allowed the company to meet practically all its financial needs and even to yield a profit of 200,000,000 francs and to distribute a considerable dividend of 230 francs per share. On closer inspection, however, the internal social and cultural developments described above were weighing heavily. Indeed FN made few investments, and salaries as well as social insurance contributions already absorbed 60% of its sales. The burden of labor costs seriously encumbered the profitability of the company, while the increase in direct and indirect remunerations as well as inflation eradicated all prospects of a reversal of this trend. By June 1965, the hourly labor cost had increased by nearly 50% in three and a half years.
On average, FN exported 90% of its total production, and not far off 100% of its arms production. In 1965, for the first time, the business volume of hunting and sporting

(so called "commercial arms") surpassed that of military equipment, which was based on government orders.

Two thirds of these civilian products were channelled through the Browning Company to the American market. But already sales prices in the United States made the less well-off buyers turn towards the Japanese guns, which were making their first appearance. How high should prices rise with respect to the customer's budget ?

That was the vexing question which would occupy FN and its principals from then on, given rising cost prices. The company would soon discover that craftsmanship, which abounded in that region, was to become a luxury, and that high labor costs compelled it to restrict its business to top quality products.

As far as the military arms division was concerned, FN's post-war expertise in the field of small arms earned the company an exceptional reputation: mainly the LAR, MAG and small-caliber ammunition ensured it the de facto monopoly in many markets in the non-communist world. FN also occupied a privileged position in the domain of aeronautics, as the sole engine manufacturer in the Benelux countries, and saw its position strengthened with the successful J79 contract as the privileged Belgian subcontractor. However, there appeared the problem of maintaining a balanced strategy in an essentially military domain which is subject to cyclic and variable international orders.

The other activities at Pré-Madame for their part helped to stabilise the work volume rather than to ensure profitability under variable, if not at times illusive, circumstances.

René Laloux receiving HRH Prince Philip of Edinburgh and HRH The Prince of Liège, on behalf of FN, at the Arms Museum in Liège (1966). An important personality in the arms industry, René Laloux (1895 – 1981) was the son of one of the first managers of FN, on which he too would impose his mark and his authority.

FN in an ever changing world

The sixties will always be known to us – and also to posterity – as the 'golden sixties'. But they were also the years of revolt against Western society, during which the twentieth century, albeit reluctantly, broke with the ways of life and thinking of the previous century. This did not come about without difficulty. Consider all the events which shook the world, from the first man-made satellite, Sputnik I which was launched on 4 October 1957, until the first hesitant steps of Armstrong and Aldrin on the moon (20 July 1969) : the assassination of Kennedy, the war in Vietnam, the decolonization, May '68, the "Prague Spring".... Not to mention the amenities of everyday life : automobiles for all, television, air travel, the first computers !

New challenges

FN was confronted with this new world, armed with a solid legacy, which however did not necessarily prepare the company to face it successfully on all fronts, neither more nor less than other industrial enterprises of the same kind and age. In fact, during this period the company showed an unsettling decline in profitability. In 1971, annual profit, which in 1965 still attained 200,000,000, dropped into the red, thereby forcing the company to cease the distribution of dividends, which had become something of a fixed institution since 1949. The FN share, which in 1968 had attained 4,900 francs, varied between 800 and 1,900 francs six years later.

So what had happened ? These mediocre results were mainly due to a stagnation in sales, which remained between 3,200 and 4,200 million francs, and the spectacular increase in labor costs.

The fall in sales resulted from a combination of factors, such as the end of the expansion in the post-war 7.62 caliber armament and the new conditions and terms of payment in government contracts, the appearance on the scene of tough competitors in the area of lower priced sporting arms, the relative mediocrity of the orders which succeeded those of the J79, and also the disappearance of the military vehicles division.

On the other hand, in the context of inflation and the unprecedented prosperity which Belgium enjoyed at that time, the level of remuneration at FN quadrupled in the period between 1960 and 1975. This ascending spiral, which was particularly noticeable in highly paid activities such as these, would continue during the seventies and was to constitute a major challenge for the Herstal company. In this way, salaries and social security contributions accounted for 66% of the sales in 1970, while in 1974 this figure would reach 85% ! Under these conditions, the level of productivity was one of the lowest in the entire Belgian metallurgical industry, with a sales figure per capita of less than 400,000 francs.

Also, the suspension of the dollar's convertibility into gold (1971) destabilized exchange rates, introducing a new element of uncertainty into international trade transactions.

New techniques

Disoriented by the loss of certain traditional activities, FN felt its way and carefully introduced new products. Its financial policy was aimed at immediate profit rather than the deferred benefits of research and investment. However, necessity makes the law. The general management judged the situation with acumen and attempted, with the means at their disposal, to free itself from the pincer movement caused by sales and salaries. The first aspect was difficult to cure in a stagnant market. The second was just as unpliable, as the handicraft character of certain products and social pressure opposed the consequences of mechanization. The sixties, particularly the second half, witnessed a substantial modernisation of the production and even the management apparatus. So it was thanks to the introduction of new techniques that FN, which had always been a leader in the application of modern technology, experienced a fresh start. It marked the beginning of the end of "granddad's factory".

The most striking innovations it adopted deserve to be mentioned, as they marked an epoch in the industrial area of Liège and even at a national level. Precision forging and electro-erosion were introduced in 1960, as well as machines with digital control, although these were developed chiefly towards the end of the decade. The arms industry, which up to that time had always been a sacrosanct domain of handicraft, started to acquire a new outlook: in 1966, a machine to measure the straightness of the barrels; in 1967 an automatic phosphating line; in 1968, mechanical fitting of the action-frame on certain guns with over-and-under barrels; transfer machines, mechanical polishing of the gun stocks and machines with automatic feed in 1969; automatic electrolytic bluing and polishing of barrels for the Auto-5 in 1970. Pré-Madame adopted electrochemical machine finishing in 1969, while the division for thermal treatments was modernized in 1971. In 1968, an industrial workshop for lost pattern casting, which had been designed on an experimental basis four years earlier, was inaugurated. The ammunition factory too was being updated, with the mass production of hunting shells with plastic cases (1968) and the progressive introduction of automation, which six years later would result in a doubling in the production capacity. In 1973, Zutendaal too started to innovate, with the installation of an automated system for the production of lead styphnate and the carousel fitting of primers. From the beginning of the sixties, computers started to occupy a place in the company. By 1966, FN had three computers: a Gamma 10 for Company Accounting, an IBM 1401 for the Turbojet division and an IBM 1620 (bought in 1962) at the test-bed in Liers. The latter was replaced in 1968 by a time-sharing system, quite innovative for its time. In the meantime, a decision was made to equip the company with a Computer Center, which was set up in 1968 and inaugurated two years later under the name "Centre de Traitement de l'Information" (C.T.I.).

This, then, was a period of numerous technical innovations. They most visibly affected the older departments, as at Pré-Madame, which has always been open to advanced techniques, modernisation of the equipment was easily carried through as new orders came in. However, the coexistence of many different types of activities at FN was to prove more than ever beneficial in the sense of a more or less intentional, but often decisive osmosis of techniques employed in the different branches. From then on, the company embarked more swiftly than ever before on a process of modernisation, which at that time was still unusual, but which now is quite common, as it constitutes a part of the industrial "struggle for life".

In the medium term this new policy was to have considerable effects. Directly noticeable was the need for new financial resources to meet investments which were starting to exceed the relatively low level of the previous years. This need partly accounts for the negative financial results at the beginning of the seventies. On the other hand, the reduction in the workforce, which had started then, is a result of the increased mechanization which, in the long run, was to reduce the labor cost to a more bearable level.

Before concluding this overview of the general progress of the company, we will point out that in 1972 other important decisions had been made, the innovatory character of which will form the subject of the following chapter and which are explained by the logic of a new attitude elaborated and assumed by a new team of managers. It concerns the participation of 36% in the capital of the Beretta company, the famous Italian arms manufacturer, the plan for the construction - in the following year - of an arms assembly plant in Portugal, and finally the setting up of the Dumont-Sclaigneaux company in Andenne, in which FN was a majority shareholder, for activities in the domain of industrial hydraulics.

Expansion of the sporting arms and ammunition activities

The hunting and sporting arms activities were developed considerably in a leisure industry market which was in full swing. In spite of the competition from Spain and Italy in smoothbore guns, Germany and Finland in rifles, and also from Japan for ordinary quality products, the increase in demand, especially in the United States of America, encouraged FN to make head way.
Undoubtedly, the cessation of production of the latest version of the Anson shotgun might be regretted (1968), which proved too costly to produce because of the handicraft involved, as well as of the double automatic (1971), an original weapon which never met with the expected commercial success.

On the other hand, the Browning Over-and-under became the world-wide hunting shotgun par excellence in the category of semi-manufactured products. FN diversified its various models : game shooting, trap and skeet shooting, sporting shooting. The gun was made more luxurious by the introduction of numerous variants and by an increased elaboration of the ornamental engravings, in which the local arms engravers were able to give free rein to their traditional talents.
For this purpose, and so as to meet the special orders from Browning, in 1977 FN set up a workshop in Faurieux, situated close to the main factory.
The Auto-5, the showpiece par excellence in the range of sporting arms since 1903 and a reference unit for FN's production standards, met with wide acclaim. In 1967, the release of the two millionth Auto-5 to be manufactured at Herstal after the war was celebrated.

Cutting of a cluster of parts obtained by precision casting. Introduced in 1964 precision casting or lost wax

process allowed FN to produce, both for its own needs and for outside customers, parts for mechanical construction in general, and for the aeronautical, nuclear and arms industries in particular.

A number of new products started to appear which no longer owed anything to the personal inventions of John Moses Browning, who had died long before: a .22 rifle, the 'T-Bolt', fitted with an unusual but effective transverse bolt (1965), and in particular the BAR (Browning Automatic Rifle) in the following year, whose name, for reasons of repute, is reminiscent of the famous automatic rifle from which, however, this original weapon differs in all respects. It was designed by Bruce Browning, John Moses' grandson. The appearance of the BAR attracted much attention in the world of hunting, where up to that moment it was considered hazardous, even impossible, to design an automatic rifle which at the same time would be sufficiently powerful and accurate for hunting big game. The success of this weapon, which had become a classic, has never failed since. Nearly forty years after its creation, in its current version it remains technically the best of its class, a class in which it has a 50% market share. It has also become the reference product of the Herstal Group's Portuguese subsidiary.

In the domain of handguns, the old 10/22 pistol was replaced in 1970 by an adapted version, the 130 model. Meanwhile, Browning had developed two versions of the old lever-action rifles, with which "the West was won", in .22 and in hunting caliber. But seeing the increase in tenders with the arms manufacturers of Herstal, he offered the manufacturing license to someone else.

This warning shot was to be an omen: it was considered not only as the far-off signal of a tidal wave coming from the Far East, but also as an indication of the harmful effect of cost prices on the old commercial transatlantic relations, which had been cemented long before.

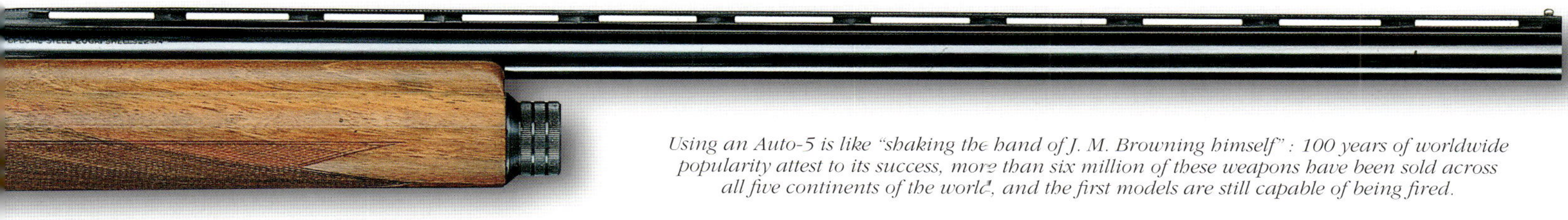

Using an Auto-5 is like "shaking the hand of J. M. Browning himself" : 100 years of worldwide popularity attest to its success, more than six million of these weapons have been sold across all five continents of the world, and the first models are still capable of being fired.

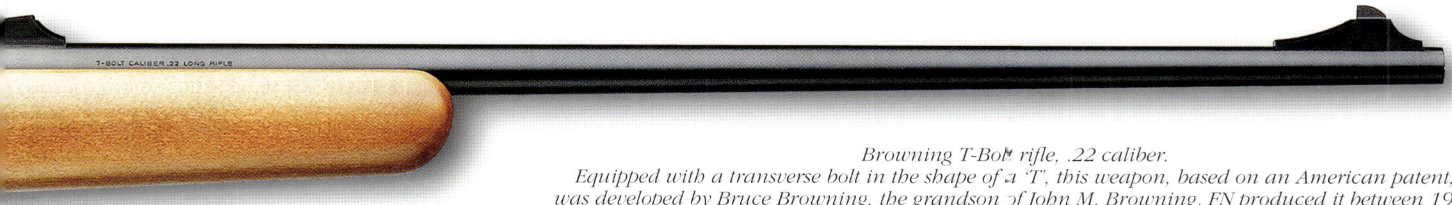

Browning T-Bolt rifle, .22 caliber.
Equipped with a transverse bolt in the shape of a 'T', this weapon, based on an American patent, was developed by Bruce Browning, the grandson of John M. Browning. FN produced it between 1965 and 1974, marketed it for a second period between 1978 and 1981 and has resumed its production since 2006.

Browning BAR MKI. The BAR, also invented by Bruce Browning, was an inspiration in hunting arms manufacturing in the 1960s. In its different versions, it remains today one of the most respected semi-automatic hunting rifles, and the biggest selling.

Echoing the traditions of the nineteenth century, in particular to those popularized by Winchester, the Browning BLR Long Action rifle reintroduced the trigger-guard lever weapon, but with more modern techniques (rotating bolt, rectilinear breech block motion) and new materials (light alloy frame) and also incorporating an extremely elegant finish.

In fact, the reduction of production costs had become a top priority in a context where the traditional gun making industry, which was not particularly economical as far as time, materials and labor were concerned, had become uncompetitive. For this purpose, it was necessary to design products which directly took account of these new considerations – which was new in this profession.

The Liège over and under shotgun, developed in 1972, answered those conceptions. The fact that it was not the work of the department officially responsible for new inventions, but the sole initiative of an engineer, called J. Martial, shows to what extent old ideas remained deeply rooted in people's minds. Overwhelmed by the prestigious image of the traditional superposed shotgun, which it was feared would be ousted by this new internal competitor, the Liège won neither the unanimous conviction of FN's experts nor that of the clientele. Nevertheless, this new arm, which comprised several technical innovations, constituted the first attempt to design and manufacture at Herstal a mid-range low-price rifle. Later, in 1973, and this time on good grounds, FN released a new self-loading shotgun, the B2000. This weapon, which unlike its illustrious 1903 predecessor was gas operated, presented a new style and manufacturing characteristics which allowed for low-cost mass production.

Hunting ammunition also benefited from the boom in the arms business. Production capacity was greatly increased in order to meet an increased demand, particularly from abroad, as the export share was considerably extended.

Publicity for the release of the Liège shotgun (1972).

In 1968, a manufacturing and sales agreement with the Cartoucherie Française in Paris reinforced the presence of the Légia brand in France, a country whose two million hunters represented a first-class market opening.
Between 1967 and 1974, the cardboard tubed cartridges were gradually replaced by plastic cased ones, a development which of course necessitated the introduction of new methods.

FN Browning Liège, the first hunting weapon designed by FN with the aim of simplifying production and reducing costs.

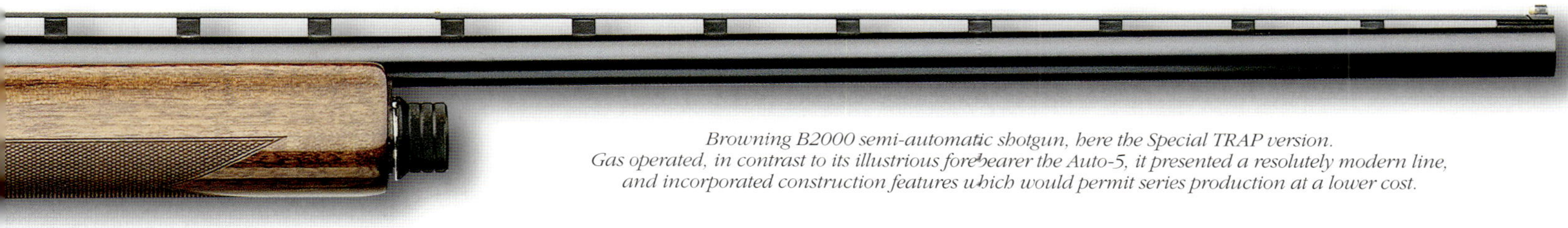

Browning B2000 semi-automatic shotgun, here the Special TRAP version. Gas operated, in contrast to its illustrious forebearer the Auto-5, it presented a resolutely modern line, and incorporated construction features which would permit series production at a lower cost.

A visit to Herstal by the Browning France team, who had taken over the Cartoucherie Française.

Browning B25 side-plate engraving Sherwood (M2) model, 12 caliber. Fine English grained engraving, edged with rock-work creating a pearled border ornamented with acanthus leaves. On the side-plate, in a medallion, flights of partridge, pheasant and duck. On the bottom-plate of the receiver there is a rabbit framed by detailed English engraving. Three piece fore-end.

Browning B25 side-plate engraving Oxford model (M1), 12 caliber. Fine English grained engraving, the tracery work interspersed with polished areas. Rock-work producing a pearled border ornamented with acanthus leaves. The axis of the hinge is decorated with a delicate English-style rose.

The Browning 'over-and-under' has become the hunting shotgun par excellence throughout the world. FN has diversified its design and also made it more luxurious by introducing a number of variations, embellishing it with ornamental engraving and endowing it with superior finishes, like the 'mirror' polish on the internal parts above.

Between two calibers

In the military domain, the period was dominated by the success of the LAR, MAG and HP, which were then at the height of their glory. Particularly the first, which a Canadian expert rightly qualified in his book about the weapon as "The Free World's Right Arm", was at that time undoubtedly the most widespread in all the countries of the non-communist world. As to the MAG, from 1969 FN trod new ground when it was asked to work out practical formulas to adapt this weapon for various types of combat vehicles.

This new approach, which allowed the Herstal company to apply its know-how to the production of more and more elaborate items, resulted in various interesting applications. But already recent experiences in combat situations had shown that the conventional calibers, particularly NATO's 7.62, proved too powerful for close combat. In the mid sixties, under the management of Ernest Vervier and his successor Maurice Bourlet, FN decided to develop a new weapon capable of firing 5.56 ammunition (.223 Remington), which was to become standard from the Vietnam War onwards. A Light Automatic Carbine, the LAC, was the outcome of this research. Production started in 1968. Not long after, the development of a new light machine gun of the same caliber, the MINIMI, got under way. The latter offered the most promising prospects, because even if various nations, the United States being the foremost, had already designed or adopted a special small-caliber infantry rifle, there still existed no squad automatic weapon in this class.

HP pistol cal. 9 mm

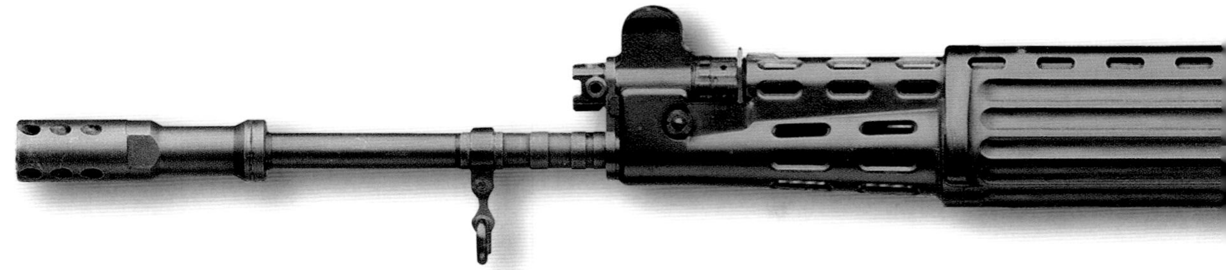

20, 30 and 40 mm FN ammunition. The production of medium calibers, which started up before the war, was accentuated and modernized since the seventies.

The level of activity in the military cartridge division was very high, with a range of products which varied from the smallest conventional calibers to the 40 mm. There was an increased tendency towards automation of production, with special attention for the mid-range calibers, which was to increase still further, as the company already foresaw the competition of new up-and-coming manufacturers in the area of small caliber ammunition, which called for less sophisticated specializations and production apparatus. This tendency to sophisticate the products was to typify FN increasingly, as in general it was aiming to exploit its wide-ranging know-how and move towards activities which involved a higher added value on account of their production costs.

The sixties witnessed experiments with and generalization of self-propelled missiles. For this purpose, FN set up in 1966 a test-bed for rocket engines in Zutendaal. Four years earlier, the company had signed an agreement with the British Aircraft Corporation with a view to joint research into a trajectory correction system for anti-tank missiles, called ATLAS (Anti Tank Laser Assisted System). The project was subsequently abandoned in 1972 after the British Ministry of Defense decided to suspend funding.

Light Automatic Carbine (LAC) 5.56 mm caliber. FN was keen to design a new arm capable of firing 5.56 mm ammunition (.223 Remington), which had become standard since the Vietnam War. Its production began in 1968.

MAG coaxial model

From 1969, a new avenue was opened up for FN when it was asked to design practical ways of adapting the MAG to various types of combat vehicles. This original approach enabled the Herstal company to apply its expertise to increasingly complex products.

There are several configurations of the MAG for internal mounting in armored vehicles. Variants with and without solenoid firing systems, short trigger frame, extended barrel and cocking cable, allow the weapon to be adapted to the turret configuration.

The MAG with spade grips is used as a door gun in helicopters.
This configuration is ideal when the gun is pintle mounted, and the user needs to rapidly swing the weapon through a wide arc.

Mounted as coaxial guns in main battle tanks and infantry fighting vehicles, or as main guns in small turrets, the MAG has been adapted to many different vehicles. The outstanding reliability and accuracy of the MAG made it the natural choice for the US Army for the M1 Abrams tank.

End of the high-tonnage vehicles, successes in aeronautics

The activities at Pré-Madame, which were already marked by the recent closure of the motorcycle division, experienced a swift development, of which the aeronautics division was to reap the best fruits. 1966 marked the end of the high-tonnage vehicles, which for too short a time had been the pride and hope of FN, especially considering the potential this domain offered. Too limited as it was to a domestic market which was dominated by military orders, on which the national industry could not always count, this division terminated its activities with the production of armored vehicles for the Belgian gendarmerie (1967). Only the AS24, that ingenious airborne tricycle with low-pressure tires, extended that tradition of all-terrain vehicles, which had started before the war. However, competition was tough, and commercial success was to remain mediocre.

The Matagri division for agricultural equipment - mainly aimed at a Belgian clientele and also at the French market through its subsidiary at Neuilly - emphasized the production of serial pipeline milking equipment, which was to replace gradually, at least in the larger farming establishments, the first individual milking machines. The more stringent inspection criteria for dairy products had, around 1970, also allowed for the development of a range of milk refrigerator tanks under the slogan 'Cold Farming'. The vogue for frozen products had opened fresh and interesting outlets in the less exclusively agricultural domain of deep-freezers. But competition from abroad, the working costs of a commercial network which had nothing in common with FN's usual channels, and finally the overheads which such a large company had invested in so specific a service, finally put an end in 1975 to that effort in diversification.

It is also worth mentioning the suspension of sales of Mitia tools with tungsten carbide plates, a legacy from the contacts established in Britain during the war.

There was also the end of the Hawk contract and FN's withdrawal from the activities of MMN, which was to be taken over by the company Franco-Belge de Fabrication de Combustible (1973). In the meantime, the knitting-machines division went through both good and bad periods, as competition became tougher. In 1973, these activities were centralized at FN Bruges.
As far as the ELDO scheme was concerned, supplies began in 1963 but were suspended three years later. The low sales and the idea of subcontracting did not excite much enthusiasm in these first hesitant steps into the space age, although FN was to occupy a significant place in that space age later on. Nevertheless, this exercise in international co-operation did have the merit of initiating several of the company's engineers in advanced technology. Other projects did not get further than the research stage, however, and were often pushed aside because of high working costs and the uncertainty about the market situation; examples were the fuel cell (1962-1970), and later the small gas turbine FN-Rateau (1971-1979) or the Jarret rectilinear engine (1973-1981).

On the other hand, the collaboration with the Boeing company (later with Caterpillar) which had led to the foundation of FN Boeing Turbines S. A. (dissolved in 1968), seemed quite promising in the sector of turbo-generators of 250 to 500 hp. FN had manufactured several hundred of these engines, more specifically those intended for the turbo-compressor group used for starting up aircraft engines, and as back-up engines for the Swedish S Tanks.

In addition, many turbines were repaired at Herstal as part of the after-sale service.

In 1968 there came an end to the supply of J79 turbines for the F104 G, apart from the spare parts and maintenance works which, logically enough, followed.
It was an exceptional contract, both in respect of its international scale and the technological expertise it had allowed FN to acquire in manufacturing (1,300 sets of parts) and assembly (374 engines).

FN armored car, version with machine gun (1967). Ordered by the Belgian gendarmerie, this type of vehicle was the last of the 'heavyweights' to be manufactured by FN.

Assembly of a kit of the Rolls-Royce Tyne turbo-propeller. This activity, along with the production of various components of the same kit, lasted from 1964 until 1973 and was relaunched in the course of the eighties.

From 1964 to 1973 we have the delivery of 750 Rolls-Royce Tyne 21 and 22 propeller jet kits intended for the Franco-German Breguet Atlantic 1150 and Transall C 160 aircraft. In addition, some 200 jet-pipes were manufactured for the Breguet Altlantic.

In 1968, FN signed a contract for the partial production and assembly, as well as the testing of 131 SNECMA Atar 9C engines for the Belgian Air force's Mirage 5.
It was certainly a deal of limited scope, considering the substantial funds required for similar production in equally small quantities, which illustrates – a perennial problem – the necessity of concerted international action in this domain.
At least this experience had the merit of putting FN in touch with a great French engine manufacturer, through whose agency some ambitious projects were to prove particularly fruitful, especially in connection with the CFM 56.

It is also worth noting that, apart from various sub-contracts with Rolls-Royce and Turbomeca, this period witnessed the development of a real maintenance and repair service for turbines supplied by Pré-Madame, as well as those of other manufacturers which were entrusted to FN by their users.

Assembly of the SNECMA Atar 9C turbo-jet, intended for the Mirage 5 fighter of the Belgian Air Force. This undertaking lay at the basis of the relations between FN and the French engine-manufacturer.

The central production division

Since the beginning of the twentieth century, FN progressively equipped itself for the production of unfinished and semi finished products to be supplied to its own production workshops. In this way the company had in the sixties a series of services, called the Central Division, which consisted of the traditional forge, the precision forge (for turbine and compressor blades), thermal and surface treatments, the traditional foundry and the last innovation which was the precision foundry. Inspired by the ancient lost wax process, this technique was industrialized during the last war in order to save time and money by manufacturing finished products which required only minimal finishing or which were ready to use. By adopting this process, FN followed the logical pattern of simplification of mechanical operations, which aimed to reduce costs.

Well equipped as it was, the Central Division worked first of all to meet the requirements of the company, but so as to increase the value of the equipment by amortizing the installation and working costs, its services were increasingly sought by outside customers.

New men, new ideas

In 1963, the Board of Directors entrusted Othon Drechsel with the general management of the company. Othon Drechsel was an engineer who had joined FN in 1929 and who had built up a career at Pré-Madame, where along with André Regout, he promoted activities in the aeronautics branch. As a result of this specialization he was open to technical innovations, and he was a shrewd manager, aware of the changes which the company should and would undergo. However, at the moment when he was entrusted with the highest position of responsibility within FN, the reign of his predecessor was not yet over, as René Laloux continued to combine the duties of Chairman and Managing Director.

The coexistence and collaboration between these two contrasting temperaments with different horizons and fields of interest had allowed the company, which was still impregnated with an omnipresent and sometimes beneficial tradition, to open itself up to technological progress and to prepare the way for adopting modern structures. These were to be the work of another generation of managers and executives, headed by Michel Vandestrick. The latter, a metallurgical engineer, who had come from the steel sector but who felt strongly attracted to management issues, was engaged by FN in 1966 to update the financial departments and to introduce analytical accounting by means of computers. In 1968 he became Company Secretary and succeeded Othon Drechsel three years later.

Towards change

In the period 1965-1973, productivity increased, thanks to mechanization. The workforce remained comparatively stable until 1970, when after three years it had fallen to 10,000. This figure is still high, even in a period which had not yet succumbed to the attractions, whether true or supposed, of 'small is beautiful'. Moreover, these people were concentrated within a limited area, as the large majority worked in Herstal and comparatively few in Zutendaal and Bruges or Ghent.

The number of executives already showed an upward trend, an indication that the need was felt to improve management skills. Independently of the FN School, which had been training the necessary qualified staff for the workshops since the interwar period, from the fifties, the need had established itself to prepare the executive staff for the new trends in management which the general evolution of the global society had encouraged as opposed to the traditional modes.

FN was not slow to act, and to that end it not only called on its own capacities but also engaged the services of outside experts. In this way, for example, from 1965 onwards its executives and foremen followed training which was provided by the Institute of Sociology of the University of Liège.

HRH Princess Paola of Belgium received at FN by Othon Drechsel, General Manager (1968). Drechsel, who came from the motorcar division of FN, was head of the Engine Division before becoming General Manager (1963) and Managing Director (1970) of the company. Another person pictured is André Watrin, former apprentice of F. Funken and master engraver.

Michel Vandestrick (1926 - 2007), Company Secretary in 1968 and Managing Director from 1971 until 1986. An engineer from the steel sector, he was the originator and great architect of FN's modernization as the 'golden sixties' drew to a close.

Aerial view of FN in the 1970s. A small town within a town, at that time over ten thousand people worked there.

"Equal pay for equal work"

The most serious problem to affect the Herstal company, quite apart from the decline in profitability, was of a social nature. Numerous disputes broke out, and work stoppages occurred frequently, often for irrelevant reasons, as in a kind of reflex.

Between 1960 and 1973, FN was hit by a number of strikes. The women's strike of 1966 was memorable in that it highlighted inadequacy in the attitudes of both the management and the union movement in the face of new situations, and because of the new trends in everyday life which were brought to the surface. This first exclusively female strike of our time, which had become famous throughout Europe and abroad – sociologists had written several works on it, while it was covered frequently in the press – originated from the industrial model which had prevailed throughout the sixties and which, undoubtedly more so at FN than anywhere else due to the high concentration of manpower, tended to devalue the worker inside the company, while society offered him the prospect, if not the possibility, of unprecedented emancipation. This distortion particularly affected certain categories of employees, who were, more so than others, confined to repetitive and soul destroying working conditions. The female workers or 'machine women' at the Grand Hall of FN, the main mechanical workshop of the Herstal factory, were in that position, just like the young workers or 'machine-kids'. They were still working under conditions which prevailed during the last century.

On 9 February 1966, seeing the vacillation of the negotiating parties, who were hampered by the problems of wage scales which a strict and swift application of the principle of equality in the salaries of men and women, as recommended by the Treaty of Rome (1957), might have generated, the women spontaneously went on strike. Adopted later, and not without trouble, by the unions, the movement progressively grew to 3,000 people, thereby leading to the lay-off of those who were not directly involved in the action. Work was not resumed until 10 May.

Equally, the management of FN was caught unawares, because it had not intended to change its traditional production methods and even less to 'reorganize the work', a formula which at that time had not yet achieved the following it was to have later. It rather chose to beat about the bush, comforted by the thought that it had sufficient resources to cope with a prolonged period of inactivity without too much trouble, and also afraid of the prospect of a drastic reorganization of the salary structure. The movement gained momentum and acquired an unprecedented notoriety. But it soon became clear that a settlement of the dispute could not be a once-and-for-all affair. Moreover, it broke out again in 1974, but here the approach was fundamentally different. Indeed, for the first time the management came up against the problem of settling a conflict which had started from the grass roots and over which the union representatives had little control. This type of situation was subsequently to have a tendency to spread. From then on, the workers' representatives were to experience a certain discrepancy between the more traditional lines of their doctrine and the new, more qualitative aspirations of their members. The slogan of the women's strike of 1966 was "Equal pay for equal work", but at the same time it demanded an improvement of working conditions. An interesting and characteristic detail is that it was not so much the strains of the venerable International which issued forth most frequently from the protesting crowds, but the words of a then popular song by the French performing artist Henri Salvador :

"Work means health, idling conserves it
Slaves to toil don't make old bones..."

Historical photographs : the female workers of FN going on strike (9 February 1966). This first women's strike caught

the news headlines in Belgium and abroad, because of its specific nature as well as the slogan "Equal pay for equal work" and the new type of demands it voiced.

A contrasting picture

There remain a few words to be said about the image of FN – up to the mid sixties – held in the minds of various people. This picture would be different depending on whether it concerns a foreigner or a Belgian. From abroad, the picture remained excellent, as it had done throughout the entire history of the company. This collective appreciation was in itself certainly justified, as it discounted the management problems and was basically merely concerned with what counted for foreign clients: the quality and reputation of the company's products and service.
On the other hand, the opinion of the Belgians was at the same time more qualified and relatively contradictory, based as it was on different criteria, and also influenced by distorting rumors.

The impression prevailed that FN was a stable and prospering company, a real national institution which should be preserved carefully. On the other hand, people started to joke about its traditional, bureaucratic and even old-fashioned style, as well as about certain obsolete working conditions. Moreover, the then fashionable attacks on the structures of Western society had made it the ideal target for opinions against the manufacturing and sale of arms, which were particularly active during that period of prosperity and full employment.
The fact that some had referred to FN as part of the Belgian 'militaro-industrial complex' was to stigmatize the Herstal company, and particularly the Liège region, as the exclusive place for 'gun merchants' thereby putting military and sporting arms, heavy and light equipment in the same box. At the same time, increasingly restrictive bills were being initiated on the possession or sale, and as a result on the manufacturing, of arms.

This climate of disguised hostility partly explains the decision taken in October 1971 by the board of FN to change the name of the company from Fabrique Nationale d'Armes de Guerre to Fabrique Nationale Herstal, thereby eliminating an unwelcome connotation and asserting a will to break out of its isolation and to modernise the company.

This task was to prove so arduous, that the Liège region, with which the Herstal company was in permanent osmosis, started to sink into an unsettling state of economic decline, as a result of the accelerated deterioration of its traditional industrial base.

Change and success

For FN the seventies marked a period of complete change, even revolution, if the basic forces which constituted the philosophy of the company had not played their regulatory, if not retarding role, by preserving part of their perenniality through a certain resistance to change.
But it is equally true that the latter proved to be of a profound and unparalleled nature in that it was conducted deliberately and systematically. This often happened in periods of recession, at a time when fundamental values were being questioned, provided the means and the will existed to turn setback into defiance.

Picking up the gauntlet

Financially, FN Herstal was in a bad way. A stagnant sales did not allow for a sufficient mobilisation of cash-flow, caught as it was between wage claims, aggravated by inflation, which absorbed up to 80% of its figure, and markets which had become more and more affected by the recession. Under such conditions, investments remained inadequate, and soon the profit margin would fall into the red. In order to ward off the danger, the new General Manager, with the support of the principal shareholder and the engagement of new or newly trained collaborators, succeeded in implementing a company scheme and growth strategy, which he was to elaborate and refine gradually in the course of the first years of his mandate. FN's recovery was in fact a spectacular one.
The credit for this success goes to the man who was its chief promoter and who made the most of the favorable conditions at his disposal. Those years were to be remembered by all those who had lived through them at FN or in its vicinity as the Michel Vandestrick years.

A new company scheme

The philosophy behind this action was based on the opinion that economic and social maters are closely interrelated. There could be no economic prosperity without an adequate social climate, in the same way that there could be no social progress without a favorable economic situation. Therefore it was necessary to act simultaneously on both these fronts. So the FN's company scheme was twofold: on the one hand, revitalizing the old Herstal company and its environment, on the other, implementing a growth strategy on a world scale and becoming an international entrepreneurial group.

Right from the start, FN was insufficiently prepared for this voluntarist managerial approach, the basic principles of which were not clearly identified and for which the promoters still lacked supporters. The recourse to various consultants from reputable international companies allowed them to accelerate the process, with the double advantage of being able to benefit from long experience and to give the credit for new structural reforms to the efforts of external and presumably impartial analysts. This assistance from outside consultants was received with mixed feelings. The traditional company mentality disapproved of it in principle, because it somewhat disavowed the talents and abilities of FN's 'ordinary' employees, who were jealous of the 'aura' which the everyday nature of their tasks did not give them, but which nevertheless exerted a certain unavowed fascination over them, simply because it made them feel uncomfortable. But whatever the case, these new considerations showed up more than ever before the close interdependence of

Michel Vandestrick (right) after the demonstration flight of the F16 Pratt & Whitney plane at the Air Show of Bourget in France (1975). The first person on the left is Guy Derom (1923-2005), former Commander of the Belgian Antarctic Expedition and public relations officer for the Engine Division. The second person from the left is Jean de Fonvent, Manager of the Engine Division and subsequently of the Aeronautical Affairs at FN.

the various factors which both constituted and influenced the life of the company. It was decided not to neglect any of them but to act on them in a coherent way.

In anticipation of what FN should and could become after 15 years of change, a manifesto was drawn up and distributed under the heading 'Prospects for 1989'. The contents do not lack audacity, considering the realities of that time : *"FN is a modern enterprise, performing and profitable. It offers prospects for growth to its workers, its shareholders, its region. It meets the demands of its clientele by the quality of its services and its products. It achieves these aims thanks to the co-operation of everybody."*

The desire to shock and to stimulate was clearly in evidence, as are so many other mobilizing slogans which were to follow. One management method was decided on: joint management by objectives, allowing for individual initiative and responsibility in a decentralized structure. The reorganization of FN, which started in 1973, was completed two years later.

Discounting a number of subsequent modifications and adjustments which took account of persons and circumstances, it sanctioned the principle of a corporate organism at the disposal of, but separate from, the operational divisions, which are market-oriented rather than product-oriented and organized in profit centers.
The new analytical accounting system and assessment according to results were to guarantee a greater degree of autonomy.

In addition, a strategic plan was worked out and implemented in 1975, with a view to establishing converging objectives for that cluster of decision-making centers which FN had progressively become.

Old and new lay-out.
The seventies were characterized by a more rational arrangement of the work stations and operations, an improvement in the working environment as well as the beginnings of automation and a reduction in the workforce.

CORREA

Releasing energies

The revival of FN and its social and cultural environment constituted the prime proposition of the new company scheme. From then on, it was to be a question of paying more attention to the expectations of the company's different partners, breaking the silence towards the outside world and between the management and the unions.
The latter challenge was not the least ambitious, particularly as it had to overcome almost a century of distrust and misunderstanding. To say that this was a success would be an overestimation. But even if the management and workers' representatives had not succeeded in becoming true 'industrial partners' (according to a fashionable euphemism), they had become at least partners in the communication process, who had perhaps come to respect each other's dignity. From that moment, the solicitude of achieving greater transparency in management dealings, particularly through the works councils and the internal press, and the desire to reach a certain consensus before taking decisions of general interest, were always part of the agenda.

In this respect, the second strike by the women workers in 1974 was also revealing. Being of the opinion that their situation had not changed sufficiently swiftly since the events of 1966, the machine operators once again walked out spontaneously, demanding urgent action on the part of the management.
By tackling all the latent problems regarding the conditions of the workers as a whole, the management succeeded within a month in defusing this new psychodrama. It set up three commissions, in which management and unions were equally represented, to look directly into all problems concerning remuneration as well as the working conditions and organization.
These meetings led to a number of profound changes in the company: abolition of piece-work, reorganization of the various functions, promotion of the women, improvements in the work place, and permanent mutual consultation in these matters. Other long term consequences of this innovative approach were: the development and implementation of a true 'lay-out' policy, in other words, an industrial structure which allows for an integration of the economic, technical and social aspects of the company in terms of working conditions and organization.

These parameters were applied not only in the restructuring of the old production halls in Herstal, but also in the new production plants which FN had set up outside its main points of anchorage (Harzé 1975) or abroad (Portugal 1973, Brazil 1978, United States 1981, etc.). At the same time, the concerns of the workforce, who were seeking an improvement in their working environment, came together with the management's intention to create smaller operational units, which would be easier to manage efficiently, and less subject to industrial unrest, than was the monolith in Herstal from which they issued. In addition, in this expansion beyond the limits of the Herstal establishment, the FN management saw a means of reducing the labor bill, at the same time moving into labor markets with surplus workers who were less influenced by political pressures and public opinion than was the case in Belgium.
Viewed with suspicion by the union leadership as a threat to local employment, these first relocations had nevertheless not yet acquired - in the current of modernist ideas which prevailed at the time - the sometimes painful connotations which globalization was to generate at the turn of the century.

Another innovation which was intended to meet the demands of the staff was the professional training scheme set up to initiate staff in new areas of employment and to prepare them more adequately for the complexity of their tasks. This was welcomed heartily by the authorities: study leave, retraining, crash courses.

FN's Training Center, which was moved to more spacious premises in 1980, opened its doors to the outside world and made efforts to co-operate with the state and private education institutes. Special attention was paid to the economic education of the workers and their representatives, with a view to familiarizing them with the new international situation to which FN's fate was linked. In this respect, the study trips which the works council of Herstal made to Sweden (1974) and the United States (1979) are particularly interesting.

'Machine woman' in the Great Hall. In 1974, being of the opinion that their conditions had not improved quickly enough since the events of 1966, the women machine operators had once again walked out spontaneously, demanding immediate corrective action.

The qualitative and quantitative strengthening of FN's executive framework during that period contributed considerably to the investment efforts in human resources, as well as in continual training, which from then on was carried out on the spot or at special seminars organized by specialized organizations These seminars aimed at improving technological or economic knowledge, as well as social and individual behavior. All this was to contribute to an improvement in the executives' flexibility and capacity for taking decisions. FN's considerable participation in the setting up and management of the Foremen Training and Improvement Center and of the Liège School of Management (now Business School for Advanced Management Training) contributed to a significant extent to the training of new managers which the Herstal company - and the whole region for that matter - badly needed.

Special efforts were also made in the internal communication of technical knowledge – this was already done through the Central Laboratory and its documentation center – but also of management sciences, which were then in full development.

All those successive theories – never absolute or entirely transposable but in any case complementary – which had enriched the thought of the managers in the course of the last thirty years have found a varied response at FN : from the (always present) "Peter's principle" to the "Just in time" along with "In search of Excellence" and the "American Challenge", "The Japanese Challenge", a touch of "Wall Street" and "L'Imprécateur", to end up with the "Chaos Management" according to Tom Peters, so that the culture of the company should fall in line with that of our time...

FN's Training Center, set up at a new site in 1980, prepared the personnel for the new economic and social challenges and, of course, in new techniques.

A deliberate civic approach

By opening its doors to the outside world, the company sought to tempt and possibly seduce its various business partners. It succeeded by offering them the material resources it had regained and the dynamism which they needed. Particularly with respect to the area of Liège, and of Wallonia in general, which had become more and more adversely affected by the loss of its traditional industries and the decline of its prestige, FN had played the role of beacon and even of locomotive.

A great number of economic, social and even cultural initiatives were supported by, and even originated from, FN. More particularly in the context of its redeployment strategy, FN was aiming at a true 'regional diversification' by setting up new plants in Wallonia: Faurieux, Harzé, Gasquy, Hauts-Sarts, Milmort, Sclaigneaux... and by participating in the recovery or setting up of companies, such as Sonaca, Raskin Liège, Belairbus... The important positions held by Michel Vandestrick at the head of or within the Confederation of Industry in Wallonia, Fabrimetal, the Confederation of Belgian Industry, and many other management organisms gave a considerable impulse to modern concepts and doctrines, at the same time giving the Herstal company a favorable image.

This propulsive effect yielded spectacular results at times. For instance, in 1983 FN took the initiative for and organized a series of events entitled 'Synergium Japan-Europe' - consisting chiefly of seminars and exhibitions - at which the privileged partner, the then far away Japan, was intended to promote the emulation of its regional competitors.

Contacts with the media were encouraged and even to some extent institutionalized. Rarely had the increasing number of public relations initiatives have such a catalytic effect. Privileged relations were also established with financial and stock exchange circles, the support of the banks as well as that of the shareholders having become of prime importance with respect to the growing need for capital experienced by the company from then on.

Moreover, FN was to be particularly attentive to government initiatives and to contacts with the various officials, who would be prepared to endorse and materially support those initiatives which created employment or development.

Gradually a new notion was taking shape: that of the social, or rather civic responsibility of the company.

The rebellious and "consumerist" Western society, in its eagerness to question the establishment and the institutions, was asking the large industrial corporations to justify their reason for existence in terms of their service to the community, rather than the strict satisfaction of their own interests.

The FN stand at "Synexpo" (Liège, 1983). FN has often taken the initiative for economic, social and cultural promotion. This exhibition was part of the series of 'Synergium' events, which aimed to make the region aware of Japanese dynamism.

FN was prepared to take up the gauntlet. It had its 'social balance' drawn up so that nobody should be ignorant of its contribution to the general welfare, it was published in 1977 along with the company's annual report. It reported that the Liège area owed 3% of its gross regional product and more than 3% of its working population to FN, and also that 3% of the working population active in the Belgian metallurgical industry were employed by FN.
The company dealt with 5,300 suppliers, half of whom were in Belgium. It held about 100 patents, registered in 25 countries, and employed more than 200 people for research and development of new products. Year on year, for every four jobs at FN a fifth was created in the country.

It would be fruitless to try to measure the tangible results of FN's civic 'messianism' during this period. To varying degrees, some members of the company's management team saw in it a way of reconciling their natural penchant for modernism and generosity with the necessity to display the technical and commercial power of the company, thereby exorcising any negative connotations linked to their basic business activity.

This driving role, certainly costly and to some extent lateral, proved to be beneficial for the image of FN and of the region – something which was no doubt necessary in order to stimulate the reinvigoration of internal productive resources.
This could only serve as a temporary brake on the structural decline of the Liège basin, however, which would be tested more severely than ever over the course of the following decade. With the benefit of hindsight, however, it should be recognized that by promoting diverse industrial initiatives within its walls – some of which became independent and survive to this day – FN has contributed significantly to the creation of new business activities, generating an estimated 2,000 jobs since the seventies. These jobs, financed by company equity and to a lesser extent government grants, are not related to the arms manufacturing sector.
That being said, in large part it was the profits from military outlets which had provided the investment for this diversification. After the event, the economist could certainly question the wisdom of spending, during the same period, an estimated 400 to 500 million euros in such a peripheral fashion. Would it not have been better to devote funds to investments which would strengthen the company's core activities ?
But the historian would retort that the most coldly rational calculations often clash with more contingent considerations, forced on decision-makers by the immediate situation. In any event, this rightly flattering picture found its match in activities of an unequalled dynamism and in new, diversified and specialized products.
It was thought that FN's growth strategy should be based on the exploitation of its strong points, those of a world leader in the field of precision mechanics with a worldwide commercial network. This proposition was something of a gamble, as it rested on certain assumptions. But could FN's expertise be maintained, even by buying out specialist companies, in all fields of this multifaceted sector? And was its sales network suited to the marketing of widely differentiated products ? Finally – and most importantly – would profitability meet the new ambitions ? The following period was to show the relevance of these fears, all too easily dismissed in the enthusiasm of a company in transition.

So, capitalizing on traditional products, such as arms, ammunition and aero engines, the company sought to diversify its products, at the same time exploring new areas, intensifying its trading activities and extending its supply routes.

Woman worker of the Great Hall, a figure symbolizing the social changes occurring at the time (1980).

Defense and security

The Defense and Security division (subsequently called Defense and Security Branch : BDS) achieved enormous success with the final development of a 5.56 mm caliber weapon system: the FNC carbine, the Minimi and the SS109 ammunition, standardized by NATO, as was FN's 7.62. The ingenuity as well as the exceptional operational quality of the Belgian weapons was once again vindicated when the United States officially adopted the MAG in 1976 and the new small-caliber machine gun four years later.
The well-known stringency of the tests, to which all equipment is submitted across the Atlantic before it is adopted by the Army, gave the arms manufacturers of Herstal an additional degree of prestige which also explained the commercial success of this military equipment in other countries.

The renewed success of the Browning .50 machine gun, which FN improved by fitting it with a quick-change barrel, started in 1978. That same year saw the release of a new product which was destined to meet with great success. This was the "pod", a container which could be fitted onto planes and light helicopters and which held one or two machine guns which were remotely operated by the aircraft pilot.

The BDS also turned its attention to the engineering sector, exploiting its know-how in the design and construction of "turn-key" weapon arsenals and workshops, according to the requirements of the user. The sale of complementary equipment outside the division's product range also expanded, notably thanks to collaboration with the Beretta Company, in which FN had taken a substantial share in 1972.
A Police division was set up to meet the requirements of that particular clientele with equipment which was akin - but not identical - to that used in the army.

Increasingly obliged to transfer production of equipment from Herstal to its client countries, FN set up a subsidiary plant in South America : FN Do Brazil (1978), and in North America : FN Manufacturing (1981). In addition, an agreement was concluded with the Australian Mayne Nickless and the CIG (Center d'Informatique Générale) to set up two common subsidiaries in Belgium: Security and Technology Systems and Security and Technology Services, with a view to the design, installation and maintenance of security systems for persons and property (1981).

Within a few years, under the impetus of Bernard Regout, General Manager of the Defense and Security Branch, FN as manufacturer of military arms and ammunition had considerably extended its range of products and skills, thus reinforcing its pre-eminent position with the armed forces of nearly 100 countries.

A lightweight and compact weapon, the FNC assault rifle is designed for mechanized and airborne units. It is extremely accurate up to 450 m with a mechanical sight and up to 600 m with an optical sight. Its firing system enables single shot, three shot or continuous firing.

"Destruction" testing the Minimi in arctic temperatures.
The Minimi gained a considerable reputation when the United States army, well known for the stringency of its quality criteria, adopted it in 1980.
This innovative weapon filled a gap in the market for a 5.56 mm squad automatic weapon.

Minimi US M249 used by the US army.

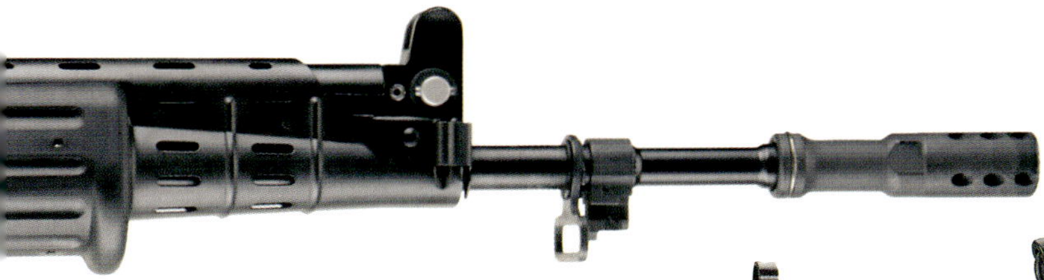

Minimi M249 SAW machine gun used by American forces.

The Minimi proved to be unusually reliable under all combat conditions. It is a true machine gun capable of delivering sustained fire. It can be used on a gun carriage, a 360° tripod or on vehicles. The lifetime of its principal parts exceeds fifty thousand rounds.

A Belgian "Blue Helmet" in operation in Africa, armed with an FNC rifle.

LAR Heavy rifle.

Photographic study of a MAG resting on a bed of 7.62 mm cartridges. At this time, FN was the only company producing both the arm and its ammunition, which gave it an unprecedented competitive advantage.

FAL cal.7,62

FN SNIPER rifle for sharpshooters. NATO 7.62 x 51 mm caliber (1978).

FN Browning 12.7 x 99 mm heavy machine gun, Q. C. B. version (Quick Change Barrel). Manufactured since the inter-war period, this famous machine gun regained acclaim after FN had improved its design (Quick Change Barrel system) and its ammunition was diversified, especially with the APEI ammunition (Armor Piercing, Explosive, Incendiary).

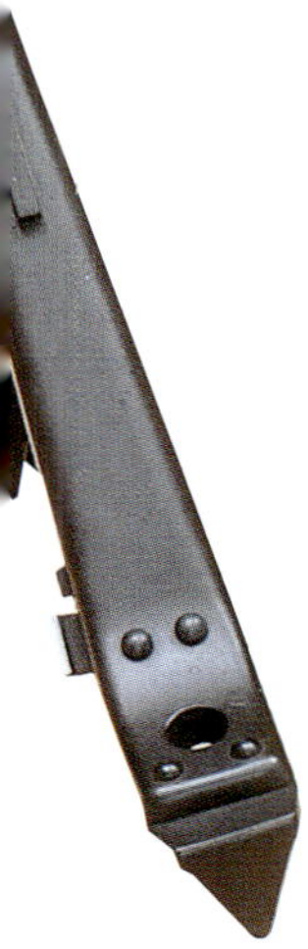

The .50 ammunition range
M33 - M17 - APM8 - APIM8 - APITM20 - APEI - Blank

The FN Browning heavy machine gun Q.C.B. 12,7 x 99 mm mounted on armored vehicle.

FN submachine gun produced under Beretta license. Cal. 9 mm Parabellum.

Photographic study illustrating the range of ammunition,

FN Twin Machine-gun Pod (TMP). This pod, which could be fitted onto planes and helicopters, was equipped with machine guns and a camera, which were remotely controlled by the pilot. The pod illustrates the sophistication of the products and the know-how of the Defense and Security Branch.

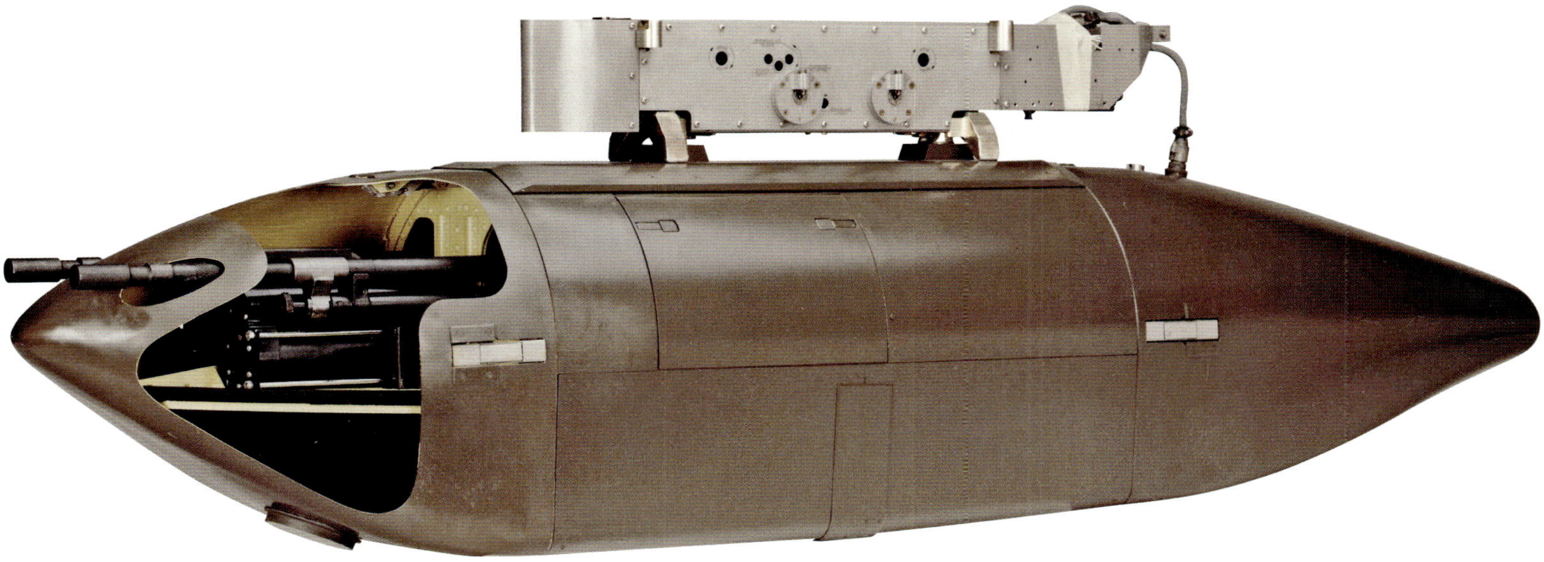

Twin Machine-gun Pod on BELL 206 helicopter.

for a slide show presentation of the Defense and Security branch (1980).

FN do Brazil. This former subsidiary, set up in 1978, allowed FN to implant its manufacturing activities in South America and to cover more directly, from this production unit, various markets on the American continent.

FN Manufacturing Inc. (FNMI). This large factory producing armaments for the American army was established by FN in 1977 in Columbia, the capital of South Carolina.

Aerial view of the FN factory in Zutendaal in the Province of Limburg (northern Belgium).

FN Sports

In the meantime, the sporting arms and ammunition division had a face lift too. Gradually the opinion gained ground that these traditional products, the market for which was in decline, fitted into a larger context, which at first was considered complementary, that of active leisure pursuits. It was decided to launch into this area, and the new division FN Sports enthusiastically took on the challenge. But in view of rocketing production costs, it was forced to reorganize production, either by abandoning the production of arms which had become totally unprofitable (such as the Auto-5, the Mauser and .22 caliber rifles) or by moving it abroad (B2000, BAR). On the other hand, the traditional principle of the over-and-under shotgun was pursued, with new models, such as the Special Sporting in 1974 and the Super Trap in 1976.

The renewed success of these great classics in the hands of experienced shooters in top competitions only served to vindicate their quality. An Express over-and-under rifle of the same design and introduced in the sixties met with wide acclaim among hunters. Furthermore the 'Liège' and the subsequent versions of this weapon confirmed the possibility of turning aside from the, albeit prestigious, beaten track in arms designing, as well as to recover some of the clientele put off by the prices which had grown well beyond their means.

In approaching the markets, mainly by taking direct control of its representatives in Belgium and abroad, FN Sports began to realize that the clientele, under the influence of an aggressive competition, had become irreversibly more attracted to commercial arguments inspired by consumer products, such as prices and novelty.
The arms and ammunition introduced during these years took account of this, in the sense that emphasis was laid on variety rather than novelty, which however did not preclude originality or exclusivity.

Another, from then on dominant, characteristic of arms production was a more marked collaboration and integration between the various sources of creativity and production : Browning USA and its own suppliers, with FN Herstal, and the latter with its Italian partner Beretta or in occasional partnerships.
These various collaborations were reflected in new types of pistols such as the .22 International (1975) and Challenger II (1976) and III (1982) pistols, a double-action pistol, the DA 140 (1978), as well as various large-caliber or .22 rifles during that same year, and a new automatic shotgun named after its year of issue, the B80.

The bi-caliber (.357 Magnum and 9 mm Parabellum) Barracuda revolver (1976) was the first handgun with cylinder introduced by FN-Browning, but this venture only met with mediocre commercial success, and production ceased in 1987.

Barracuda revolver.
The only revolver in FN-Browning's handgun production (1976 to 1987), the Barracuda did not meet with the commercial success anticipated.

DA 140 Pistol. The Double Action pistol (DA) was introduced in 1978. Of foreign design, its sales were very limited and production ceased in the 1980s.

Browning B25 "Ducks Unlimited"

Browning B25 Hunting Special, 12 gauge, tulip fore-end, curved stock, "Langeais" engraving (B2) : English engraved motifs on the barrel wings and on the top of the receiver. Pearled trim. Finely engraved pheasants, grouse and young partridge framed by fine English engraving. Delicate roses. Grey tempered.
In the late 1970s, FN-Browning named the different types of engravings on the over-and-under models after fancy designations, intended to emphasise the luxury nature of these partly handmade weapons.

Browning B25 Special Skeet, 12 gauge, half fore-end beaver tail, pistol stock, 'Birmingham' engraving C3 : fine engraved motifs of acanthus leaves on the barrel wings. Borders on shells and on the internal face of the action-frame underlined with a pearled trim. Fine 'Embroidery' type English engraving. Grey tempered. Walnut, 'teardrop' refiling, oil-finished.

Browning B25 Special Trap, 12 gauge, half fore-end beaver tail, pistol stock with rubber anti-recoil pad. 'Trianon' engraving D5 : very beautiful engraved motifs of garlands of acanthus leaves and flowers on the barrel wings. Borders on shells and on the internal face of the action-frame.
Elaborate engraving in Louis XVI style, finely engraved. Grey tempered. Specially selected wood, beautifully ribbon or flame marbled, 'teardrop' refiling, natural finish.

Browning B25 Special Trap 150, 12 gauge, half fore-end beaver tail, pistol stock with rubber anti-recoil pad, floating ventilated rib. 'Palatinat' engraving A1 : engraving of garlands of acanthus leaves on the barrel wings and the top of the receiver. Trim with scrolls and roses. Blued finish.

Browning B25 Sporting Special, 12 gauge, tulip fore-end, curved stock. 'Carnaby' engraving D2 : motifs of pearled trims and trims with scrolls on the barrel wings. Borders on shells and on the internal face of the receiver underlined with a beautiful pearled trim. Very fine English engraving of bouquets. Grey finish. Specially selected wood, beautifully ribbon or flame marbled, 'teardrop' refiling, natural finish.

Browning Express rifle, 9.3 x 74 caliber. Custom-made. Model with side-plates decorated with engravings depicting boar hunting scenes and framed by deep engravings. Pistol stock made from oil-finished walnut, finely checkered, 'teardrop' refiling, with three-piece tulip fore-end.

BAR MKI rifle. During the 1970s, in order to maintain its competitiveness, FN Sport reorganized its production across several sites, particularly for the BAR which from then on was produced at FN Viana, near Porto in Portugal.

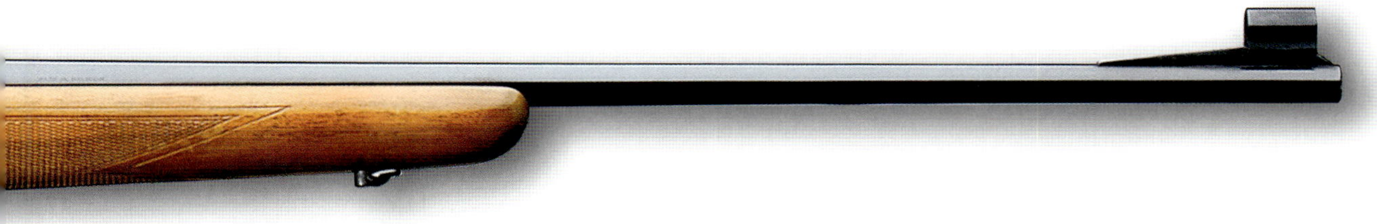

Mauser hunting rifle, produced in all the standard big game calibers.
Its production started in 1922 and it was very similar to the military model; during the 1960s this weapon was redesigned as a luxury sporting model. Faced with a surge in manufacturing costs, with some sadness FN was obliged to end its production in 1978.

Browning Auto-5 Hunting model, 12 gauge Magnum.
Another victim of production reorganizations, this legendary shotgun, which had always been produced at Herstal, was subcontracted to Miroku in Japan in 1976.

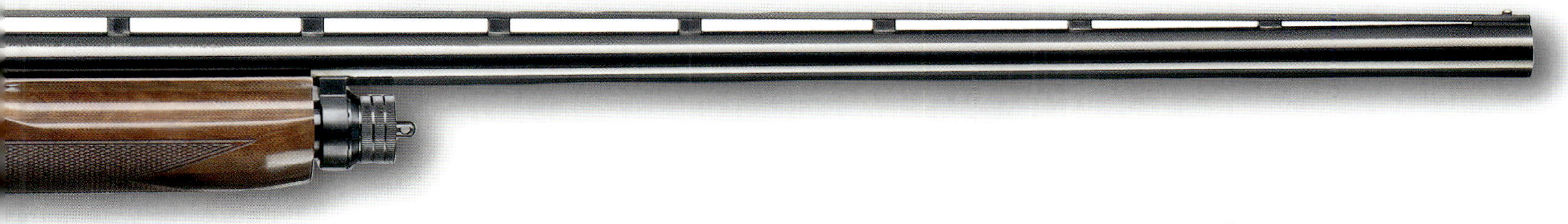

Browning BPS Hunting, 12 gauge. Created in the United States in 1977, the Browning Pump Action Shotgun was also subcontracted to Miroku, just like the Auto-5.

Harmon Williams, Sr. guided the Browning firm through difficult times with great wisdom and foresight during the 1970s.

The take-over of Browning

The major event of the decade in that sector was undoubtedly the take-over of Browning USA by FN, following a take-over bid concluded during the summer of 1977. In fact, in the previous few years its transatlantic partner had ended FN's seventy-five year monopoly on the supply of arms from abroad.
This was because, keen to take a larger share of the market, it had extended the range of its sporting arms beyond those supplied by Herstal. It turned mainly to the Japanese manufacturer Miroku, who supplied it first of all (1969) with the B-SS side-by-side shotgun (the Browning Side-by-Side), just at a time when FN had withdrawn its own Anson model.

Next came the Citori (1970), which was gradually to replace John Moses Browning's traditional over-and-under model. Then three years later the BPS pump action shotgun (Browning Pump Action Shotgun). Other models would follow, the result of the collaboration between the Morgan designers and their Japanese counterparts, thereby reinforcing Browning's brand image and presence on the world markets. At the time of the takeover bid in 1977, Browning was suffering financial and legal difficulties, which Herstal took advantage of - in a way some denounced as hostile - to curb what it saw as a loss of exclusivity.
This seizure, however sudden and unexpected it may have appeared, nevertheless constituted a considered policy of direct investment in the markets by a manufacturer, in this case haunted by the prospect of losing an essential outlet upon which depended its survival and access to which was controlled by a single client who had become rather less regular.

In this way, FN and Browning had become one entity, with their former relationship of subordination now reversed. So the parent company was obliged to assume technical and commercial responsibilities which had become considerably increased by its power of decision over a foreign company, which contributed its know-how and its approximately 10,000 sales outlets in a continent representing half of the world market in sporting equipment.

Beyond the immediate preoccupations which at the time dictated the actions of Herstal's management, the acquisition of Browning would prove to be paramount for the company's future. In terms of design, production and marketing, the group's strategy for sporting arms could then really be deployed on a global scale.

This was borne out by the significant increase in sales figures, although at an internal level there was only a slow gestation of the potential synergies before this promising merger, conceived and executed in 1977, was fully to bear fruit.

View from the entrance and aerial view of the Browning installations at Morgan, near Salt Lake City, Utah.

The Brownings : a genius of invention,

The greatest ever inventor of firearms was undoubtedly an American – John Moses Browning. However, he began his professional activity in an unusual location, very far from the Connecticut Valley which was the cradle of American arms manufacturing. Because of family circumstances, he was born in the still wild and mysterious Far West where his parents had emigrated in the mid nineteenth century. His father Jonathan Browning (1805-1879), a Mormon patriarch who was to acquire three wives and 22 children, was a gun maker who had set himself up in Ogden, close to Salt Lake City, in 1852. The small workshop he ran mainly undertook repair work, using the modest equipment which such an isolated region offered at that time, but supported nevertheless by a community of pioneers who needed to protect themselves in a hostile environment. He was a skilled and very creative craftsman. Before going into the West he had invented two repeating rifles, one with a 'harmonica' magazine, the other with a cylinder.

From a very early age, his son John Moses (1855-1926) demonstrated a penchant for mechanics and in particular for gun making. In addition to his personal skills in this field, and in spite of a rudimentary, largely self-taught, training he also benefited from circumstances which were exceptionally favorable to the development of his talents. The second half of the nineteenth century and the beginning of the twentieth were particularly characterized by unprecedented developments in gun making technology, such as the change from muzzle loading to breech loading, single shot to repeating and automatic weapons, the abandonment of black powder in favour of smokeless powders, the appearance of self-loading handguns, the replacement of mechanical with automatic machine guns, the widespread use of rim- or central-firing cartridges, the diversification of calibers offered.

In a few decades, John Moses Browning had taken the major points of all these inventions and from them created completely new models which were simple, practical and lent themselves to mass production.
It is astonishing to think that the man who once modestly described his talent as "a drop of genius in a barrel of sweat" took out at his own expense 128 patents on 80 different weapons. Many are still in force today, or were copied later for a multitude of new versions. Assisted by his brother Matt (Matthew Sandefur Browning) who was the businessman of the team, and by another skilled craftsman, his half-brother Ed (Jonathan Edward Browning), he soon moved into the limelight after the death of his father.

John Moses Browning's parents.

John M. Browning at eighteen. He was already in charge of his father's shop. Everybody called him "John Mose".

John Moses Browning at the time of his first inventions.

He first worked his way up as a manufacturer and retailer, and soon caught the attention of the large gun manufacturers of the East coast, among them Winchester, Remington and Colt. These companies hurried to acquire the rights to his inventions to industrialize and market them... even to monopolise them so as to prevent their competitors profiting from them.

But the United States alone proved to be too small for John Moses Browning's creativity. In 1897, he signed an agreement with the Fabrique Nationale in Herstal, which was then part of one of the largest arms manufacturing consortium in Europe.
This made Belgium the second country in the world (after Switzerland) to adopt an automatic regulation pistol, and the word 'Browning' entered into common speech to denote a hand gun of this type. Shortly afterwards, Browning crossed the Atlantic for the first time and came to Herstal to present his automatic hunting shotgun. He was to return on another sixty occasions, and drew his last breath there in 1926 during a routine visit.
Encountering Browning would be an exceptional opportunity for the future destiny of Herstal. In both the civilian and military sectors this collaboration, unchallenged for over a century, would bring Herstal a succession of new inventions to commercialize and develop and gave it access to international outlets, particularly to the massive United States market.

By the same token, the Belgian manufacturer's prestige and resources, more recently extended to the Herstal group's entire international empire, would enhance the Browning brand and give it world class status.
The rise of the Browning family, numerous as it was, did not stop with the death of the great inventor. Several members of the two generations which followed became famous either for the invention and development of new weapons or for their management of the Browning company. This was set up in Ogden (and later in nearby Morgan) to market an ever-growing range of products, from hunting to outdoor sports, not forgetting the exploitation of the military patents. These men, though, were managers and technicians, a very different breed from the men of the heroic times when the West was won.

Trained in educational establishments, some to a very high level, their profile is that of the twentieth century technocrat, not the self-taught entrepreneur of the previous century. This does not detract from the credit due to them, as a number of inventions were their work. And it should also be said that business affairs were much more complex for them – at international level – than for their forerunners.

four generations of talented gun makers

The Browning Brothers Factory, Ogden, Utah Territory, about 1882.
(Left to right : Sam, George, John Moses, Matt and Ed Browning, with Frank Rushton, an English gun maker).

One of John Moses' sons, Val Allen Browning (1895-1994) perfected the machine rifle, the .30 and .50 machine guns and the over-and-under hunting shotgun invented by his father, created new models such as the 'double automatic' and the BAR rifle (in collaboration with his son Bruce) and managed the family business.

He also lent his name to varied corporate patronage programs and philanthropic works. His older brother John Browning, known as Jack (1880-1964) had often plowed the same furrow as his father.
He had learned French and German and studied at Berlin University. A poet and writer in his free time, he held the reins of the Browning Arms Company for several years. A nephew of the great inventor, Marriner A. Browning, and his son Matt S. Browning were also involved in the management of the family business.

Finally one of Val Allen's sons, Bruce Warren Browning, was instrumental in bringing the BAR, Nomad, Medalist and T-Bolt designs to the world marketplace.

His brother, John Val, born in Liège in 1925 and died in 2003, would be the last in the Browning line to hold the post of Managing Director.

So, for a hundred and fifty years the members of this family created, produced and commercialized some of the best hunting and army regulation arms.

Lieutenant Val A. Browning, in 1917, demonstrating the use of the Browning Automatic Rifle invented by his father.

Matthew Sandefur Browning (1859-1923).

Val Allen Browning (1895-1994).

Bruce Warren Browning (1928).

Formal portrait of John M. Browning with model 1895 machine gun.

The four Utah shooting champions in 1892. The "Four B's" are from left to right : G.L. Becker, John M. Browning, A.P. Bigelow and Matthew S. Browning.

The first extension of FN Sports' activities was oriented towards the setting up and running of clay pigeon or target shooting clubs in Belgium. Here the division in fact saw a means of distributing, if not popularizing in the more favorable sense, sports which paradoxically were not very widespread in a country known for its arms production, and which were regarded with distrust by those who merely associated them with criminal affairs.

The S.P.I. (Sports Promotion International), which was set up in 1974 to co-ordinate this promotion, had, for lack of clear profitability motive, the merit of shedding a more favorable light on the shooting sport in the context of peaceful and responsible training.
This concept was from then on also developed, in consultation with the users and the authorities, by an originally corporative organization, which was set up two years later : the European Institute for Hunting and Sporting Arms.

The sports equipment developed by the FN Group during the 1970s and 1980s earned the Browning brand worldwide recognition at a number of championships and competitions.

Target shooting range at its inauguration (1980). This range, located in Battice near Liège, illustrated the effort FN put into popularizing the well understood practice of sporting shooting, and extending its activities into an area which to some degree complemented its core products.

But the great change in what up to that moment had been the exclusive domain of hunting and sporting arms and ammunition lies in the commercialization, development and manufacturing of products henceforth qualified as "non gun".
Following the very old example of Browning in the United States, and soon encouraged in that direction by the take-over of the American company, FN Sports extended its range of products in 1976 with sportswear and especially golf clubs.

Developed at Herstal with the human and technical resources of FN, particularly its ballistics laboratory, these golf clubs soon met with great success, thanks to their original design, which distinguished them from their competitors.

Tennis rackets were released the following year, in view of the infatuation at that time with a sport which was being popularized, with television highlighting the champions and spreading the image.

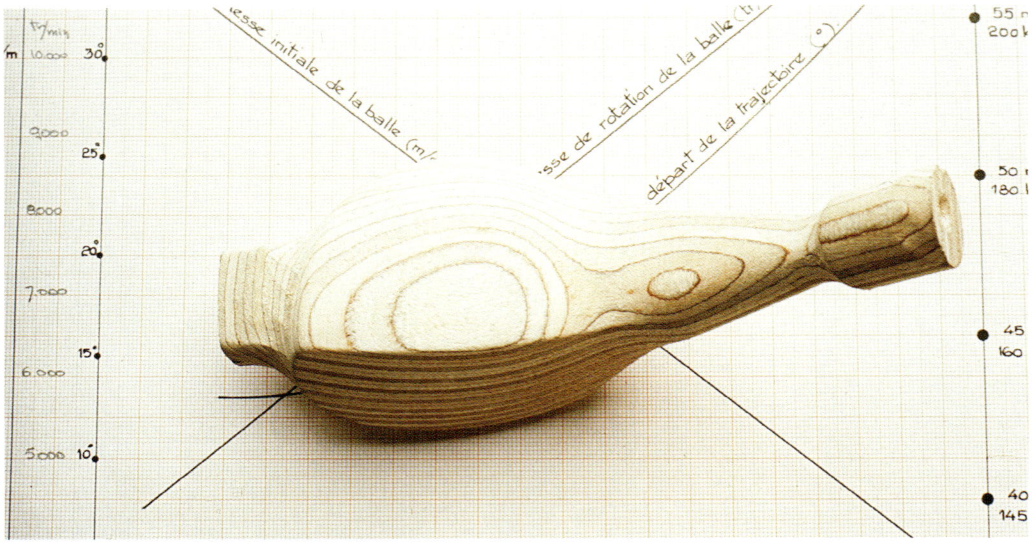

The first Browning golf clubs.
Aiming to take advantage of the prestige of the Browning brand in order to move into the leisure industry, FN applied its technical resources to the manufacturing of golf clubs of an unusual design.

PLAY
BROWNING®
GOLF
BROWNING1
JOSE-MARIA CANIZARES
RONAN RAFFERTY
JOHN WILKINSON
GERY WATINE

The structure of the first Browning tennis rackets incorporated materials generally used in aeronautics, such as aluminium and 'honeycomb', characterized by their lightness and high degree of rigidity.

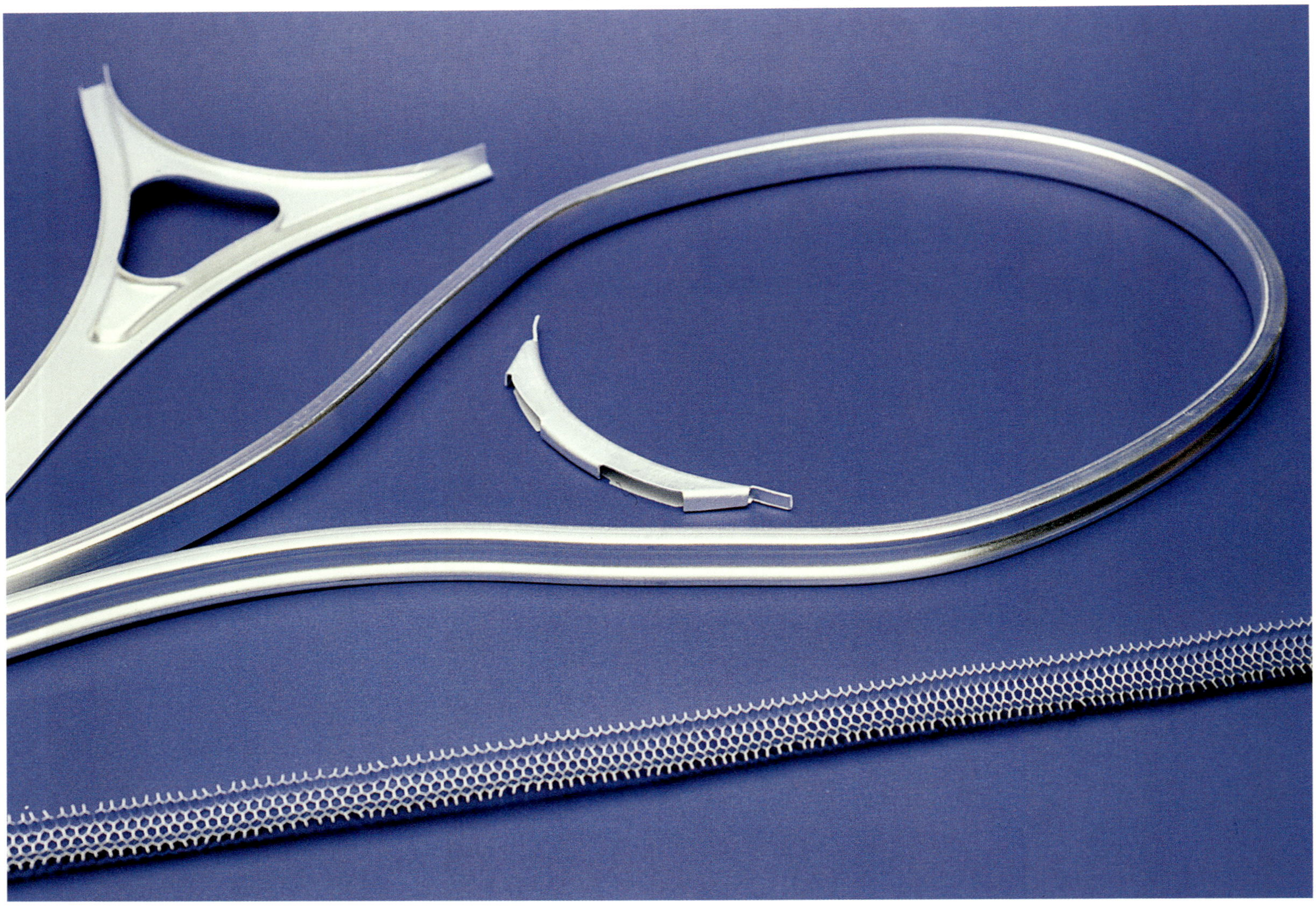

Inspired particularly by the techniques and new materials in their own aeronautical fabrications, in 1978 the FN Group launched itself energetically into the production of tennis rackets, tennis being an expanding and widespread sport at the time.

Production of Browning tennis rackets at FN Bruges. After the production of lorries and knitting machines was closed down, FN Bruges equipped itself for the manufacturing of the first type of Browning tennis rackets with metal frame and honeycomb structure. This structure chiefly helped absorb the impact of the ball on the racket, thus reducing the risk of "tenis elbow".

In 1978, the take-over of the French company Lerc, which specialized in composite materials, cleared the way for new techniques which were applied successfully in the manufacturing of fishing-rods, which FN Sports added to its range of products, as well as second-generation tennis-racket frames.

1979 saw the release of bows, which already featured in the American product range of Browning and which were supplied to Europe.

In 1980, the sail board appeared. With an extensive commercial network at its disposal, as well as various worldwide sources of supply, FN Sports had become a multinational group, enjoying considerable potential and prestige.

Browning hunting bows. Using modern materials and equipped with cams facilitating the tensioning of the bow string, this bow – one of the many models in the Browning range – allows arrows to be shot with a great degree of accuracy and considerable penetrating power, with relatively little effort on the part of the archer.

Aerial view of the factory of the LERC company in Valenciennes in France, specialized in the manufacturing of composite materials.

Browning sail board. At the beginning of the eighties, Browning manufactured and commercialized various models of sail boards.

BROWNING
BROWNING

Browning fishing rods. As a natural complement to its hunting equipment, fishing tackle was added to the Browning product range in the late seventies.

Aeronautics and space

CFM 56 turbo-jet. This engine was developed jointly by General Electric and SNECMA. FN collaborates on the European side of the project.

The period was auspicious and decisive for what, in the context of the general restructuring of FN, was henceforth called the Engine Division. The production apparatus of that unit was spectacularly extended in order to meet what was to be called for the second time in Belgium (since the formula was first applied to the J79 contract) the 'contract of the century' : the co-production at an international level of the General Dynamics F16 fighter, which was adopted by Belgium, Denmark, Norway and the Netherlands, and also offered to the United States Air Force as well as to other countries.

In 1976, FN was entrusted with the partial manufacturing, assembly and testing of the F100 engine of Pratt and Whitney, intended for the European users of that aircraft. The company seized this opportunity to occupy a favorable position in the aerospace sector and embarked on the construction of a new factory, which was to be the most modern of its kind in Europe. The works were progressed rapidly. They commenced in April 1977, were completed in the summer of the following year, and the staff were transferred forthwith from the old workshops of Pré-Madame, which was then used for other purposes.
In the meantime, the test bench of Liers was adapted and modernized in collaboration with SABENA. On 29 September 1978, FN delivered the first F100 engine. The whole business had mobilized a great deal of energy but needed to overcome some measure of reservation. Initial investments exceeded 3,000 million francs, in fact double that amount if all the additional expenses which were subsequently needed to complete the project are included.

A large number of personnel were taken on and trained for the new tasks which they were expected to perform. Employees from other departments of the company who had been laid off were also retrained for that purpose. The new factory of Milmort was undoubtedly one of the most spectacular and advanced industrial achievements of that decade in the whole country, not to speak of the acquisition of know-how required for the manufacturing of the American engine, which constituted a real "technological barrier" which the Herstal engineers needed to break through for that occasion.

Without entering into the detail of the business deals concluded by the Engine Division during that period, there appears clearly a general tendency towards a certain balance between military and civilian orders and especially towards co-production, rather than the sole production under license practiced during the first years. So in addition to the production and maintenance of Snecma-Turbomeca Larzac engines intended for the Alphajet training jets ordered by the Belgian Air force in 1978 in the framework of a compensation agreement, and the resumption of the Transall and Breguet Atlantic orders, there were two important initiatives which held more promise for the future.
The first was in 1972, in the context of an agreement signed with SNECMA for the release of a generation of 10 ton thrust engines, the CFM56, to be retrofitted in a number of military and also civil aircraft.

Parts of the Viking engine for the Ariane launcher, manufactured by FN. Since 1978, FN was subcontractor and subsequently codevelopper of particular components of the engine for the Société Européenne de Propulsion. Inset: the Ariane LOI launcher taking off on 24 December 1979.

For this purpose, FN participated in the joint venture CFM International, which included the French engine manufacturer and General Electric, and which gave FN the right not only to supply subsections of the engine, but especially to codevelop the engine. This program, allowed the Engine Division not only to reinforce its business strategy in the civilian sector, but also to occupy a position, albeit modest, in the restricted circle of industrial co-operation in the domain of aeronautics.

The second initiative definitively confirmed FN's involvement in the domain of aerospace. Since 1975, the Société Européenne de Propulsion (SEP) subcontracted to FN the manufacturing of parts of the Viking engine for the successful Ariane launcher. This agreement subsequently developed into a participation by the Engine Division in the design of specific components of later versions of the engine.

With three decades of economic ups and downs behind us, we can today view these 1970s developments – exhaustively publicized – with a degree of composure. They were the work of Herstal, which paid a heavy price. By placing unprecedented importance on an activity which it had created in the post war years to obviate a scarcity of orders for arms, FN committed itself – both excessively but at the same time inadequately – in a field where the enormous size of the projects and the investment requirements, the cyclical and deferred nature of economic repercussions, risked taking it beyond its means. This is finally what happened with the affiliation and then out-and-out sale of FN Moteurs. Be that as it may, FN has left a legacy to the Walloon region, and consequently to Belgium, a durable driver for development in the aeronautical and space sectors which is one of the most promising areas in advanced technology.

F16 General Dynamics aircraft of the Belgian Air Force. For these planes which were also destined for Denmark, Norway and the Netherlands, FN was entrusted with the partial production, assembly and testing of the Pratt & Whitney F100 engine, thus benefiting along with other Belgian companies from what was called the "contract of the century".

Aerial view of the jet engines test-bed. Installed in 1950 in the disused fortress of Liers, near Herstal, they were continually modernized and extended – mainly on the occasion of the F100 contract – in order to test more and more powerful and sophisticated engines.

Aerial view of the Engine Division. On the occasion of the F100 contract, FN built a modern factory on a rural site, not far from its headquarters, which was started up in 1978.

Formétal

In 1975, the Central Production Division was reorganized and acquired a fully independent status as FN Formétal (FNF). The following period witnessed an expansion of the activities of this division, both in the context of internal supplies – particularly the turbine blades ordered by the Engine Division – and for an increasingly select foreign clientele, among which France and the United States occupied an important place.

The industries of aviation, nuclear energy and energy in general showed increasing needs which FN could help to satisfy. It was precisely to satisfy the demand in the domain of aeronautics - not exclusively but nevertheless predominantly - that a new, ultra-modern and American-designed workshop for lost pattern casting was set up on the industrial estate of Hauts-Sarts in Herstal: FMP Aero (1977). It allowed FN, where osmosis between the different divisions was more than ever before being encouraged, to acquire expertise in a series of advanced techniques which were to earn it a solid reputation as a manufacturer of turbine blades, still enhanced by the collaboration between DM and FNF as well as by the activities of precision forging. Traditional casting however experienced a serious decline, which was to result eventually in this division being closed down at the beginning of the eighties.

Since the sixties, FN tried for technological and economic reasons to break away from the old methods of machine finishing by milling.
The precision casting of intricate parts particularly for aircraft, and more recently the application of composite materials, reflects this trend.

Pouring molten metal into the mould of a cluster.

Robotic coating of the clusters.

Assembly of the clusters for wax parts.

Wheels, aircraft turbine manifolds and other precision cast parts.

An international industrial group

In 1974 FN set up a Directorate for Subcontracting, Engineering and Joint Venturing (DSEP). The aim was to look for outside subcontractors to support, or partially replace, integrated workshops in the best possible economic conditions on the one hand, to apply the know-how of the company to the design and installation of 'ready-to-use' mechanical workshops on the other, and third to seize each opportunity to extend the domain of the parent company by acquiring a share in other ventures for complementary activities.

The first aspect aimed at stimulating the internal departments by letting them compete with their counterparts in the market and also at breaking down FN's traditional autocratic mentality. This was achieved quite well, and it also helped to develop more commercial and global concepts of the products rather than the purely manufacturing considerations of before.

The second aspect capitalized on FN's own needs with respect to new industrial space as well as on the demand of developing countries for technological know-how. DSEP was successfully entrusted with new outside establishments : Viana, Harzé, the Engine Division, FN Do Brazil, FMP Aero, FN Manufacturing ...

It also launched into industrial co-operation abroad, in general in the framework of joint ventures, led by several companies under the sponsorship of interested governments. Since 1975 the department of Industrial Engineering and Co-operation of the DSEP (subsequently called Industrial Services Overseas) succeeded in signing several important contracts for training, technical assistance and industrial maintenance with Bangladesh, Iraq, Algeria, Indonesia ...

The third aspect of course reflected the general strategy of the Herstal company: entrepreneurial growth on all fronts, with the advantage of a rich industrial know-how, and with a transnational foundation. It was an effective way of harnessing the creativity of the company's region of origin, of achieving a growth in sales – as we shall see below – and also of breaking through the initial geographical boundaries with their social constraints.

The competition from abroad, which was becoming more and more worrying, and the higher labor costs applied in certain countries, supported the management's theory that – for economic reasons – for each collaborator employed locally a complementary activity must be carried out by another person abroad. In 1978, FN asserted publicly its quality as an "international industrial group" and for the first time published the consolidated accounts of its empire. Three years later, it held a share in the capital of more than 60 companies, 37 of which are subsidiaries.

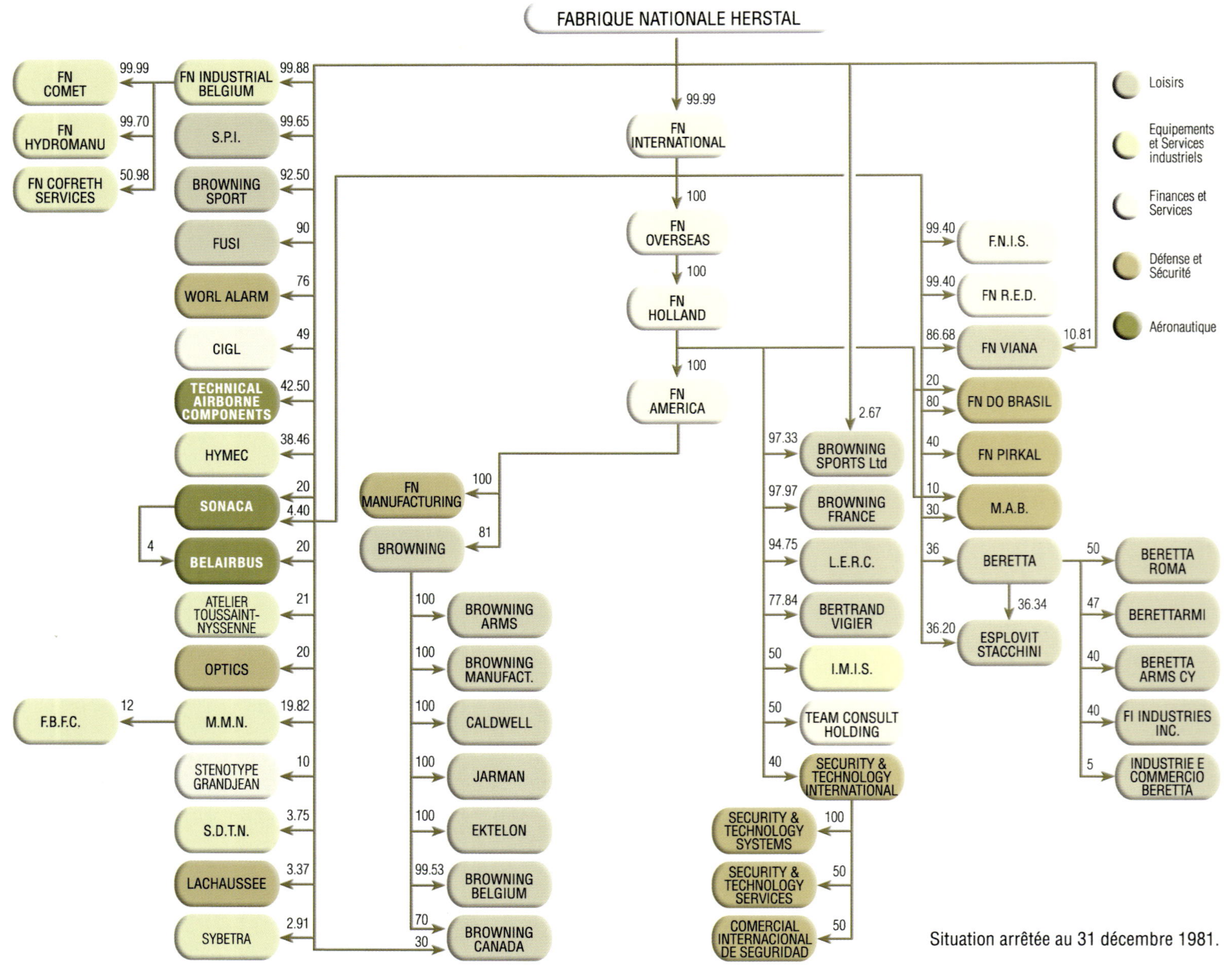

Structure of the FN group in 1981. This diagram shows the extent and complexity of the company's network of subsidiaries at the time of its greatest expansion.

'Robert', a member of FN Manufacturing's personnel. Located in the heart of the Deep South of the United States, this factory has a first class labor force and has gained orders from a demanding customer, the American government.

Genuine products

In 1979, DSEP joined with a division which had been set up four years earlier, the Industrial Equipment and Services Division, or FN Industry as it was called from 1980. To start off with, this unit had taken over two activities which were not to survive for long as a result of competition : the division of agricultural equipment and electrical household appliances, as well as that of knitting machines. On the other hand, it soon associated with another branch, which was to prove more promising in the long run: industrial oleo hydraulics.

Taking advantage of the possibilities presented by the decentralized structure of FN, this division, within a few years, was to constitute a conglomerate of smaller companies, gradually extending its activities into very diversified sectors. From industrial cooperation and hydraulics to industrial maintenance and services (including those for the parent company), hydraulic platforms, recycling of plastic waste, automation, perirobotics and robotics itself. FN Industry, which dealt mainly with businessmen and the authorities, developed service-intensive activities, releasing original products such as the Comet platforms, well known with the fire brigades and the public works sector, and the "plasticizer", developed on the basis of a patent of CRIF (Centre de Recherche Scientifique et Technique de l'Industrie des Fabrications Métalliques).

Generally speaking, FN had, within a matter of a few years, established a balance in the division of its activities, thus ensuring a spreading of the risk due to the cyclic nature of some traditional orders, and at the same time somewhat attenuating the controversial connotation attached to its traditional strategic focuses.
More so than before, we find both civilian and military, government and private clientele, industrial and consumer goods, products and services.

PALL oleo hydraulic filters distributed in Belgium by FN Hydromanu.

Restoration of the "Perron" of Liège by means of the Comet hydraulic platform.
Surely one of the best products of the FN Industry division, this platform was widely adopted, both in Belgium and abroad, by fire brigades and public works companies.
The monument pictured here is the old symbol of the city of Liège, of which the stylized picture serves as a hallmark for the firearms produced and sold in Belgium.

COMET
COME

Comet
ECT-125

FN Comet aerial lifts delivered to the Namur Fire Brigade.

FN Comet telescopic aerial lift in action on the gantries of the cable car of the Citadel of Namur.

In 1974 the Herstal Group began looking for a core product for the civilian industrial market, and one of the focuses of the logic of strategic diversification for FN Industry rested on the following question : in the future, what outstanding, new MECHANICAL equipment might provide the answer to the needs of an emerging economic sector in the decade to come?

Ecology and solid waste recycling being an obvious choice (and being already involved in hydraulics in the field of urban waste collection with their compacting skips) FN Industry began the search for complementary and emerging areas of activity, with the help of the Group's Central R&D department.
Equipment which was original for the time, such as combustible methane producing biodigestors of organic waste, aeraulic solid waste separators by density, and thermoplastic materials recycling machinery, was tested in collaboration with various inventors and research centers. In this context, FN Industry purchased from CRIF the operating license for Professor Patfoort's plastificator. This short screw plastificator could remelt a physical emulsion of the different families of polymer present in sorted urban waste.

And secondary recycled products could be produced from the reinjectable granules. These presented the characteristics of wood, cardboard or terracotta, but with the advantage of being easily shaped, and having the physical characteristics of thermoplastics.

At that time selective collection of plastics did not exist and FN Industry was a trailblazer, developing in collaboration with the municipality of Liège a prototype for a selective collection wagon for paper, plastic and the remaining household waste.
Belgium's first ever selective collections took place in Liège as part of EC funded research under the INTRADEL plastics recycling module.

It very soon became apparent that the concept should be sold in the form of a turnkey package solving the full recycling cycle of urban plastics waste, finished products included.

The first export sale of this complete system was the 'Plastic Recycling RHEO' module for 6,000 tons per year, bought by the city of Shanghai in China. From the plastic waste collected, this module could produce enough recycled bicycle handles to equip the two-and-a-half million bicycles manufactured annually in the Shanghai region.

In our regions, the selective collection of plastics only became widespread 20 years later, as the result of the FOST PLUS program and European directives.

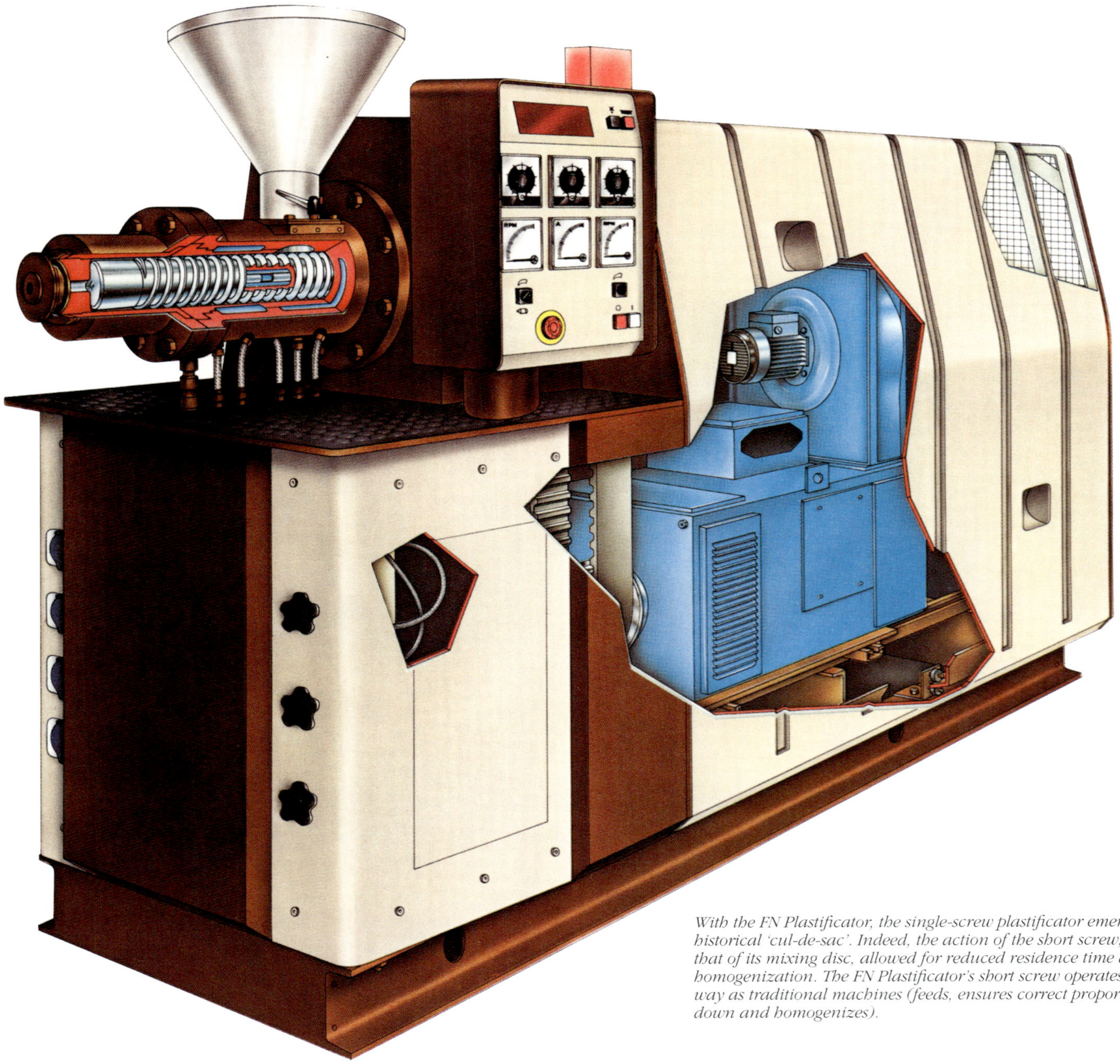

With the FN Plastificator, the single-screw plastificator emerged from its historical 'cul-de-sac'. Indeed, the action of the short screw, added to that of its mixing disc, allowed for reduced residence time and improved homogenization. The FN Plastificator's short screw operates in the same way as traditional machines (feeds, ensures correct proportions, melts down and homogenizes).

Various items made from recycled plastic.

Automation of the handling of powders at FN's Zutendaal factory, made by the Perirobotics department of FN Industry, Bruges.

Installation of automatic palletization at the "Sucreries de Tirlemont" (Belgium - production of sugar), made by the Perirobotics department of FN Industry, Bruges.

Robotic system for sampling the melts, made for the Cockerill steel company by the Perirobotics department of FN Industry, Bruges.

Promotional playing card decorated with FN Industry's colors (1985).

"Success story"

The seventies witnessed the American decision of August 1971 to suspend the convertibility of the dollar, as well as the oil crisis of 1973, which announced the end of the era of growth which the world had known after the war.
The fall of the dollar and its subsequent fluctuations were to form part of the world economic scene, with their attendant disturbances and uncertainty, which have certainly not made policy planning easier for multinational groups such as FN.

Nevertheless the latter was able to benefit from favorable conditions which have facilitated and, to a certain extent, explained its high volume of business: the abundance of 'petrodollars', which many regular clients had, and the demand for Western products, services and know-how by the most well-off developing countries. Indeed, it was above all temporary economic circumstances which confirmed some of FN's managers, as well as a good proportion of observers, in their opinion that the reorganization and diversification had brought about a durable structural improvement.

From 1974 FN's sales climbed to a spectacular height. By 1980 it had quadrupled (in present monetary terms) up to more than 16,000 million francs, of which 2,000 million as a result of trading. For the whole group that figure reached 20,000 million.

Bernard Regout, Director of the Defense and Security Division, which was the primary generator of financial resources during the period of modernization of FN.

Computer Aided Design. Computer technology applied to the designing of new products has, since the eighties, revolutionized the company's research and design departments.

The share price, which continued to plummet since 1968, recovered during that same period and reached its climax in 1980, without however equalling the record value it had reached towards the end of the sixties. The shareholder could once again receive a dividend from 1976.

These brilliant successes did however conceal a profitability which was insufficient to meet the needs of an industry in full change. The modernization of the production apparatus, which had commenced a few years earlier, pursued its course : renewal of the machinery, adoption of complex automated equipment - which at FN had vindicated since 1975 the expression "technological revolution", especially in the arms division - , the increased use of computers (gradually including Computer Aided Designing (C.A.D.), the finite element method and modal analysis), new social provisions, setting up of plants in Belgium and abroad.
The annual figure of the investments, which at the beginning of the decade was limited to 150 million francs, had increased tenfold since 1976. Moreover, inflation led to a huge cost increase : from 1967 to 1977, if the retail price index had doubled, the average labor cost had no less than quadrupled.
Under these conditions, the company's growth strategy contained a high level of risk, considering the very limited resources. Already in 1978, there was concern that the long term debts had already surpassed the company's capital within a matter of three years. In 1980, there was talk of a financial crisis, along with a call for recapitalization. As for the company's profit, once again in the black since 1975, this rapidly reached its peak within two years – more than 200 million – only to fall once more into the red in 1980.

A company with a human face

The overall workforce of FN remained more or less stable during this period. In 1980 there were still 10,000 employees. In the meantime, the structure of the workforce had substantially changed as a result of the reduction in the number of blue-collar workers and the increase in the number of white-collar workers. This was a logical consequence of the mechanization on the one hand, and higher qualification on the other, as well as the aim of reducing the general labor cost. While in 1974 the proportion of labor cost to sales almost reached 80%, this figure dropped to about 48.5% six years later. However, the burden of labor cost was still considered too high. In 1976, following a general tendency, a bridge pension scheme was implemented as a poor palliative for a structural crisis which had become extremely painful for the managers in the Western world.

Throughout the seventies, FN was aiming to become that 'company with a human face', about which Jean-Louis Servan-Schreiber once wrote a best seller.

While within the limits of the system the company was trying to remedy the social diseases inherent in the traditional working methods, it certainly succeeded in making room for consultation and participation by the personnel, attempting to realize that impossible dream of reconciling man's management with his basic aspiration. Over and beyond fluctuations in the economic situation, which frequently give an industrial concern other things to think about than considerations of that nature, it is to be hoped that the lesson will not be forgotten.
Relations between the management and the different categories of staff had in the meantime developed significantly. Power and authority had also lost a lot of their synonymity.
Both industrial and union leaders had substituted persuasion and seduction for command. On the other hand, the men who were submitted to their authority were no longer responding as faithfully as before to their expectations or behest. The relations between the industrial parties were however not necessarily facilitated by this new situation.
Because even if the unions have remained the privileged communicators with the management and the necessary link in

Browning super de Luxe superposed shotgun for the occasion of the millennium of the former principality of Liège (1980).
The decoration of this piece, from a special series of ten, is the work of Louis Vrancken, who was trained at the engravers' workshop at FN.

The engraving evokes the countryside and the traditional crafts of the Liège region. The inlaid work on the side-plate represents the emblem of the Prince Bishops of Liège, the crosier and mitre. On the bottom-plate of the receiver we can see the Perron of Liège and the coat of arms of the Blacksmiths Trade, where gunsmiths used to gather.

negotiation and application of collective labor agreements designed to keep the industrial peace, they too found it difficult to lead people in a society which had become more and more rebellious. Since the eighties, industrial relations started to place more emphasis on participation in the organization of the work process and in the qualitative management, instead of the traditional notions of information and supervision.
In the expansionist context of FN, the managers of the company opted for conciliatory relations with the workers' representatives, so as not to hamper the pursuit of the strategic objectives. Salary increases were considerable. By 1975, they had reached the annual record of 26.5%.
But if the dialogue between the industrial interlocutors had smoothed over a great number of difficulties and defused some conflicts, the old tendency to go on strike still persisted. FN was still hit by a number of strikes during these years. Prosperity, be it only apparent, was, in the Liège region, a luxury for which a high price had to be paid.

In the period from 1974-1984, FN was certainly one of the "sure values" in Belgian public opinion. Following the logic of the company's revalorization with respect to its environment, the successes of the FN group were systematically broadcast and spread by the media. Industrial investments, the new installations, the modern equipment, were all factors which helped to create a flattering image which the staff, the region and the country in general were in need of.

At the end of this period of change, FN appeared a modern, dynamic and prosperous enterprise, at the forefront of technology and industrial relations, and, in addition, generous.

The recession

Side slips

In the beginning, nothing changed. In spite of the economic recession which badly hit the industrialized world, especially Europe, and which gave a full load to a structurally impaired Wallonia, everyone thought that FN would once again escape the tough laws of general economics. Sales reached unprecedented levels: 24,000 million Belgian francs for the parent company and 34,000 million for the whole group in 1985. However, there were unsettling side-slips. The results of Herstal showed a loss in 1980, but the impression prevailed that it concerned just a simple accident on the way. The figure showed a profit in 1981, but barely maintained an equilibrium until 1985, with a new drop in 1984. The consolidated results, the share price and the dividend showed the same trend.

The euphoria of the boom towards the end of the seventies and the extraordinary leap forward on the technical, commercial and social level which the company had achieved within less than ten years obscured the fragility and snags of the system.

The prosperity of the moment depended on the good times which the military divisions of FN experienced and which allowed the company to rebalance the budgets of the various sections. Moreover this guarantee, which by nature fluctuated in the absence of a significant stable market, could not suffice to ensure the viability and profitability of the other sections. Soon the BDS was no longer able to compensate for the losses of the other divisions. FN had opted for risk and audacity instead of carefully consolidating its financial position by means of the resources created by the most profitable of its activities. Moreover, in changing into an international group, its structure had grown heavy, while its finances, already encumbered by high investments, were still aggravated by excessive working costs. The level of debt had become unbearable : 26,000 million in 1985, with financial costs of more than 1,800 million. FN had entirely lost its ability to rely on its own resources. In full economic crisis, the company's funds were themselves under threat.

The time of financial "constructions"...

The eighties at FN were characterized by a serious shortage of capital. The Herstal group attempted to tackle that painful problem with the financial co-operation of its principal shareholder, the Société Générale de Belgique, private and State banks, as well as the governments of the Walloon Region and of Belgium. It was a time of financial 'constructions', in other words technical pipelines intended to let a maximum amount of capital flow to the company by exploiting all the possibilities of the capital market, taking into account the usual parameters: the discount rate, exchange rate risks, taxes, economic recovery laws, overdue payments, etc. A first increase of capital in 1980 amounted to 435 million francs. A convertible and subordinate debenture loan followed in 1984, another in 1985. In that same year, a subsidiary was set up, the S.A. FN Co-ordination Center, with a view to establishing for the Belgian companies of the group long term credits necessary for research and investment.

Also in 1985, the company Technofin S.A. was set up with the joint co-operation of the private and public sectors, to finance FN's participation in the development and production of the Pratt and Whitney 4000 engine. A while later, in February 1986, the FN Aéronautique S.A. was founded, a reconversion company with mixed capital, destined for the acquisition of production material for this engine and for the CFM56, both for civil applications. At the end of the year, FN Aero-Finance S.A. was founded to ensure the funding of floating assets for particular aeronautics programs.

These constructions, however difficult and necessary they may have been, did not suffice to help the company out of its chronic lack of resources and liquidity. Internal reorganization was required, as well as a regrouping of the company's activities.

... and reorganization

At the beginning of the eighties FN's expansion strategy continued. In 1980, still concerned with diversification to complement its Defense division, the company concluded a co-operation agreement with the Australian Company Mayne Nickless. In 1981, in collaboration with the Société Nationale d'Industrialisation and Aerotech, FN set up the company Technical Airborne Components S.A. for the manufacturing of connecting rods for the aeronautics industry.

FN Sports, on its part, continued to expand or set up in Belgium, France, the Federal Republic of Germany and Italy. By the end of 1982, this division opted for prestige and assumed the name of Browning.

The BDS set up offices in Africa, Latin America, Asia, and started up the factory in Columbia, South Carolina.

FN Industry also extended its activities: treatment of waste, recycling of plastic material, escalators, perirobotics, expansion abroad, transfer of technology in the framework of co-operation agreements.

Section of the CFM56 jet engine for civilian use, developed jointly by General Electric and Snecma.

FN Comet aerial lift in action, fighting a fire in a factory warehouse in Namur.

FN Comet aerial lift belonging to Liège City Council's Planting department.

Robotic installation for positioning and welding the roof onto the R5 body, made for Renault Vilvoorde by the Perirobotics department of FN Industry, Bruges.

Albert Diehl, General Manager and subsequently Managing Director of FN. A former manager of Rhone-Poulenc, A. Diehl was engaged at FN as consultant and deputy to the General Manager whom he succeeded in 1986. He gave his name to the recovery plan which allowed the company to revise its strategy and its activities.

The blow came in 1983, at a moment when the low point seemed to have been reached and the company management, reflecting on their optimistic talk during the good years, started to express their gravest concern about the way the company was going. Emergency committees were set up in order to make drastic cuts, review their staff policy in the light of the adverse economic situation and redirect the company strategy. Not long afterwards, in late 1984, Albert Diehl, former manager of Rhone-Poulenc, was entrusted with identifying the priority measures to be taken in order to improve the operation and profitability of FN. Initially a delegate to the general manager, this French consultant would eventually succeed him at the head of the company in 1986 and give his name to a plan of recovery which he conceived and developed with an obstinacy required in crisis situations.
Diversification had not yielded the expected results. In spite of various technical or commercial complementary activities, some of the company's activities constituted in some cases too disparate a whole, where synergy remained theoretical rather than real. Profitability rates proved very irregular, more especially as the dispersion of efforts did not guarantee adequate investments being agreed to for the new branches necessary to acquire an important position on the markets.

It was too much and too little at the same time. Moreover, in spite of the efforts made to increase the human resources in the company, there was still a shortage of skilled staff in certain sectors. The notion of the FN-group itself created a problem, because such a gigantic structure, created so suddenly, did not help effective integration. It was decided to regroup the activities along the traditional lines by bringing them back to their essential features. This operation was conducted gradually and deliberately from 1986 onwards.
These cuts were not effected without some measure of resistance, pain or bitterness, more especially as this loss in activities was reflected in a considerable decline in employment.

The era of redundancies

The automation and computerization of the production and management, the reduction in activities, the collaboration with decentralized or subcontracting workshops, the competition from foreign countries which benefited from transfers of technology, and the economic recession, all account for the excess of personnel with which FN had been faced for some years now. In contrast with the last decade, during which the social pressure on the one hand and the need for human resources on the other, had allowed the Herstal company to maintain the level of employment, the decade now approaching its end had not been that lucky or has not had that clemency. In 1981, the group employed 12,631 people, 10,004 of whom at FN Herstal, while in 1985 this had fallen to 10,721 and 8,228 respectively. In addition, unemployment started to affect a considerable proportion of the staff.
The bridge pension system resulted in an increased financial burden for the company, which was forced to pay the immediate price for a mid-term lightening of its structure. It also led to a loss in know-how, particularly in the handicraft jobs. So from a large enterprise in the traditional sense, FN had become an industrial group of smaller businesses.
Up to 1985, the year in which the Belgian media started to question FN's invulnerability, the recent uneasiness felt by the management did not really convince the workforce in general of the gravity of the situation. The old reflex of social Defense underlay the reputedly deliberate remarks from the hierarchy. Moreover, in spite of the improved relations between the industrial representives, the confrontational tendencies remained very much alive, since it was hard to believe that the company's resources could dry up that easily.
This was shown clearly in the strike of March 1986, which some called suicidal, and which lasted a month. The two unions came face to face, with the Christian union displaying a more moderate attitude, while the Socialist union formed a hard core, attached to the exemplary merits of disputes which had Herstal for their battle-ground.
The approach of the executive staff was more qualified with respect to the basic problems which affected their company. But they were apt to be severe in their opinions - particularly in the management-executive consultation committee, which was set up in 1980 - regarding a management voice which they considered to be too optimistic.

The industrial ballots of 1987, which confirmed the growing consciousness of the executive staff, and which for the first time in Belgium allowed them to choose their own representatives, gave, at FN, 90% of the votes to a specialized professional organization which was distinct from those selected by the other categories of staff.

As far as labor organization was concerned, the recent years are marked by the increasing importance of continual and alternated training (school /factory) and retraining, as well as of quality and progress circles with, more recently, a tendency to set up 'Production Units', small operational cells based on the principle of autonomous planning and control as well as on the multi-faceted character of the tasks. 'Industrial development' was the slogan, from now on aiming at supporting change in the company and to help the staff be a part of it. Certainly technical progress has not ceased to increase : computerization and robotics are gaining ground everywhere, new materials are gradually replacing the traditional ones, while 'flexible' workshop and project management have become instilled.

After an agreement was concluded with the French company CNIM, FN Industry obtained in 1981 an order for escalators for the Brussels underground. They were constructed at FN Bruges, where considerable investments had been made.

Escalators for the Brussels underground, Jacques Brel station.

Old products and new releases

The period under discussion here was marked in the Defense and Security Branch by the mass production of the FNC carbine and especially of the Minimi. The latter met with spectacular success when the American Army decided in 1980 to adopt it: its production at the factory in Columbia under the name of M249, along with the MAG (M240), which was the subject of a supplementary contract in 1984, lasted the best part of the eighties.

Canada joined the circle of countries using those machine guns in the framework of the compensations linked to a contract for military vehicles.

In spite of an urgent desire to turn more towards medium-caliber ammunition, for which a modern production unit was built in 1982, the success of the smaller calibers, particularly the 5.56 cartridge, was considerable : the adoption by NATO of this second standardized caliber – after the 7.62 of the fifties – was accompanied by the decision to take the SS109 ammunition of FN as a reference.

New models of 'pods' were also released, along with a .50 caliber armor piercing, explosive and incendiary cartridge (APEI). In the meantime, the staff of the Engineering and Cooperation Department helped to spread the technology of military products from Herstal throughout the entire world. In addition, the R&D department of BDS started to develop a heavy machine gun, called the BRG15.

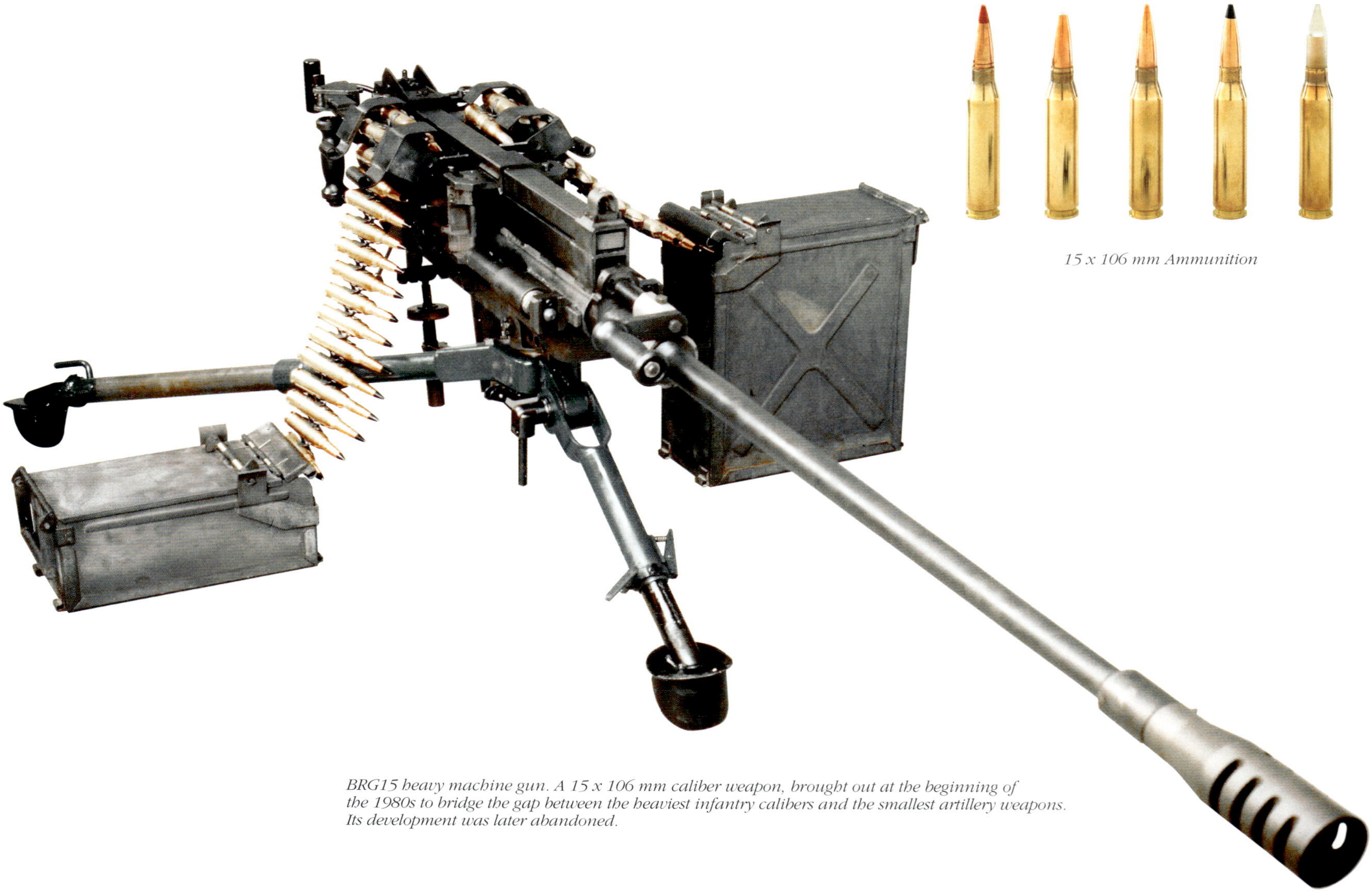

15 x 106 mm Ammunition

BRG15 heavy machine gun. A 15 x 106 mm caliber weapon, brought out at the beginning of the 1980s to bridge the gap between the heaviest infantry calibers and the smallest artillery weapons. Its development was later abandoned.

FNC cal. 5.56 mm. Arabia Model (1985).

7.62 mm ammunition NATO

FN 5.56 SS109 cartridges.

Buildings at the cartridge factory, built in 1982 at the Pré-Madame site.

Range of 40 mm L70 and Drill shell ammunition (1986).

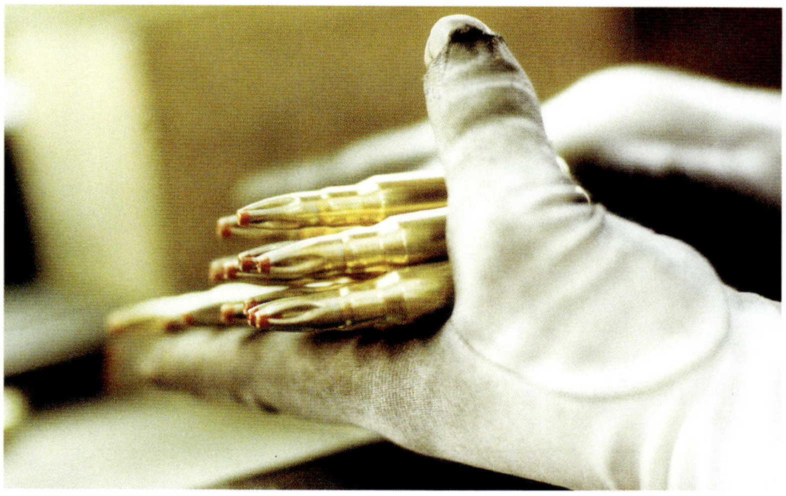

Firing a telescopic grenade from the FNC carbine, 5.56 caliber.
FN, which continued to concentrate jointly on its activities in the domain of arms and ammunition, developed various models of carbines, machine guns and cartridges of 5.56 caliber.
The telescopic grenade combined two original concepts : the telescopic structure (reduced bulkiness) and the 'bullet thru' system (flexibility and safety in use).

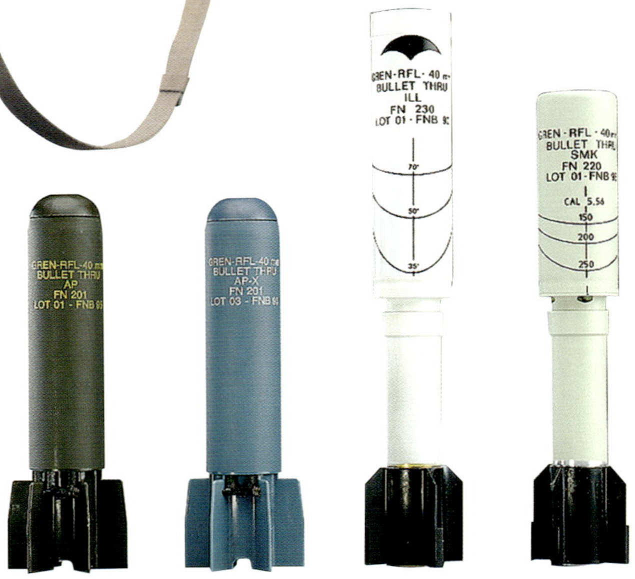

Firing of the 12.7 mm APEI ammunition : Armor Piercing, Explosive, Incendiary.

Mobile workshop for maintenance and repair of arms. The Defense and Security Branch has developed prefab workshops in containers which can be moved to different theaters of military operations. (1985)

During the 1980s, new models of pod were brought out, such as this unit, the Pod MA AFSA HMP-MRL (Heavy Machine Gun Pod - Multiple Rockets Launcher) (1986).

Demonstration firing by a helicopter equipped with FN HMP-MRL (Heavy Machine Gun Pod - Multiple Rockets Launcher) pods (1987).

FN HMP-MRL (Heavy Machine Gun Pod - Multiple Rockets Launcher) pods on the MBB-BO 105 (1985).

New production centers

For Browning, the period was characterized by the abandonment of a certain number of side activities, as well as by the redefining of a range of products which could be either internally contracted or subcontracted. In this respect, the share of Japan in the manufacturing of sporting arms increased, as did that of Viana in assembly and finishing.

In 1973 FN, already facing problems of competitiveness related to the cost of labour, had decided to open a production facility in Viana do Castelo, 40 miles from Porto in northern Portugal, an essentially agricultural region with a large labour force keen to take on industrial jobs. So the Fábrica de Armas e Artigos de Caça S.A.R.L. was formed with share capital of BEF 4,000,000.

At the time, the aim was clearly that the Herstal engineers should retain control of design and industrialization and that qualified workers continue to produce reputedly complex hunting weapons, but that finishing and assembly – both costly in terms of labour – should be carried out externally under direct control rather than on a subcontracted basis. Indeed, the cost of one man-hour in Portugal at that time was three times lower than in Belgium (and still is today). From late 1973, the entity employed 84 people and its training school, run by Herstal technicians, taught engraving, polishing, surface treatment and assembly. In the ensuing period, the company developed with the flow of the many product designs originating at Herstal : the B2000, the BAR, the B80 in collaboration with Beretta, the A500 semi-automatic, the GP 9mm and .40 SW handguns, the Gold semi-automatic and the BDA among others.

FN Viana industrial plants in Portugal.

Members of the personnel of FN Viana.

A number of arms were updated as shown by various types of over-and-under shotguns, derived from the Liège design, the success of the Express rifle, the diversification of HP models, the resumed production of "Baby" 6.35 pistols from 1980 to 1983 at the Manufacture d'armes de Bayonne (France) – the development of a range of lever action rifles in hunting or .22 caliber, reviving nostalgia for the old classics of the Old West. There is also the mass production of the B80 and, for the modern "Rambo"-type rivals to Robin Hood, a permanently extended range of high performance bows.

In the "non gun" business, we find a renewal in the range of fishing rods, golf clubs and sail boards, where the technological progress in the use and application of composite materials is significant.

Browning cartridges for clay pigeon shooting (1983). Originally commercialized under the name of FN and subsequently under that of Browning, the hunting and sporting cartridges were first manufactured at FN, later by specialized subcontractors.

An outstanding Browning B25 over-and-under. Inlaid work with 24-carat gold, inspired by Chinese art, by José Baerten (1985).

The "Legend" range of Browning golf clubs.

Browning Renaissance HP pistol, old silver nickel finish, walnut grip.

Browning I1 HP pistol.
Fine 24 carat gold pin-striping inlaid and plained to the level of the metal.
Blued finish. Walnut grips.

Browning L1 HP pistol, Louis XVI style presented in grayed finish.

Browning Vigilante HP Pistol with fixed sights, blued and polished, walnut grips finely checkered.

Special, varied versions of the Browning 9 mm HP pistol. The HP pistol, also known as the P35, is J. M. Browning's final legacy. It has probably been the most widely used handgun in the world as it was adopted by armies, police forces and other law enforcement agencies in over 70 countries.

Browning telescopic fishing rods. These fishing rods combine solidity with flexibility. These characteristics enable fish of a good size to be caught, even using very fine lines. The carbon tip absorbs the strike and then the struggle of the fish.

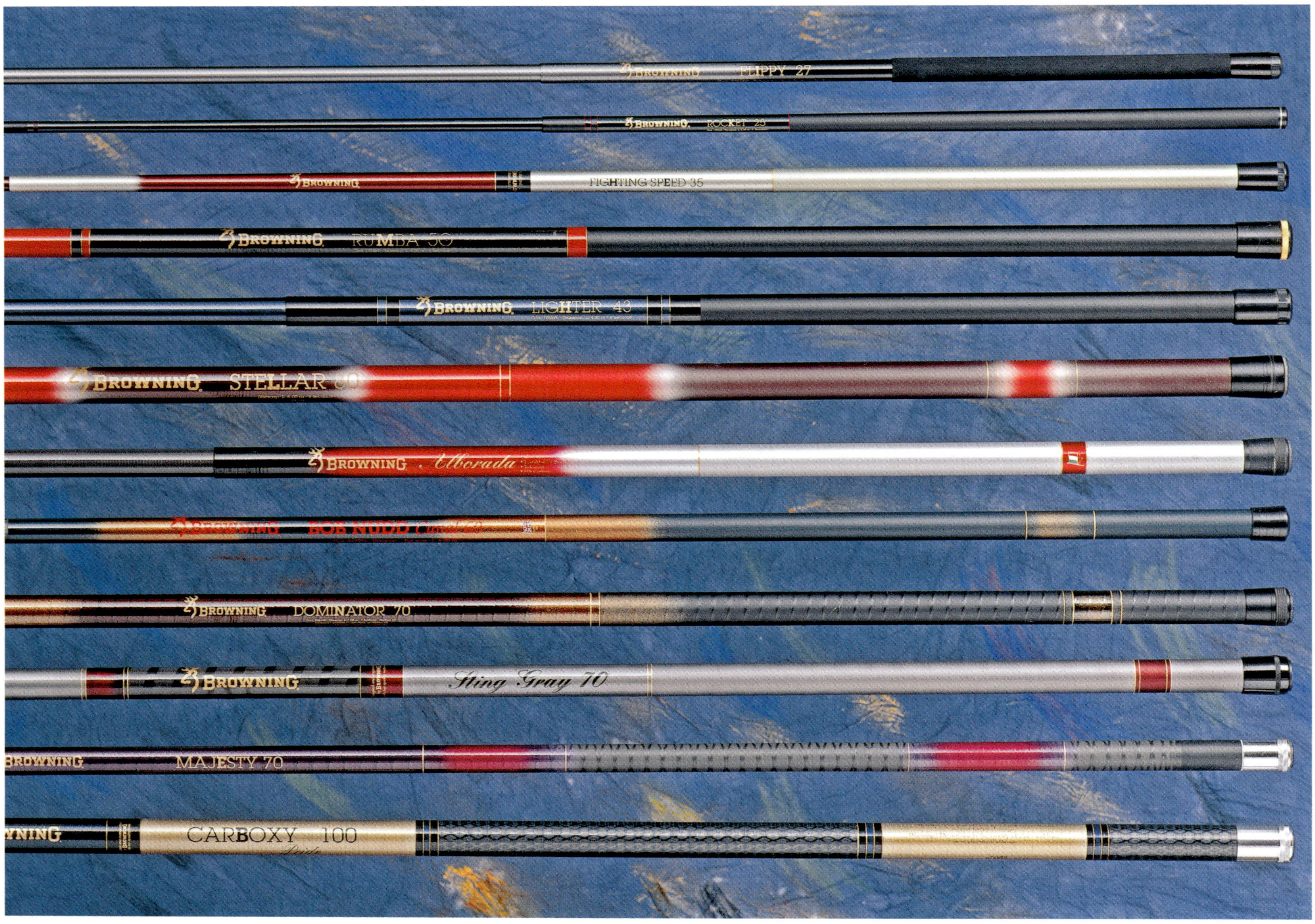

Browning 9500 and 8500 fixed drum reels with a carbon structure. The 9500 is ambidextrous, has an extra-sized line-guide in ceramic graphite and is mounted on a self-lubricating bearing. The oscillating motion of the spool causes the line to be reeled in crossed coils, thus permitting smoother unreeling when casting.

During the 1980s and 1990s, Browning supplied equipment to the best fishermen and earned a number of successes in national and international competitions.

Bob Nudd (UK), World Champion 1990-1991.

Jean-Pierre Fougeat (F), World Champion 1978 and 1988.

M. Van den Eynde (B).

World Pole Fishing Championship at Maribor, Yugoslavia (1990).

The Engine Division, for its part, was going through a difficult period, which coincided with a worldwide recession from 1981 to 1983 in the aeronautics industry.
The continuation of the F100 contract and its supplement, awarded in 1984, fortunately provided some kind of safety net, which, however, was inadequate in view of the incomplete and hazardous nature of the other activities, among which the place given to civil aircraft is counteracted by the stagnation in business.
In the meantime, the supply of "kits" for the Tyne engine was resumed while the depot-maintenance and delivery of spare parts for older engines still carried on.
In 1983, FN secured the production under license of parts and assemblies of the Pratt and Whitney JT9D, version 7R4, intended for the Boeing 747 and 767-200 and Airbus A300 and A310 aircraft.
From the following year, it took part in the design of the TM333 engine of Turbomeca for the Dauphin helicopter and its military version, the Panther. Meanwhile, the collaboration with SNECMA was continued for the manufacturing of the lubrication module of the CFM56-2 and -3, the latter destined for the new versions of the twin-engine Boeing 737 (1982).
In addition, FN was negotiating for participation in the development and production of the Pratt and Whitney 4000 engine for heavy carrier aircraft, for which the contract was to be signed in 1985. Finally, as an extension to the European space program, the Engine Division continued to manufacturing butterfly valves and control units for the different versions of the Viking engine for the Ariane launcher. It was equally associated with the development of the Ariane-5 program.
As it became involved in various projects as subcontractor or co-manufacturer, the Engine Division obviously increased its expertise in the increasingly advanced domain of aeronautical technology.

F100 engine.

Pratt & Whitney JT9D-7R4 jet engine. In 1983, the FN Engine Division was involved in the manufacturing of specific components of this civil aero engine, intended for the Boeing and Airbus aircraft.

F100 jet engine at the Liers test bench. Left page, top : view of the control center for engines being tested.

CFM 56 engine lubricating module

Main Ariane III valves.

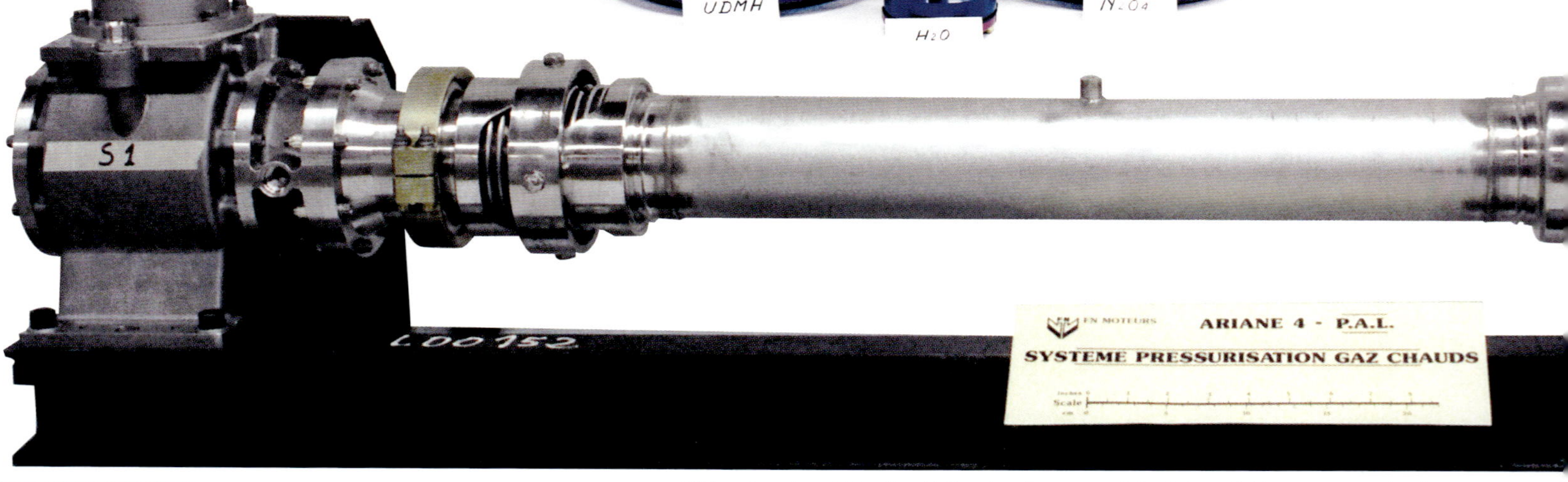

Viking 6 engine. Extra liquid thruster, first stage of the Ariane 4. Hot gas pressurization system.

Ultraviolet inspection of a valve body for the Ariane rocket (1981).

The crisis

In 1986, with decreasing sales, FN showed a loss of more than 3,000 million francs, and in the following year the loss still reached 1,800 million. During these two years, the Defense and Security Branch saw its income decrease by more than 30%. It was a consequence of the worldwide production overcapacity in the area of small arms, as well as cuts in military budgets, particularly by some regular oil-producing clients. Moreover, the devaluation of the American currency formed an additional handicap for countries in the dollar zone, which the division always used to trade with. This disadvantage also affected the competitiveness of the Engine Division. Moreover, the latter did not have sufficient expertise in the new aeronautical programs it became involved in. And in spite of the recovery of business, especially in civil aviation, its profitability remained precarious in the absence of the additional security of a proper industrial policy, both at a national and international level.

In general, FN was encumbered by very high production costs and overheads, its productivity declined and it found it difficult to adapt to a contracting market.

With finances stretched, inadequate capital, shareholders and moneylenders becoming less and less sure of its fate, the company was getting into a difficult position.

The recovery plan

Commenced in 1986, the recovery plan was a global and radical one. Considering its urgency, it was especially the aspect of cost cutting which was felt from the start. The reduction of overhead expenses, the staunching of the hemorrhages linked to the diversification and redundancy of staff (for the whole FN group, from 10,721 end of 1985 to 8,402 end of 1987) allowed for a substantial reduction of the structural costs. They also led to a change in the culture of the company - as well as its image - from an industrial well-being supported by the prosperity displayed not long ago, to a rigorous management of a company in peril.

But the rescue plan proved more ambitious. It contained two aspects, a financial and an industrial one. The first was developed in 1987 and the beginning of 1988, and called on private banks as well as on the government. It appeared under several guises: recapitalization, review of the debt, additional loans, extension of payment deadlines by the Inland Revenue and the Department of Social Security, as well as assistance from the Guarantee Fund. All this provided the company with the means it needed to recover, at least provisionally, with a view to its reorganization.

From the industrial point of view, the plan aimed to regroup the company's activities to the 'traditional strong points' or 'basic trades' of FN.

These were mainly military and civilian arms and ammunition, a striking coincidence with the strategic focuses retained after the recession of the thirties.

From 1986, Browning started to abandon its activities which related to golf clubs, sail boards and tennis rackets, in order to concentrate on hunting and sporting arms and ammunition, as well as fishing equipment. On 18 December 1986, the Browning division became a subsidiary, showing a profit that same year, and turning into a limited liability company distinct from FN, and heading its own international group.

Moreover, following agreements concluded between December 1987 and April 1988, Browning was to acquire a share in the American Company USRAC, which produced and distributed the famous Winchester rifles, as well as the exclusive worldwide distribution of these arms outside the United States and Canada. This union of two famous brands gave Browning a vast market coverage, new possibilities for development and industrialization, as well as a more secure guarantee against the risks of the dollar.
Subsequently, in December 1988, FN became the sole owner of Browning USA, after having bought up all the minority shares in that company.

In any event, the buy-out of Winchester was part of a strategy for the reinforcement of the hunting and sporting sectors as compared to "non gun" accessories, the proliferation of which was beginning to be viewed as a fragmentation of the company's core activities.

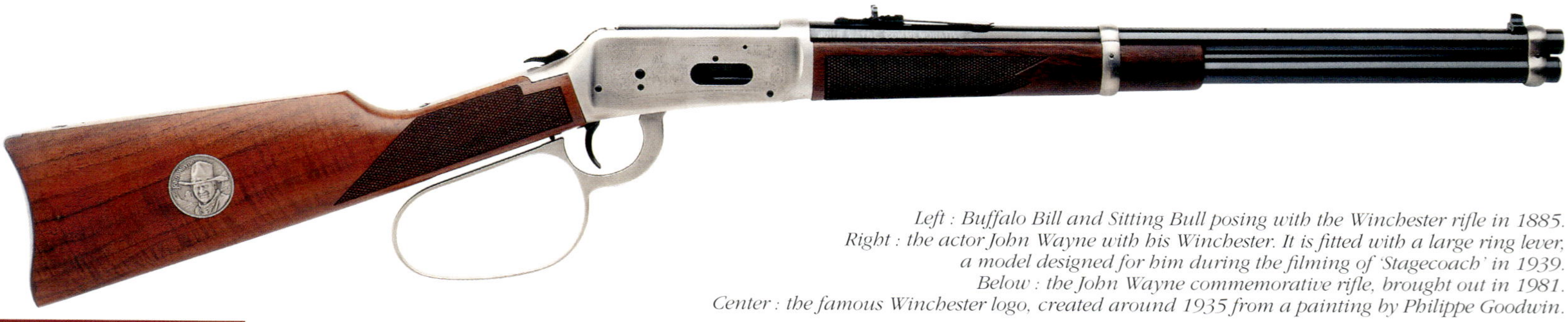

Left : Buffalo Bill and Sitting Bull posing with the Winchester rifle in 1885.
Right : the actor John Wayne with his Winchester. It is fitted with a large ring lever, a model designed for him during the filming of 'Stagecoach' in 1939.
Below : the John Wayne commemorative rifle, brought out in 1981.
Center : the famous Winchester logo, created around 1935 from a painting by Philippe Goodwin.

Rifle commemorating the 125th anniversary of the establishment of the Winchester company (1866-1991). Deep engraving by E. Bontacchio (Italy).

In 1986, the Formétal division was dismantled and its production sent to to FN Engines and the Defense and Security Branch. Later on, FMP Aero was closed down.
FN Industry succeeded in maintaining its field of activity, with various successes, but proved unsuccessful in the robotics branch. Paradoxically, in 1985, it achieved a record sales of 1,100 million francs, but two years later this division, too, did not escape the process of transfer and being broken up into subsidiaries.
Thus the industrial co-operation with the Third World was entrusted to a new company, FNDI, a joint subsidiary of FN and D'Ieteren, one of Belgium's largest car importers.

In the meantime, the Engine division was also being turned into a subsidiary company : on 17 December 1987, FN Moteurs S.A. was born. Two years later 51% of its shares was sold to the French company SNECMA, already a long-standing partner.

Industrial activity also concerned the pursuit of a better joint exploitation of resources and activities in military and civilian arms manufacturing, which constituted a special trump card for the old Herstal company, but which had been exploited in different ways in the course of time, depending on the circumstances.
So the campaign of cost reduction was pursued unremittingly and at the same time served to instill into the company a rigorous perception of the value and precariousness of the resources and of their management.

The development of new products remained crucial for the company, particularly in this difficult period when resource creation was more important than ever. There is more to this type of manufacturing than design and making a new product profitable, however.
For the time being, FN could not anticipate profiting from any innovations. This fundamental activity was nevertheless pursued with varying degrees of success.
A notable innovation, for the Defense and Security Branch, was a new telescopic grenade, designed to be fired with any type of infantry munitions (1987).
But this period also saw the development of a new-generation machine pistol, the P90, the benefits of which were to make themselves felt later. Its design was based on the principle that battle units who were not required to engage combat using light arms (tank and vehicle crews, engineers, artillery and signals corps etc.) do not need a classic weapon, unsuitable in terms of performance and weight.

However, these soldiers, and even some law enforcement personnel, do need an efficient short range weapon, light and compact with a high firing capacity.

The P90 (the "90" taken from 1990, the year of its release) caused an immediate sensation in specialist circles because of the innovative way these imperatives were met – the use of composite materials, ergonomic shape, ambidextrous handling, high firing rate, its lightweight and highly efficient ammunition. Technically, the P90 has proved to be one of the most original inventions in its field of the late 1980s.

It should not be forgotten that in the meantime the Minimi was meeting with wide acclaim, as noted above.

FN Minimi machine gun.

First prototypes for the P90 (1986).

Not only did the P90 fire a revolutionary and original ammunition, but it also marked the advent of two significant technological events in the history of gun making at FN.
First of all, its patent specified the placement of the 50-round magazine along the barrel and axis of the weapon, while all the machine pistols of the time were still encumbered, either laterally or vertically, by their long magazines.
Furthermore, through the power of its spring, the P90 magazine itself ensured the rotation and exit of the ammunition into the gun's chamber.
Also, for the first time in FN's history, the development team chose to put their faith in the possibilities offered by technical polymers.
This choice placed FN Herstal definitively at the forefront of High Tech technologies.
More than 50% of the new arms later developed by FN Herstal would be made of composite materials.

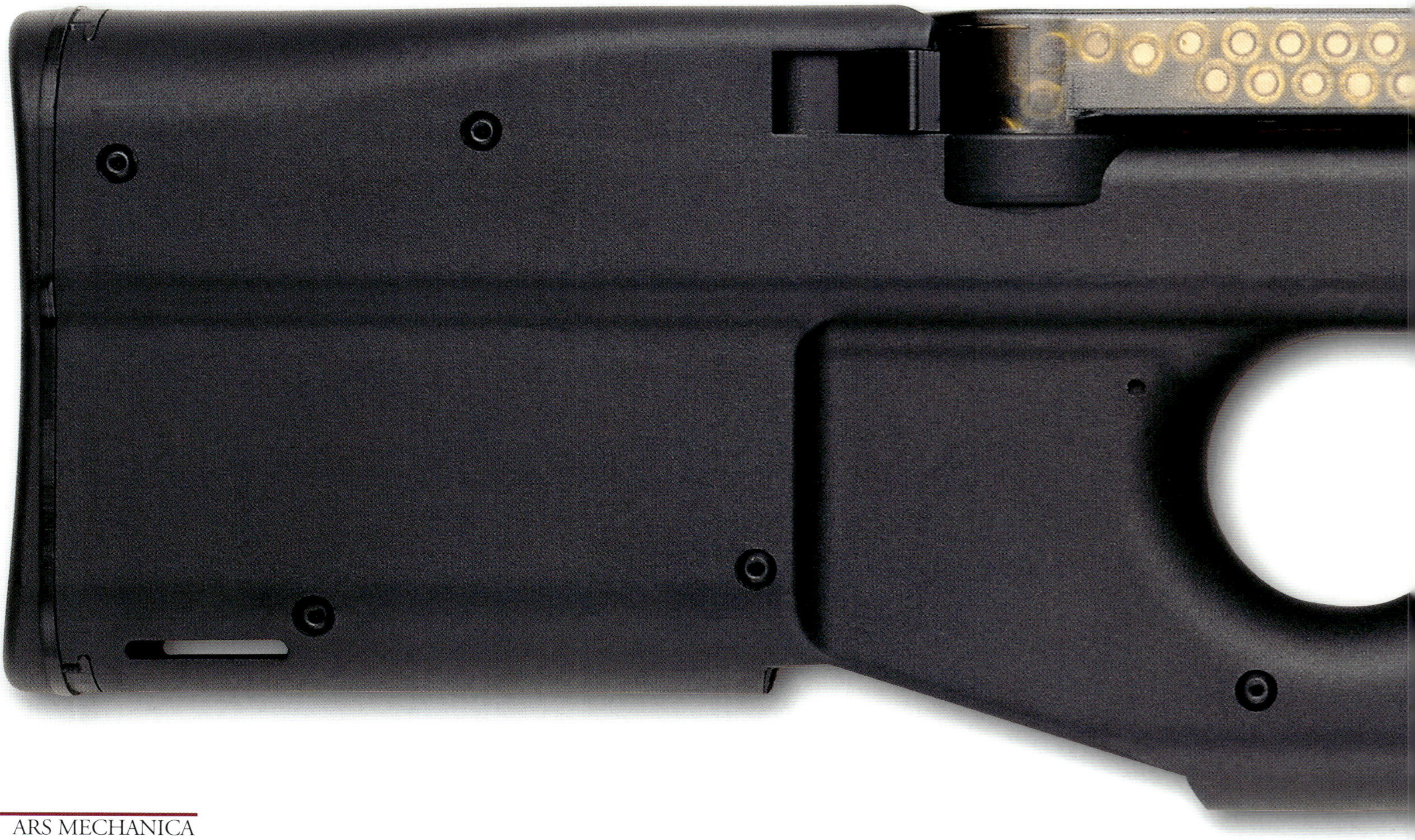

Submachine gun P90.
Developed by FN in the 80', this futuristic weapon of 5.7 x 28 mm caliber, combines the advantages of an ergonomic approach, a great firing capacity and an amazing performance in short-range combat.

At Browning, efforts were directed towards diversifying the classic range with the over-and-under B125, GTI, B127, B135 and the BAR MKII, bringing out new products such as the A500 self-loading shotgun, or developing innovative projects: the Browning 'double mode' pistol (BDM), for example, equipped with a function selector (single or double action).

The advanced design of the BDM (Browning Double Mode) allowed the user to choose between the standard Double Action Pistol mode or the Revolver mode. With the selector at mode P, the BDM functions just like any other double action pistol. With the selector at position R, the BDM functions like a revolver, and each shot is like the first shot fired. The advantage of the Revolver mode is in having a constant trigger pull from the first to the last shot, which enables the shooter to increase his accuracy significantly.

Photograph taken with Pascal Delaroche (World Sporting Champion), for a publicity campaign at the launch of the Browning GTI, a sporting version of the B325 shotgun. In a fiercely competitive market, the company always positioned the brand as being that of a leading manufacturer, anxious to meet the needs of its customers. Here, the GTI label, borrowed from the automotive sector, offered a series of extra, performance-enhancing items of equipment : 'palm swing' stock, larger fore-end, three adjustable trigger positions, ventilated side ribs and decorated in red and black, very fashionable at the time.

The campaign wording went as follows : "Understanding shooting is first of all to understand the motivation, the experience and the demands of shooters. The result of this approach is the Browning GTI, the gun made with the shooter in mind. Discover the new GTIs !

Developed at the same time as the GTI, the Browning B125 Trap F1 has also borrowed its name from the automotive field, but this time from prestigious Formula 1 racing. Upper barrel stiffened by the sighting rib and free-floating lower barrel. Each of the three triggers provided can be adjusted to suit the shooter's preferences. Adjustable forend pressure. Special checkering on the stock and ergonomic grip. Curved anti-recoil pad.

Browning A500 semi-automatic shotgun.
Designed and produced industrially at Herstal since 1987, the A500 has a short barrel recoil operation system, linked to the rotating bolt with four locking lugs. With the A500, the Browning engineers have drastically reduced the number of parts which ordinarily make up a semi-automatic shotgun. Well before building the first prototypes, computer modelling enabled it to be tested down to the very last detail. The prototypes were then subjected to tests to guarantee very tight manufacturing tolerances.

BAR MKII, Battue model. The perfected model of the BAR, called the Mark II, was introduced in 1993.

BAR MKII, Affût model. Custom version.

BAR MKII, Affût model. Custom version.

In the field of aeronautics, we will mention the start in 1986 and the production of parts for the General Electric GE110 engine, intended for the U.S. Air force's F16 aircraft.

In the following year, the extension of the CFM56 program led the factory to undertake the development and production of various parts of the -5A version for the Airbus 320. In that connection, a co-operation agreement was concluded in December 1988 with SNECMA for the -5C version, for the long range four-engined Airbus A340.

In general, the internationalization of the aeronautical industry and the often laborious sharing out of responsibilities and tasks at an international level, are in the case of FN intensified by the internal problems related to the federal structure of Belgium and which form part of the national scene.

Pratt & Whitney 4000 jet engine on the testing-stand. Since 1987, FN Moteurs has taken part in the development and production of this engine, thanks to funding by the public and private sectors.

New challenges

The drop in sales - which fell from some 34 billion Belgian francs in 1985 to 24 billion in 1987 - and the costs of restructuring, unfortunately cancelled out the effects of the recovery plan. Several factors explain this unfavorable situation, in spite of the efforts which had been made to correct it.
First of all, there was the reduction in military orders, then the negative influence of the American exchange rate, both on the consolidation of subsidiaries in the United States and on the competitiveness of FN Moteurs and BDS.
As for the expenses incurred by the reductions in manpower and the capital losses linked to business closures or sales, these amounted to 4.7 billion between 1985 and 1988. If the burden of debt is taken into account too, it had to be recognized that the FN Group was well on its way to a desperate situation. The logical conclusion to which the company managers, the workers' representatives and the well-informed public opinion arrived, was that it was necessary to recapitalize FN. The company therefore turned to its principal shareholder, The Société Générale de Belgique, which had for a long time been the Deus ex machina of business and finance, to whom had been readily assigned, rightly or wrongly - but often rightly - a predominant role in the management of the national economy. So, at the end of 1988, after some considerable effort, the 'Great House in Brussels' became FN's majority shareholder, while at about the same time it was itself taken over by the French group Suez, in the midst of much media comment.
At first receptive to its subsidiary's restructuring plan, the Générale was to perform a U-turn one year later: from then on, light arms would no longer be considered a focus for development.

In plain terms, this amounted to the abandonment of the Herstal company by its most important shareholder after a guardianship of more than 88 years. This decision represented a major turning point in FN's story, already over a century old.

And it is why the post 1989 period constituted a new chapter in this story.

Entrance doorway to the headquarters of Herstal SA (1890). The bas-reliefs depict the heraldic lion of Belgium and a military arm and hunting shotgun, symbolizing both the company's geographic anchorage and its basic products.

Requiem for the Société Générale de Belgique

The disappearance in 1988 of the Société Générale de Belgique as an independent Belgian holding company represented such a financial - even political and cultural - upheaval that it is difficult, with the distance of time, to appreciate its scale. It is not easy to discern the role played by this principal shareholder in the destiny of FN over seven decades - often behind-the-scenes, sometimes openly, but always with a paternally protective approach. Of course, its role was a vital one during the purchase of "enemy assets" in 1919, and again during the great depression of the 1930s. In more recent times, the influence the Générale had on the Herstal company, as on every facet of the Belgian economy, remains albeit often vaguely in the collective memory.

Founded in 1822 at the initiative of the King of the Netherlands, who also reigned over Belgium at that time, the "Générale" had a profound effect on the industrialization and prosperity of the country.
As a promoter of the industrial revolution, of Belgian expansion into China, Russia and the Congo; as grand master of the mining and manufacturing industries, in construction and public works, in transport, energy and financial services; also close - and for good reason - to power at the highest level, the holding company was omnipresent, if not omnipotent, as a national economic player. Consequently from the nineteenth century onwards, companies outside its control criticized its haughty dominance. More recently, economic analysts questioned its predilection for the traditional sectors at the expense of more innovative branches. For their part, trade union organizations considered it the dark bastion of a shady and manipulative capitalism. Be that as it may, in the second half of the last century the holding company's management still appeared highly autocratic, with a governor at its head holding all the decorum and privileges of old-style authority. However, behind this regal facade opposing forces confronted one another within the Board of Directors.
This can be clearly detected in the fate reserved for FN in the last years.

For a long time, the Générale considered the Herstal company to be moderately profitable, in a strategically important sector in which it felt its presence was necessary. Managed competently - and with a certain paternalism - by the Laloux family and its allies, representing besides a not inconsiderable share holding until the early 1970s, FN had been a jewel in the crown of the "old lady" of Brussels since 1919. However, the holding company was criticized for keeping the Herstal gun maker in a chronic state of under capitalization, little suited to the entrepreneurial dynamism and modernization called for by society in general, and the company itself.

Around this time, however, attitudes changed. Under the public eye, the connection with arms manufacturing was an embarrassment for the principal shareholder, having become even more dominant as the historic managers of Herstal had in the meantime realized the majority of their holdings.

Without ousting these managers, behind whom Brussels could hide in the eyes of critics of the "military-industrial complex" in Belgium, through the intervention of Michel Vandestrick, it did try to change FN's image by orienting it towards diversification; diversification which, with some financial efforts and after a period of gestation, was also naturally expected to generate new profitability.
Anxious that Herstal should retain the support of the holding company, on whose Management Committee they also sat, two successive FN Chairmen, André Dubuisson and Marcel Goblet, attempted for their part to reinforce the Générale's current viewpoint, at the risk of embellishing somewhat the potential of a diversified approach.
In this way, they succeeded for a time in appeasing the prejudices of certain of their Rue Royale colleagues.
This "makeover" took several years, as is necessary for setting up any new industrial enterprise. Although it succeeded in changing the company's image, and an improvement on its early Taylorism, the economic hopes for it were disappointed. Paradoxically, it was FN's traditional sector - military weapons, the least reputable in the view of some - which wiped out the losses incurred by diversification and saved the group from bankruptcy.

Conscious of this constant factor, some of the principal shareholder's Management Committee opted in the end to refocus activities on the armaments sector, with the prospect of having to approve new investments as part of an inevitable restructuring. Albert Diehl's mission fell into this context. The watchword "diversification" would now be replaced by "a return to core activities", and this was to prevail in the long term. But the new path was not an easy one. Because the unfavorable economic climate, the worldwide depression in the armaments industry and the difficulties inherent in unilaterally abandoning industrial or trade activities declared to be marginal, as well as social pressure, meant that it took some considerable time to achieve the success we know today.
In the meantime, other Société Générale managers were won over to a more radical attitude to Herstal. They declared their wish, purely and simply, to rid themselves of a financial and politically incorrect burden, particularly in the region of Liège where economic and social difficulties were little to their liking. As we have seen, this trend ultimately carried the day, putting an end to a marriage of seventy years' standing; a marriage in which - in a very different climate - all the affection nurtured over long years of cohabitation had finally disappeared.

Back to the core business

Super de luxe Browning shotgun, commemorating the centenary of FN. Special series of ten pieces. The engraving, chasing and inlaid work, by Sophie Purgal, of the cooperative society Les Graveurs de Herstal, evoke the foundation of FN and its vocation in the arms business. The tradition and activity of the engravers' workshop at FN, which was closed down in 1987, lived on for a while, in that new independent workshop, situated not far from the headquarters.

The new order

Jacques Germay, General Manager (1989-1990).

In 1989, FN celebrated its 100 year anniversary. This was an occasion for the company to commemorate the creative phases of its long history, through a prestigious event which included every branch which had contributed to the company's name.
A century which has indeed forged a legendary company, a company which has always been at the forefront of technology, known and respected the world over.

After a decade of diversification into activities which have not always produced the expected results, whether – like the aeronautics division – they exceeded the company's investment capabilities or whether they were clearly outside the company's area of expertise, the company has pursued a radical policy of refocusing on its core business activity : arms.
Designing, manufacturing and selling throughout the world innovative and high quality small arms, ammunition and accessories for both civilian and military markets : this was once again its essential vocation.

At the time, no-one could foresee the scale and speed of the global turmoil that year was to see: the fall of the soviet regime and the break-up of the USSR heralded the end of the Cold War, the Iron Curtain and the Berlin Wall.
It could be said that a 'short twentieth century' came to a close between the summer of 1989 and the autumn of 1990. Opened in Liège in 1914 to the sound of the guns of the Great War, it closed with the peaceful fireworks celebrating German reunification and the Peace Treaty which ended the division of Europe.

In the euphoria of the moment, some were already forecasting the spread of liberty, democracy and market economics throughout the world, which would bring shared peace and prosperity.
All the more so as advances in the applied sciences, information technology, space, telecommunications and biotechnology appeared to encourage great hopes of generalized development.

In 1989 FN celebrated its centenary year. To the great joy of everybody, many "venerable ones" attended the celebrations organised to mark this momentous occasion.

Fabrique Nationale

FN

9950.U

A gathering of FN's vintage cars in front of the Management Office building during the celebration of the company's Centenary (1989).

A development which would emerge from three poles :
– First, the American giant, the only world superpower in all fields.
– Next, Europe, which via the Communities would move towards a political Union beyond the formation of a large market and pursue a common project.
– Finally East Asia, where, along with Japan and South Korea (both new economic beacons in South-East Asia) China is slowly waking up to capitalism, while retaining its own specific political regime.

Consequently, some people were raising questions linked to "the end of history" – to what extent is an armed force legitimate or even necessary ? – but it was not long before such humanitarian optimism had to be seriously qualified.
Beyond European co-operation, the old demons had not gone away : adventurism, war mongering, ethnic superiority pushed to genocide and spectacular terrorism, sometimes cloaked in religious integrationism and hate of "the other"; the unrestrained race for control of the world's resources, by any means. Progressively such issues moved to front stage.
International violence then, widely reflected in media images, but violence close to home too, in our inner cities.
A reduction in educational curbs, easy escalation to acts of aggression, a decline in values of solidarity, illicit trades run by international organized crime, the difficulties in defining and controlling policies on immigration and integration, in practicing the theory of a multiracial society, the slowness of the police and legal institutions to adapt, especially as regards inter-European and international co-operation... all elements which perpetuated a diffuse feeling of insecurity.

Faced with this state of affairs, it had to be recognized that now more than ever an effective police force was vital, alongside preventive diplomatic and social measures. This might be at international level with military units placed, more often than not, under the sponsorship of the UN or NATO, or at national level with forces of law and order having adequate training, coordination and equipment.

Consequently, in the global defense industry, of which FN was a part, a dual phenomenon could be noted. First, throughout the 90s the industry would be subject to a massive market crisis, the result of huge restructurings.
On the other hand, those like FN who had a tradition of continuous technical and marketing adaptation, with research and investment resources and who were sensitive to market developments, could ride the tides of chaos in this new world and seize new opportunities for redeployment which would guarantee their future. Often with the help of new shareholders... and much trial and error and many difficulties.

Would life be any simpler for FN's other flagship division, that of hunting and sporting arms ?

The hunting section would suffer a certain decline, not compensated for by interest in sporting shooting among new population groups. Even the American market would suffer fluctuations, for very specific reasons: the anticipation or the effects of new laws, restrictive judgements, anxiety about the year 2000, among others.

Beyond the legal "turns of the screw", the crises and changes in the leisure civilization, there were still sales prospects and market share to be had in this market. Provided of course that the combination of tradition and innovation, quality and profitability, attractive prices and adequate margins could be maintained, particularly by using suitable distribution and after-sales service tools.

Also, it would have made little sense for FN to neglect a sector which complemented the military arms sector so well, for reasons of technical synergy (R&D and industrial processes), activity cycles, and also economies of scale linked to the requirements of global markets.

Furthermore, how could the mythical name of John Moses Browning have been put aside, or the intelligent interrelationship between the hunting and sporting activity on the one hand and the defense and security forces on the other ? As early as the 1890s, FN had invented this 'dual' approach concept.

The historic headquarters of the Herstal Group's Senior Management. Recently renovated, the building dates back to 1906.

Members of the personnel of FNMI.

Aerial view of FNMI.

New lines of action

From 1989 onwards, FN's activity centered on a few key themes :

- First, adaptation to and even anticipation of current changes in the two complementary core businesses: on the one hand, military and security force armaments, and hunting and sporting arms on the other.
- Then, restructuring the company's industrial and commercial tools in line with new potentials, requiring a more diverse range of techniques and products, smaller production runs, and a greater degree of flexibility.
- Finally, strengthening the links between the Group's entities, distributed as they were throughout the world, and the creation at global level of organized co-operation networks, particularly for marketing, research, development and industrialization of products. Thus, without renouncing the virtues of a firm anchorage at Herstal, the concrete organization of the Herstal Group at global level was pursued.

These developments came within the framework of huge downsizing operation, and therefore significant restructuring, linked to the contraction of the market.

A play in three acts

The period 1989 to 2005 was marked by three major stages in FN's development :

- the withdrawal of the Société Générale de Belgique in favor of the French group Giat Industries, which culminated in the creation of the Herstal Group, designated as Giat Industries' global small arms division.
- the episodes in the life of this Group within the French giant.
- after the withdrawal of Giat, entirely focused on other priorities, the handover to the Government of the Walloon Region.

Controlling cartridges

Assembly of the HP pistol (1999).

A "FaN" member of the personnel of FNMI.

Refocusing

The moment of truth

In the spring of 1988, the Société Générale de Belgique (SGB) came under the control of the French group Suez after an epic stock market battle. This coincided with a radical refocusing on the part of FN on its 'core business', a refocusing which had significant social and financial consequences.

To finance the restructuring in progress at FN, in 1988 and 1989 SGB turned to the capital markets and proceeded with a sizeable recapitalization of its subsidiary. At the culmination of these transactions, the 'Old Lady of the Rue Royale' held three quarters of FN's shares for the first time.

At the same time, sale of control of FN Moteurs, affiliated for the purpose, to the French group SNECMA was going ahead. This sale took effect in late 1989.

Refocused on aeronautics and space, FN Moteurs would soon take on its own existence under the name Techspace Aéro, and its stunning successes would make it a beacon of Walloon industry in a high technology sector.

Thus, through its efforts and considerable financial investment, FN has contributed to providing Belgium with a world class aerospace tool. State financial assistance notwithstanding, it is probably the only company in the world to have assumed such a risk to its cash flow. Moreover, the consequent shortfall in investment in its core business would result in a lack of armament products for the future.

The sale or liquidation of the component parts of the FN Industry section, as well as the precision forge and the Formétal foundry, was also under way. The latter, taken over by several of its managers, would continue its activities in collaboration with a French partner. Sadly, the same could not be said for the precision forge, which had to throw in the towel in 1992.

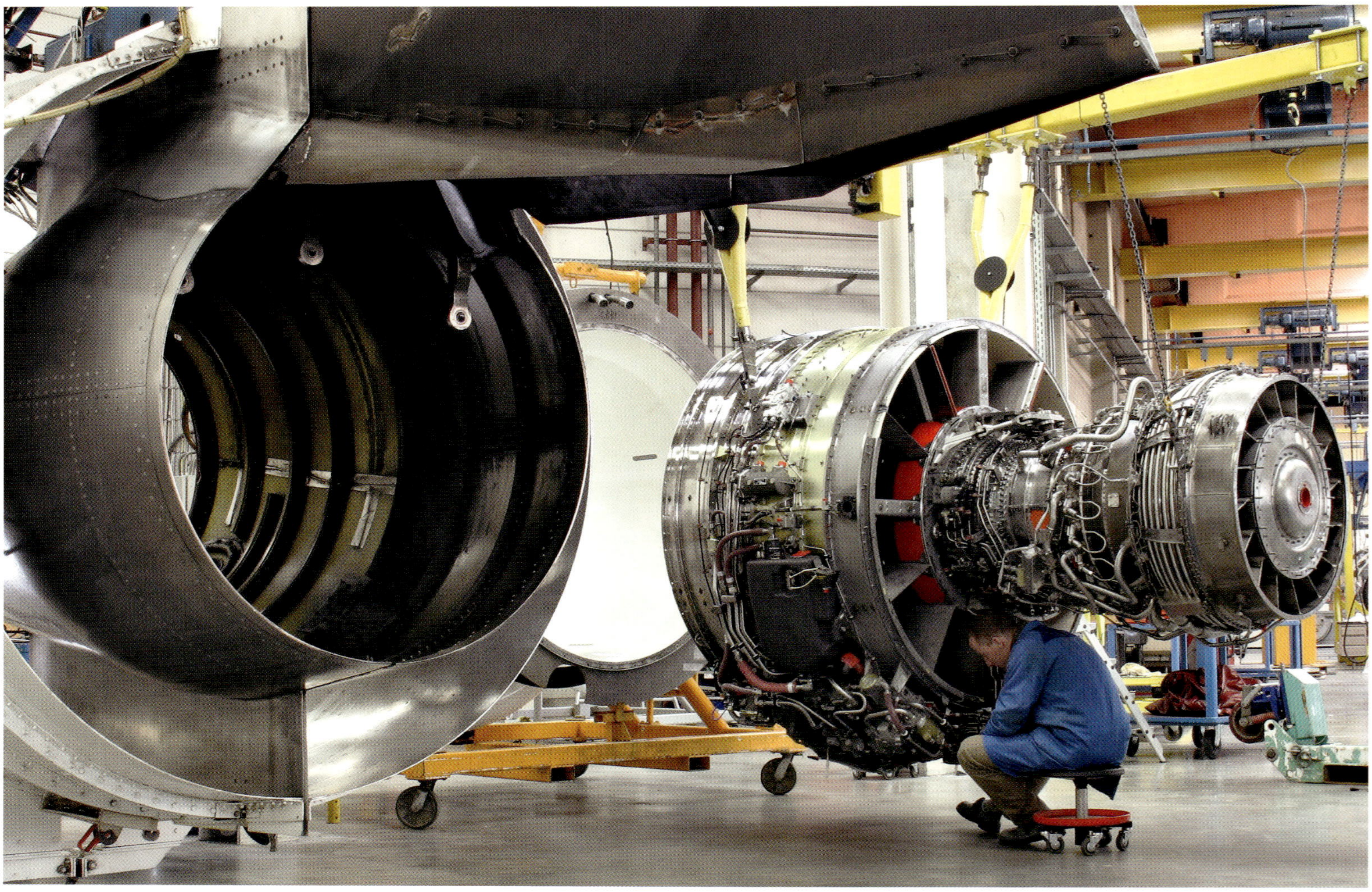

FN Moteurs, sold in late 1989, was to pursue its own story under the name Techspace Aéro, a story of stunning successes as a beacon of Walloon industry in the high tech aeronautics and space sector. FN therefore contributed, thanks to a considerable financial injection, to providing Belgium with a world-class aerospace tool. It is probably the only company in the world to have assumed such a risk to its cash flow.

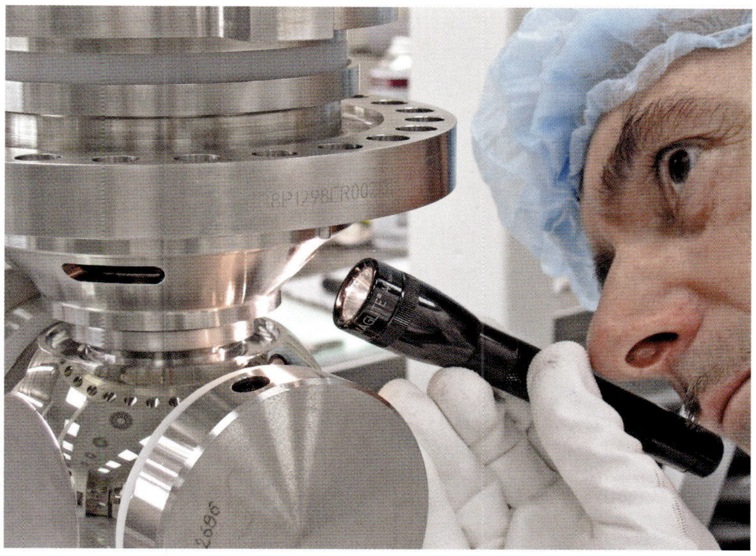

Now a member of the international industrial group SAFRAN, Techspace Aéro continues to develop around four centers of excellence: civilian and military aeronautic propulsion (low pressure compressors), special aeronautic and space equipment, maintenance and repair services, and finally engineering test stands for aircraft and helicopter engines.

But in late 1989, SGB decided to withdraw from FN as part of a radical revision of its portfolio of activities, largely in line with the conclusions of its new French shareholder, and gave itself a year to relinquish the company.

This decision may in part have been influenced by the situation in the military armaments sector, which after the end of the Cold War was anticipating a long and painful redeployment.

For FN in particular, it seemed that redundancies, especially at Herstal, would have to be more far-reaching than anticipated, and that adapting to the new order would cost far more than had been budgeted for. Besides, would it be possible to complete the process of adaptation and achieve lasting results without the backing of a large investor with an interest in the field of land armaments ?

During the six months or so spent searching for this new investor, followed by negotiations with Giat Industries, SGB would support the company's cash flow at arms length. During this period an additional redundancy plan was drawn up with staff representatives.

It included early retirements or severance pay and was staggered from late 1990 to early 1993. This plan reduced the workforce by almost a half, or some 1,500 retirements and redundancies, on top of those agreed in the two preceding plans. It was during this same period, the autumn of 1990, that Giat Industries gave an indication of its interest.

This new company, owned by the French Ministry of Defense, having taken over a number of French land armaments companies in addition to the former national arsenals, was keen to acquire a European and global dimension.

The acquisition of FN's business certainly gave it the opening it was looking for, in principle in a context of close co-operation, since the two companies seemed to complement one another rather than compete - Giat Industries was more involved in the medium and heavy arms market, and vehicles. It also had the advantage of considerable financial provision and credit, which would fund European acquisitions, thus enabling it to become a company of European and world dimensions.

Consequently, Giat Industries declared that if it were to take over a resized FN, the latter would become its global division for small arms defense activities, and also for hunting and sporting arms, a new departure for the French company. This division would, it seemed, enjoy a high degree of autonomy within Giat Industries, and even become a basis for other partnerships, particularly in Western Europe.

So in the autumn of 1990, all interested parties lent their support to Giat Industries' intelligent European strategic project. The project reached its conclusion with the creation by Giat Industries of new Belgian structures in late 1990, 8% controlled by the Walloon regional government. This was largely thanks to the strenuous efforts of Albert Diehl, who was appointed Chairman of the new Herstal Group, a post he held until the autumn of 1995.

From left to right: MM Schneider, Emmanuel Carré (Giat Industries), the Countess de Brion (Negociator), MM Pierre Chiquet (Chairman-Giat Industries), Albert Diehl, Jo Labye (respectively Chairman and General Manager - Herstal Group) and Jacques Gentgen (Giat Industries).

Highly mobile, easily dispatchable and with formidable fire power, the Leclerc tank is one of the stars of world armored weaponry.

AMX10 in operation with the Daguet Division in the middle of the Gulf War (1991).

Center photograph : Giat 105 LG1 artillery gun. Bottom : Giat 155 TR artillery gun.

Luchaire WASP light anti-armor system.

After the sale of the bulk of its business to Giat Industries and under the leadership of Albert Diehl, the former share quoted company, FN, would make plans to start up new activities, even activities outside arms manufacturing, and look for partners to achieve this, under the name Her-Fic (Société Herstalienne pour la Finance, l'Industrie et le Commerce).

Unfortunately, the contacts initiated for this soon came up against limitations, particularly legal ones. This block proved to be a persistent one in spite of several changes and partial restructurings.

At the beginning of 1991, to try to do something about this experience, immediately declared to be temporary to the Her-Fic shareholders, the SGB and the SRIW (Société Régionale d'Investissement de Wallonie) had sold their majority holding in FN for the symbolic one franc to a new not-for-profit company, the Armurerie Liégeoise. This was made up of Liège figures from economic, political and union backgrounds, and for 9 years supported the Her-Fic board with advice as well as action. In 1999, its members were forced to recognize that the revival of the company's activities was no longer feasible and so the company was put into liquidation.

In the meantime, the Her-Fic's continuing activity had nevertheless solved a series of transitional problems : the sale of the remainder of its Techspace Aéro shares, the sale of a cartridge engineering plant in Mulhouse - France (which Giat Industries had not wanted to take over); the repurchase at a fraction of their face value of bond debts arising from two FN loans; the dismissal of personnel not transferred to the Herstal Group - and a sizeable contribution to their retirement, early retirement or severance payments, with a large grant of 100,000,000 euros from two funds set up and funded by SGB towards the non-statutory provision.

This would allow the Herstal Group to remain focused from the outset and to concentrate on its own future, without having to pay for a financial legacy from the past.

Expanding the Group would require first of all a recapitalization and adequate financing. These were forthcoming, and enabled the company to take advantage of the first significant geostrategic development since the end of the Cold War. This was the Gulf War, a major confrontation between a broad western military alliance and Iraq, which had invaded Kuwait at the beginning of August 1990. On this occasion, Herstal was able to adapt, with the necessary political support, to the subtleties of Belgian legislation on arms export licences. This legislation would later be changed, along ever more stringent lines.

The financial consolidation of the company also enabled it to bolster its research resources, a traditional pillar which had always been jealously maintained. And to make advances with a range of innovative arms of a new caliber, the 5.7 x 22mm, with exceptional features (lightweight, easy to handle, low recoil, high fire power at close range) likely to be of interest to armed logistic or special forces, as well as to other law enforcement agencies.

In spite of initial reservations about accepting this new caliber– sometimes on industrial grounds – the quality of these new weapons, Five-seveN pistols and P90 submachine guns, would slowly and progressively ensure their presence in all markets – military, law enforcement, even hunting and sporting, with adapted ammunition.

Finally, the restructuring of the company together with the revival of military business brought about by the Gulf War, shortly followed by the FN machine gun models becoming standard issue in a number of Asian countries, would provide, in addition to the contribution of Giat Industries, the extra financial resources needed for the radical modernization of the industrial installations at the Herstal headquarters.

A notable example was the new high precision machining hall for complex parts, which was equipped with the latest digital control centers; its high degree of flexibility was perfectly adapted to the requirements of military orders.

In the same vein, the Herstal site saw the integration of the activities of Delcour, a supplier of military and hunting barrels, and which was soon completely merged with FN's barrel manufacturing section.

Not forgetting the continuous re-equipping of the research and industrialization departments, particularly in terms of computer assisted design, or high tech information technology, gradually introduced into all the company's departments.

In late 1991, all was sufficiently advanced to put in place the organizational and legal structures of the Herstal Group.
At the top of the organization chart, a holding company called Herstal S.A.. Immediately below, two subsidiaries 100% owned by the holding company - the Fabrique Nationale Nouvelle Herstal SA (which would adopt the simpler name FN Herstal S.A. in 1995) for defense activities and later law enforcement agencies, and Browning S.A., for the hunting, sporting and leisure activities. Between these two, the Canonnerie Delcour, the barrel manufacturer for both sectors. Finally, a series of industrial or sales subsidiaries in Western Europe and the United States, and privileged links with the Japanese Miroku company, manufacturer of hunting and sporting arms for Browning. Thus was a truly global group constituted, based on the banks of the Meuse, but with a strong presence in the United States, the market of reference for the Group's two fields of activity.

*Souvenir office watch, made on the occasion of the official inauguration of the modernization investment of the Herstal plant.
As the Herstal Group in 1993, the bird shown here "leaves the shadow to come into the light".*

GIAT industries

RÉGION WALLONNE

The American section

Since 1981, FN Manufacturing at Columbia, South Carolina, was devoted to produce arms for the American government, first of all with M240 and M249 machine guns, and then with M16 rifles from 1989, a contract which had been won after intense competition with the Colt group, the owner of the designs, in response to an invitation to tender issued by the American government.

In 1991 the relative importance of the United States for the Herstal Group was further increased with the takeover of the operations of a second American brand, Winchester, owned by the American conglomerate Olin.
This involved first of all the manufacturing and distribution under license – largely exclusive – of shoulder arms, controlling a new company, the US Repeating Arms Company (USRAC) located in New-Haven, Connecticut.

This idea was born of the conviction that the Browning and Winchester ranges were not in competition, but complemented one another. An idea which had been put into practice by the time Giat Industries arrived on the scene. In addition, some members of Giat Industries wanted to stake even more on Winchester, by taking over its Italian cartridge manufacturing plant at Anagni, near Rome, and by distributing, in Europe, the metal ammunition manufactured by Olin in the United States. However, in contrast to Browning, the Winchester activities, lacking in the development of new arms and in arms manufacturing expertise, were to prove to be unprofitable in the long term.

From 1991 onwards, apart from some economic fluctuations, all these elements combined to give the Herstal Group a firm foothold on either side of the Atlantic ocean. The best thing being that it had a large circle of enthusiasts demanding civilian arms but also had a loyal client in the American armed forces, an unequalled reference. On the other hand, it was also faced with a series of difficulties to manage parameters, including fluctuations in the value of the dollar, American regulations, global control of the group, a financial resource threshold. But finally the absolute strategic necessity for the Group to maintain its options for action throughout the world, centered on a European, Belgian and Walloon base.

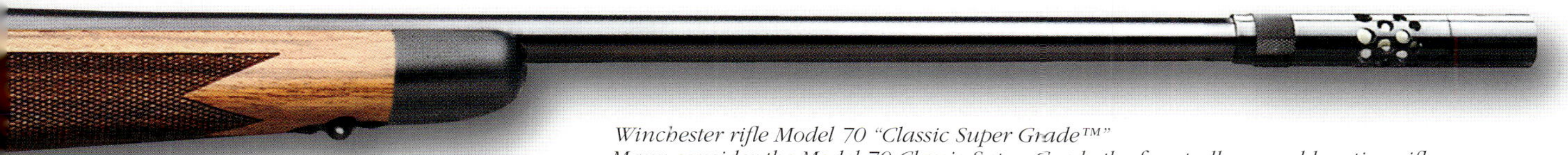

Winchester rifle Model 70 "Classic Super Grade™"
Many consider the Model 70 Classic Super Grade the finest all-around hunting rifle ever made. It features a selected walnut stock, black fore-end tip, inletted swivel bases, shadowline cheekpiece, and bases and rings. The stock itself features an ample cheekpiece. (BOSS™ optional).

Winchester Model 94 "Legacy™"
This year choose your Legacy with a 20" or 24" barrel. Both have all the famous Model 94 features you want plus a degree of finish and detailing that you'll appreciate for years. The stock is American walnut distinctively sculpted in a half pistol grip stock design. The forearm and grip are beautifully cut checkered.

View of the old Winchester factory in New Haven, as it appeared around 1889.

Members of the personnel of the U.S. Repeating Arms Company.

Albert Diehl, Chairman of the Herstal Group.

First results and outlook for the Herstal Group

The end of 1991 saw a significant recovery in the Group's consolidated sales (some € 450,000,000 in 1991) and positive operating profits in both the military and civilian branches, with a total cash flow of € 20,000,000.

This improvement was the result of a significant reduction in staff and financing costs, among others. It was also the fruit of continued customer loyalty in the company and its products, exceptional opportunities in the military markets, in the Gulf and East Asia regions in particular, and finally the ability of the civilian sector to withstand the effects of the economic recession brought on by the Gulf War.

For the immediate future, all pointed to the continuation of this slight upturn in defense activity. The hunting and sporting business, however, still had serious difficulties to overcome, principally to make the new Winchester arms and ammunition niche profitable and to return Browning (outside the US) to profitability. It continued to suffer from the lack of genuinely new models, apart for the Browning BAR MK2. This was also true for the new Winchester over-and-under model, expected on the market for 20 years.

Furthermore, it was also expecting increasingly severe legal restrictions on the purchase, possession and use of firearms by private individuals in a growing number of western countries.

However, to maintain its position in the civilian market, the Herstal Group could count on its solid market position (gained through Browning USA) and on the brand's traditional products. There was also the prospect of substantial assistance from Giat Industries, which had set up a Browning unit in Paris in order to follow progress more closely. A little too closely perhaps, in view of the Herstal Group's autonomous operating rules.

And so in 1991, after the catastrophic years 1989 and 1990, the company, in all its aspects, recovered its vigor, courage and confidence in the future.

Up to this point the FN personnel who had survived the restructuring, when faced with the various relocations and the centrist tendencies displayed by certain of Giat Industries' representatives, had harbored the fear of not seeing the specific nature and benefits of its labors recognized on the ground. It was that simple.

In this context, in November 1991 the Belgian unions and Giat Industries' senior management negotiated and signed an agreement affirming the autonomy of what was later to become the Herstal Group. As a further guarantee, the agreement provided for the presence of a director proposed by the Walloon regional government, the right of veto over strategic decisions (a surprising concession and a potential source of conflict in view of the equity financing level) and the confirmation of Albert Diehl in the post of Chairman of the Herstal Group. Diehl was tasked with entering into direct dialogue with Giat Industries' Chairman and Managing Director Pierre Chiquet, the driving force behind the strategy.

Big game hunting scene with the BAR MKII Browning rifle, fitted with a telescope sight.

The ups and down in the Herstal-Versailles marriage

Generally prosperous years

The period from 1992 to 1995 confirmed the prime position held by the company in the arms manufacturing world, and the exceptional opportunity offered by several big military contracts, in a sector which was still depressed and in turmoil.

The performance of the military section

Herstal was winning large orders, particularly for machine guns, while at the same time orders placed by the American army with FNMI, for machine guns and M16 rifles, were growing. After a decade of investment, FNMI was finally breaking even. These elements, together with the confirmation of certain traditional business volumes, guaranteed the Group's military section several prosperous years to come.

In 1992 and 1993 the financial results were largely positive, with military sales in the region of € 225,000,000, or almost half the Group's total sales. For the first time in years, FN distributed a dividend to its shareholders.

On 9 December 1993, a large modernization project for military and civilian production at the Herstal plant was inaugurated with much ceremony by the President of the Walloon regional government, Guy Spitaels. A few months later on a sunny day in June, more than 8,000 employees and their families visited the new installations at the Herstal headquarters.

The results were similar in 1994. The economic indicators for the military sector slowed down during 1995, but were still at an acceptable level – at least for the operating profit as the net result was still only slightly positive.

Top, Pierre Chiquet, Chairman and Managing Director of Giat Industries which took over the Herstal Group. With Guy Spitaels, Minister-President of the Government of the Walloon Region, during his visit to Herstal on 9 December 1993.

Georges Delcour is explaining the techniques of barrel making to Guy Spitaels.

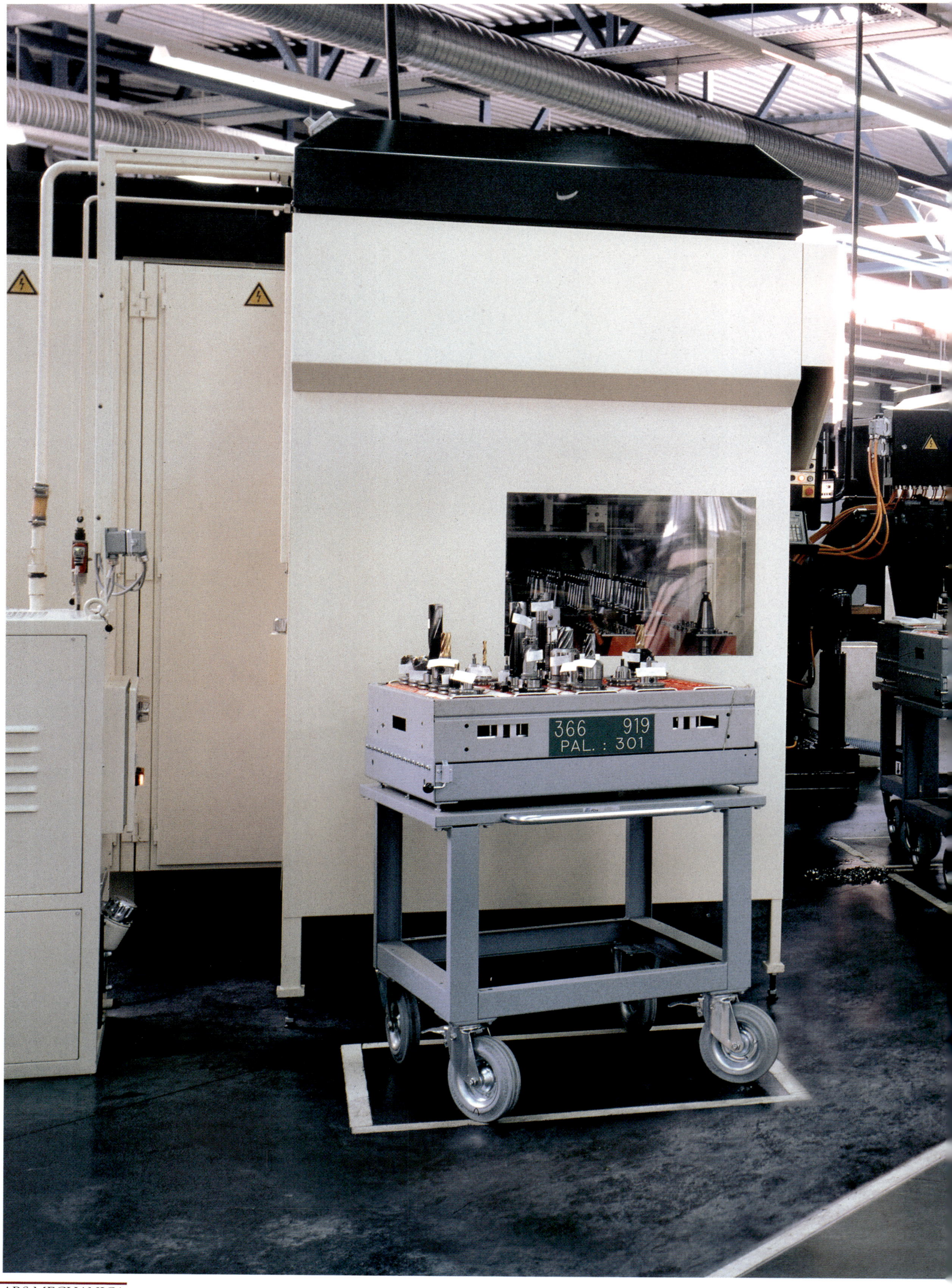
366 919
PAL. : 301

New machining hall equipped with the very latest in digitally controlled machines.

One of the many tool pallets equipping the 18 digitally controlled machining centers.

General view of the "controling office" in the machining workshop in Herstal. Totally computerized, it could manage every machine of the workshop at a distance. A totally new concept at the time.

Making of prototype parts for the research and development department at FN Herstal

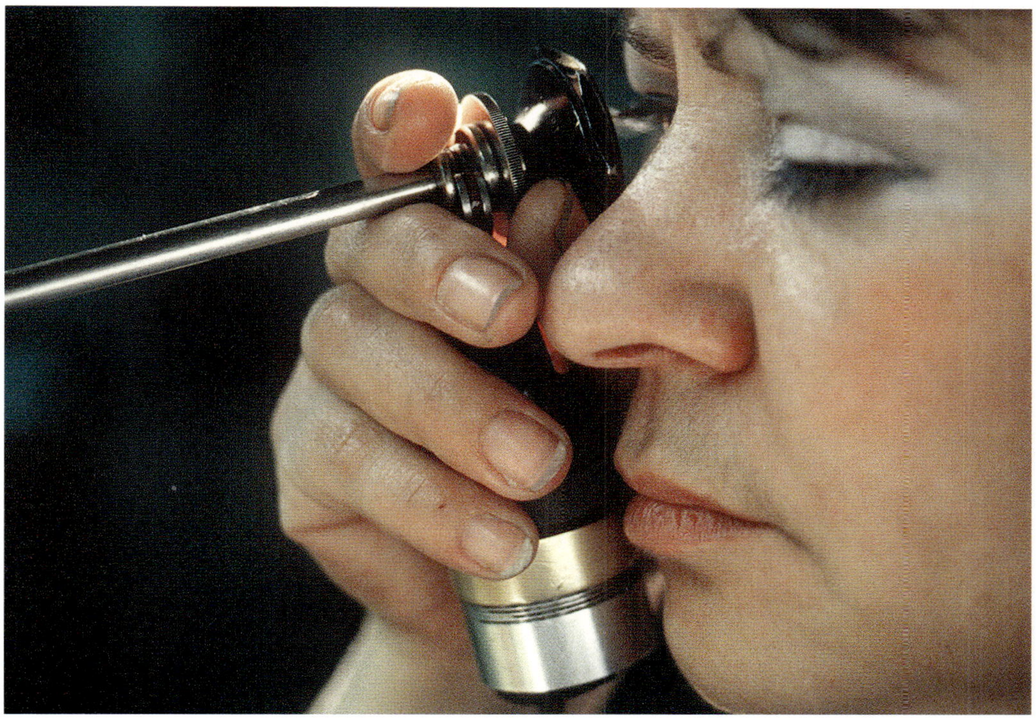

In late 1991, the Canonnerie Delcourt, a manufacturer of civilian and military barrels, was taken over by the Herstal Group and relocated, together with its entire workforce, to a new factory built at the Herstal site.

Poster designed to promote the Open Day on 11 June 1994.

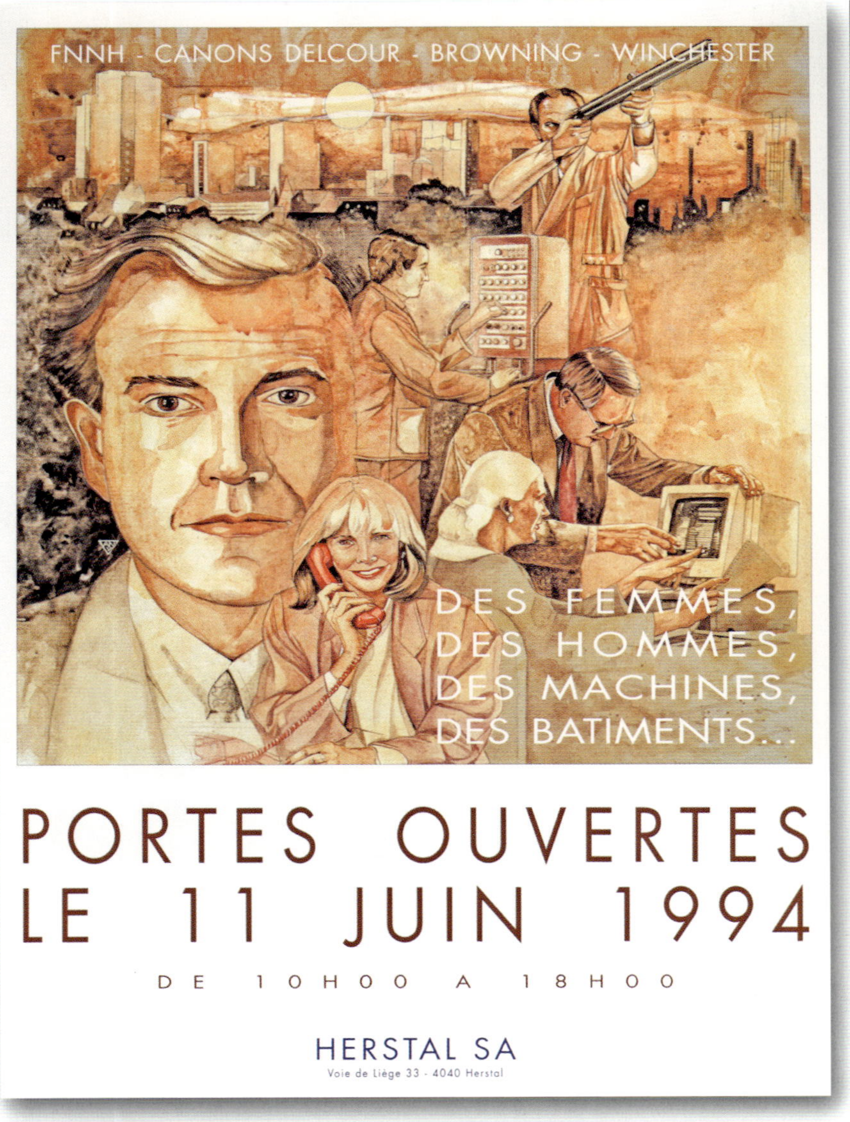

This event was a great success as more than 8,500 people accepted the invitation.

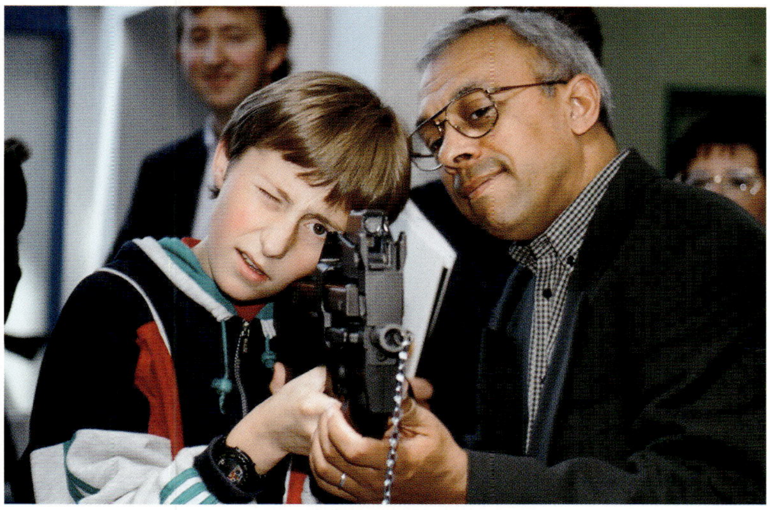

Throughout the day, employees and their families were able to visit the production facilities and find out about the product ranges, both military and civilian.

PORTES OUVERTES
LE 11 JUIN 1994

CAMPUS

chiron

GROUPE
HERSTAL

Hopes and anxieties in the civilian sector

In 1992, the results for Browning and Winchester were affected by the recession and by large losses linked to the many diversifications, which led to their sale under license, such as fishing equipment sold to the American company Zebco, golf equipment and sports rackets sold to Brg Sport, a company founded by a former FN manager. These sales of unprofitable activities were recorded as capital losses, without even the prospect of royalties or any other future return.

Finally, there was the problem of the profitability of the new Browning acquisitions – Winchester ammunition in Italy and Winchester arms in the United States – which would entail planning and financing major restructuring, including the construction of a completely new plant, at the expense of investments in equipment and products.

In 1993 the situation improved, essentially thanks to a vigorous upturn in the American market, but also thanks to significant capital increases for Browning subscribed by Giat Industries and the Walloon regional government; the latter thereby gaining a second seat on the Herstal Group's board. In 1994, another year boosted by the American market, the new manufacturing plant for Winchester arms was commissioned at New Haven, Connecticut, matching the renovation investments made at Herstal at the start of the decade.

The first half of 1995 was still buoyant. However, from the summer of that year the civilian market slumped in the United States, and although the operating profit for the year as a whole was acceptable, it was necessary once again to burden the civilian sector's net result with provisions in anticipation of the looming lean period.

Legendary American motion picture actor John Wayne was associated in the mind of millions around the world with the Winchester lever-action rifle. In this bronze statue, decorating the lobby of the new Winchester plant in New Haven, Wayne is shown holding his trademark Model 1892 lever action rifle.

Don Gobel, Chairman of Browning US. Under his leadership, a new factory for U.S. Repeating Arms was constructed in New Haven, Connecticut, during the mid-1990s.

Aerial and interior views of the new Winchester factory in New Haven. Here, the assembly hall.

1994 marked in Winchester's history the 100th anniversary of the most famous gun in its range: the model 94. In celebration, Winchester issued an exceptional commemorative limited edition which included three models: bottom, the Winchester 94 Grade I, decorated with engravings featuring Whitetail stags. Center, the High Grade model, decorated with engravings depicting mouflons and stags.

Top, the Custom model, only 94 were produced, with hand-engraved chamois and caribou, underlined by a gold trim.
The rear peep sight with micrometric adjustment is an exact replica of the Lyman n°2 model.
The three models were produced in .30-30 Win. caliber.

Advertisements for Winchester sporting shooting cartridges.

"For those who believe that you get to be World Champion three times in a row through sheer luck, here's a good luck charm for them"
Winchester, three times World Champion

Winchester won the title of World Sporting Shooting Champion in the Trap class in 1989, 1991 and 1993. Winchester was then sponsor of the Italian shooter Marco Venturini, who performed the remarkable feat of winning three times in succession, the World Championships taking place every two years.

Winchester metal cartridges for handguns.

In 1992, Browning transferred its fishing activity under license to the American company Zebco. Above, two advertisements which appeared at the same time in the French specialist press for the launch of new products in 1993.

A worrying reduction in government orders

Up to 1995, several large contracts had been able to compensate the defense activity for the crisis in the worldwide military armaments sector. But from that year the company too was to feel the effects of market contraction :

- the abolition of conscript armies in Europe and reductions in defense spending across much of the western world.
- lack of money and disorder of varying degrees of gravity in a growing number of non-western states, where in terms of armaments the cut-price stocks (with the Kalashnikov in first place) of the former Eastern Block dominated, countries for which Herstal would have been anyway to obtain an export license.
- the timidity of the peace-keeping interventions of the Western powers and international organizations in local conflicts, punctuated occasionally by atrocities and even genocide.
- the chronic absence of a common European defense policy, which had been announced in the Treaty of Maastricht creating the European Union, and its hoped-for corollary: a rationalization of the European market, even a specialization in military production on the Old Continent.

As for the area of hunting, sporting and personal defense arms, this was increasingly suffering the repercussions of restrictive measures, plus a relative lack of interest in anything which was not novel. And here too, in the second half of 1995, the contraction of its market affected the company, mainly in the United States.

Were these structural movements or a conjunction of unfavorable cycles? Both, in all probability, with the added problem of business rivals which had firm footings in their national markets and capacities they intended to exploit to the full.
It was an uphill struggle to convince a shrinking number of demanding potential customers, often already equipped with durable products, to purchase new ones. This naturally posed the problem of adding to and renewing product ranges, something which had not been sufficiently pursued up to that point.

It also led to the company turning to the neglected potential of the law enforcement market. Two approaches which could produce results only in the long term.

One thing was clear – a further reduction in the workforce would have to be contemplated, as well as keeping a tighter rein on expenses, and even investments. This was not only because the company was facing a gloomy economic outlook, but because the burden of poor decisions in Browning's recent past also weighed in the balance.

Browning, together with Olin, wanted to promote the revival of the Winchester brand. But in doing so a series of mistakes were made: the heavy financial burden of the initial buyout and successive restructurings of the Winchester cartridge factory at Anagni; the absence of a real product policy; the 1993 release of the Winchester 10-01 over-and-under shotgun. This was manufactured by the Italian company Marocchi and was based on one of their models, the "Ten O One", which had to be withdrawn from the market following a series of incidents in the United States, which, although minor, might easily have caused serious injury.

And above all, excessive investment in the new USRAC plant in New Haven, even taking into account the economic booms of the American market. This was to emerge as a problem from 1995 onwards. The company had counted on annual production and sales of 400,000 or even 500,000 Winchester arms, instead of the existing figures of 150,000-250,000 fundamentally outdated models.

To indebt the Group, under the guarantee of Giat Industries, for overrated investments was an adventurous gamble. A losing one, as it turned out, and one which led to a costly downsizing. This setback was a blow to USRAC's future, as it also faced recurrent productivity and distribution problems in the United States.

While the Group's attention was focused on Winchester, from late 1991 to 1995, not enough attention was being paid to bringing out genuinely new Browning arms.

It was another two years before a new model was successfully released : the Browning Gold semi-auto shotgun. So from the end of 1995, there were a number of areas of weakness in the Herstal Group, just at a time when it was presented with a generalized depression in the military market and a marked recession in the hunting and sporting arms market, especially in the United States.

So it seemed inescapable that the group would have to contemplate yet another adaptation – in terms of size, resources, organization and operation – to the challenges it faced, challenges which were not only related to the economic climate but were also structural.

Giat Industries was itself in a delicate position with regard to its other fields of activity, particularly in the aftermath of difficult operations and limitations in its French establishments.

The revival of the Herstal Group machine within Giat Industries would require close co-operation based on mutual trust, specific and relevant ideas, strict management of scarce resources and effective change management.

Unfortunately, these objectives were not to be achieved by the replacement of Albert Diehl at the head of the Group by Claude Elsen, a move prompted by Giat Industries. Tasked with performing a diagnosis and drawing up a plan to remedy the company's ills, Claude Elsen failed to provide a credible or coherent plan. Worse, a climate of disorder and conflict reigned within the company.

Winchester Model 94 Wrangler with American walnut stock
The Wrangler, with its 16" barrel and rapid follow-up shots is a fast handling rifle. Offered in four chamberings including the .357 Magnum cartridge, the Wrangler is equipped with a large loop lever that offers a real advantage during late season hunting when wearing gloves. Its total length is just a bit over 34 inches long – easily fitting into a backpack or a scabbard.

Three advertisements for a campaign promoting the Browning B25 and its associated products.

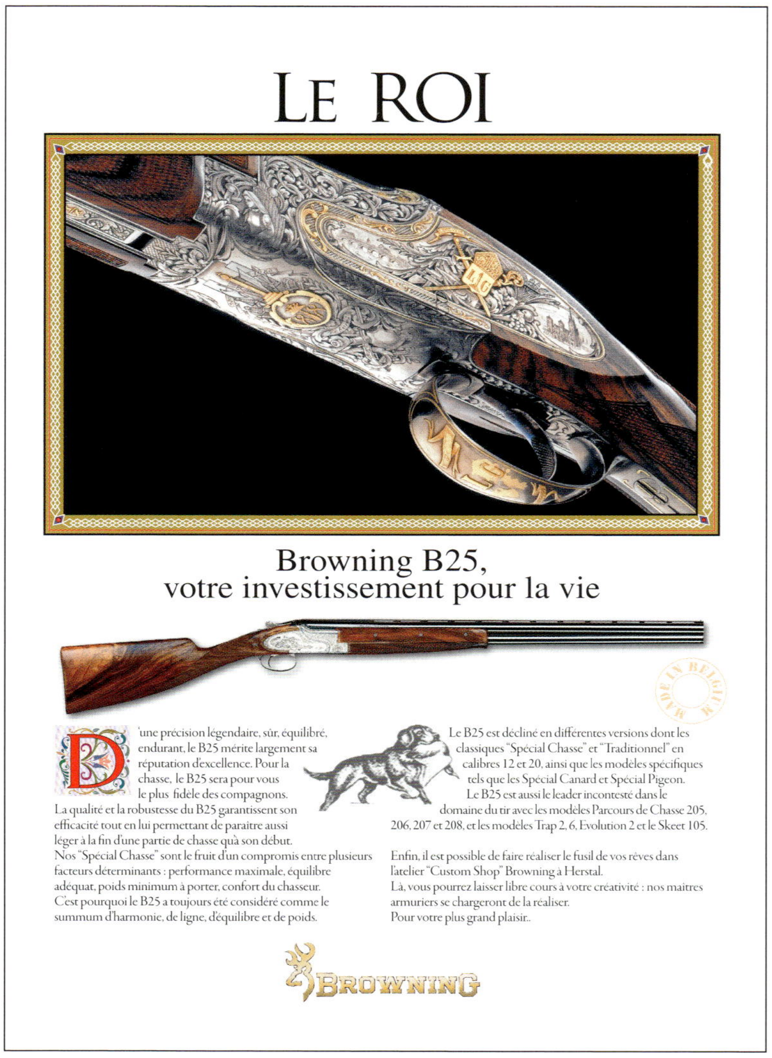

Promotional poster for Légia cartridges.

Promotional campaign for the Browning brand, as a prestige name for outdoor activities.

A difficult separation

Given these circumstances, Giat Industries, whose own financial situation, like Herstal's, was worsening on a daily basis, decided to draw a line under the experience. They announced that, not being in a position to devote more money to restructuring the Herstal Group, they were forced to seek serious buyers for it.

Jacques Gentgen, an experienced Senior Engineer who had been Giat Industries' representative on the Herstal Group's board, was then appointed General Manager for one year with the agreement of the Walloon regional government, a minority shareholder but nevertheless legally influential.

His role was to 'pour oil on troubled waters', and to produce a realistic revival plan which was to be drawn up in consultation with all the company's protagonists. Six months had been wasted, but the plan was ready and accepted in late 1996.

There remained the labor issues, including in particular several hundred redundancies at Herstal, and finding the resources to finance them.
With the help of a specialist French organization, three potential buyers were found and they came to audit the company at the beginning of 1997: a finance company, a European company with an interest in arms, and a well-known American arms manufacturer.

In the meantime, and unsurprisingly, a significant drop in sales and orders and a deterioration in both operating and net results began to appear in the Herstal Group's accounts. This negative change continued into 1997, when it reached an all-time low for the Group since 1991 – the net results suffering two years in a row as the result of big provisions for restructuring in a number of locations.

Was it for this reason that two of the three prospective buyers withdrew in short order ? In any event, in July 1997 at the end of Jacques Gentgen's term of office only the American one remained in contention, in all probability only interested in the Group's, and its American flagship's, capacity.

Giat Industries had already signed a letter of intention in its favor and the potential buyer had carried out an audit of the Herstal Group. However there remained serious concern in the economic, social, political and trade union circles of the Liège region. Being itself in a precarious industrial, commercial and financial position, was the buyer's intention not merely to take control of the Group and carve it up as quickly as possible after transferring its center of gravity to the United States ?

In the interim, from 1 July 1997, the Herstal Group was led by Philippe Tenneson. This French engineer, with a solid training and a fruitful and diverse career behind him, was Giat Industries' Managing Director at the time. He was appointed by the Chairman Loppion to put the Herstal Group's house in order, so as to guarantee its future and to take the company through the forthcoming transition, potentially a sale to the evicted buyer or a takeover by the Walloon government.

The Walloon government showed itself to be highly reluctant to accept the demands made by the potential buyer as to its financing of a new redundancy plan for the Herstal plant. It would, however, be prepared after audit to intervene financially under a debt reduction program by Giat Industries and, after a redundancy plan accepted in advance by the trades unions was financed, to take up 100% of shares in the Herstal Group.
Certainly, this buyout by the Walloon government was announced as a 'carrying operation', which would be limited to the time needed for the Group to recover its solid bases everyway, in the context of a rehabilitation plan validated and monitored by a rigorous management and by Regional representatives. A necessary condition if the group was to have any chance of negotiating and gaining respect in a satisfactory general partnership with a financial or industrial third party. In this context, the Walloon Region agreed with Philippe Tenneson that he should oversee the drawing up and control of the plan. The evicted buyer withdrew and the management of Giat Industries gave it their support.

From September 1997 things began to move quickly, at least in the basic orientations : the Group's restructuring plan was updated and an audit was carried out on behalf of the Walloon Region by the Société Régionale d'Investissement de Wallonie (SRIW); the plan was accepted by the labor organizations and was adopted under the name 'Herstal Plan 2000', covering the years 1998, 1999 and 2000 and implemented under the direction of Philippe Tenneson.

Assembly of the BDA9 pistol.

Production of the M249 at FNMI.

Finally, on 25 June 1998, all the social, legal and financial conditions having been fulfilled, Giat Industries sold its holding in the Herstal Group to the Walloon Region for a nominal sum, after reducing the Group's debt to the tune of € 200,000,000. By a subscribed capital increase, paid-up in cash, of € 62,000,000 the regional government was able to provide the financial resources necessary to pursue the Herstal Plan 2000, a large proportion of which would go to financing the redundancy scheme at Herstal in 1998 and 1999, early retirements for the most part.

Robert Collignon (left), Minister President of the Walloon Regional Government, helped in this by his right-hand man Pierre Sorveaux (right), was the main artisan of the takeover of Giat Industries' share in the Herstal Group by the Walloon Region.

From 1 July 1997, Philippe Tenneson managed the Herstal Group. This French engineer with sound training and a successful and diverse career behind him, was formerly Managing Director of Giat Industries.

An international Walloon company

The Herstal 2000 Plan

Approaching the year 2000, the Herstal Group is ready to write a new chapter of its history.
The objectives of the "Herstal 2000 Plan", which the company launches in 1998, are quite ambitious. Indeed, the Group wants to reinforce its presence on the worldwide market and to reach an optimum size, only capable of financing the research and investment in the long term and to offer partnerships which can contribute to guaranteeing its future. It also intends to introduce a more coordinated international structure, centred solely on its historic trade: the design, manufacture and commercialization of small arms for the markets of defense, law enforcement and hunting and sporting shooting. The company needs to rise to the challenge of the realities of the market, where new players show great commercial aggressiveness. It is necessary to manage innovations and costs more effectively, principally through improved synergies between the Group's entities. The entire product ranges must be renewed. The Group needs to integrate more effectively its global network of Research, Development and Industrialization, basing it within joint centres of expertise and strengthening its sales policies.
It also has to reinforce its commercial policy, to rationalize and modernize its production assets and, finally, to adapt the workforce wherever this proves necessary.
Implemented under the impetus of Philippe Tenneson, the "Herstal 2000 Plan" is intended to lead to the recovery of the company on foundations which guarantee a solid and durable international position in its two core areas of activity. Given time for its implementation, we are to see it re-establish the Group's "fundamentals" and bring it back into profitability

The Defence and Law Enforcement pole

In the context of the rapidly diminishing military market of the 90s, the result of East-West détente and worldwide economic conditions, the Herstal Group's Defence pole had to radically reposition its activity.
Initially, it was to adapt its size to the evolution of the market then very rapidly modernize its product ranges and production tools, and finally redeploy its organization with a view to consolidating its international presence.
From its base at Herstal, the nerve centre for its organization, research and sales, FN Herstal constantly strengthened the tools for its international presence so as to be even closer to its customers and so offer them a guarantee of quality.

It is also in this spirit that a new subsidiary was created in Washington, FNH USA, tasked with developing a marketing strategy in step with the needs of the US military market. Next, FNH USA's mission was expanded to encompass the development of its Law Enforcement activity in the United States. From then on, these two activities were to undergo continual development.

At the same time, large investments were allocated to FNM (FN Manufacturing Limited – formerly FNMI) in Columbia for the rationalization and modernization of the factory there. In particular, this was to significantly increase its activity, linked to the needs of the American army during the conflict in Iraq, and strengthen FN Herstal's position in the USA with the aim of making it a major light arms supplier to the American armed forces.

Advertisement (1995). At the time, FN Herstal was alone in offering such a varied and prestigious range.

The transformation of the business portfolio

Machine guns

During this period of great change for the Herstal Group, FN Herstal consolidated its position as global leader in the field of machine guns; a leadership based on flagship products, the MAG, the MINIMI and the .50. This is a niche in which the Group's mastery has long been recognized and is a major strategic focus which has been invested in continually to keep pace with users' changing needs, especially in terms of operational performance, weight reduction and simplicity of installation. Against this background, numerous variants and adaptations of basic products were to see the light of day.

FN Herstal also capitalized on certain more specific requirements, such as the need in several countries for a light machine gun like the MINIMI, but in calibre 7.62.
Originally designed to be developed in calibres 7.62 and 5.56, the MINIMI was finally launched on the market in calibre 5.56 only because of the adoption of this calibre by the armed forces of NATO. A number of armies then adopted it, most of them from NATO countries. New demand from the US Special Forces gave FN Herstal the opportunity to adapt the weapon to the higher calibre.
The MINIMI 7.62, adopted after rigorous assessment by these Special Forces under the name MK48, is expected to see significant growth on the international market.

Beyond this, one the major challenges for the company's future will be to pursue its thinking on long-term developments in support weaponry.

MK46 machine gun, developed at the request of the US "Navy Seals".

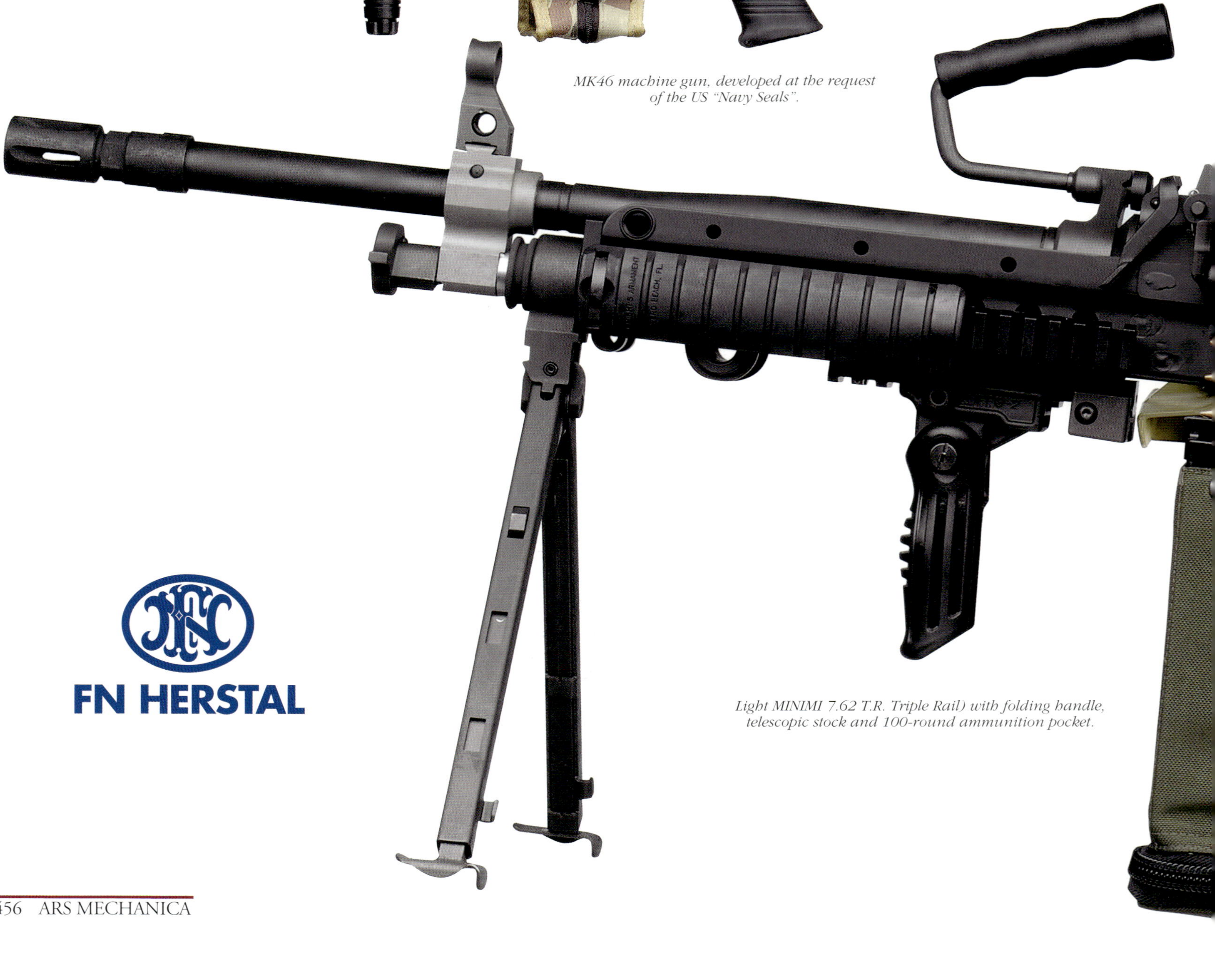

FN HERSTAL

Light MINIMI 7.62 T.R. Triple Rail) with folding handle, telescopic stock and 100-round ammunition pocket.

MINIMI 5.56 x 45mm, Para model with telescopic stock and hydraulic shock absorber.

MINIMI 7.62 in its standard version on a 360° gun carriage

Onboard weapons systems

Based on its machine guns, FN Herstal very actively pursued the development of Weapons Systems, a field in which it was consolidating its successes and skills as a designer-integrator year on year. Above and beyond armaments, it was integrating all the technologies which allow the remote operation of its weapons on subsonic aircraft, helicopters and land vehicles.

And so it developed new high-performance systems, such as the port or ramp gun pod systems, based on the M3M machine gun, whose performance and firing rate characteristics makes it one of the most efficient products on the market.

On this basis, several resounding successes have been achieved at international level, including in the USA where the concept has been standardized on several helicopters and is rapidly becoming a qualified standard under the codification GAU-21 for the FN M3M system.

M3M system for multifunction vehicles.
The system includes an M3M 12.7mm machine gun, mounted on a gun carriage absorbing the recoil and accepts 600-round ammunition containers.

Test firing of the M3M with tracer ammunition.

M3M machine gun (12.7mm caliber) integrated into the two wings of the ALX aircraft, also called the Super Tucano, produced by the Brazilian builder Embraer.

Designed around the M3M machine gun, the MRP was specially developed to be installed on the ramp of the main helicopter models on the market.

M2M machine gun (.50 Cal) mounted on a vehicle, during operations in Iraq.

M3M machine gun (12.7mm caliber) mounted on a marine gun carriage.

The Tiger (or EC-665) is a combat helicopter built by the Franco-German Eurocopter company. Here, the Tiger is equipped with two FN HMP-400 pods.

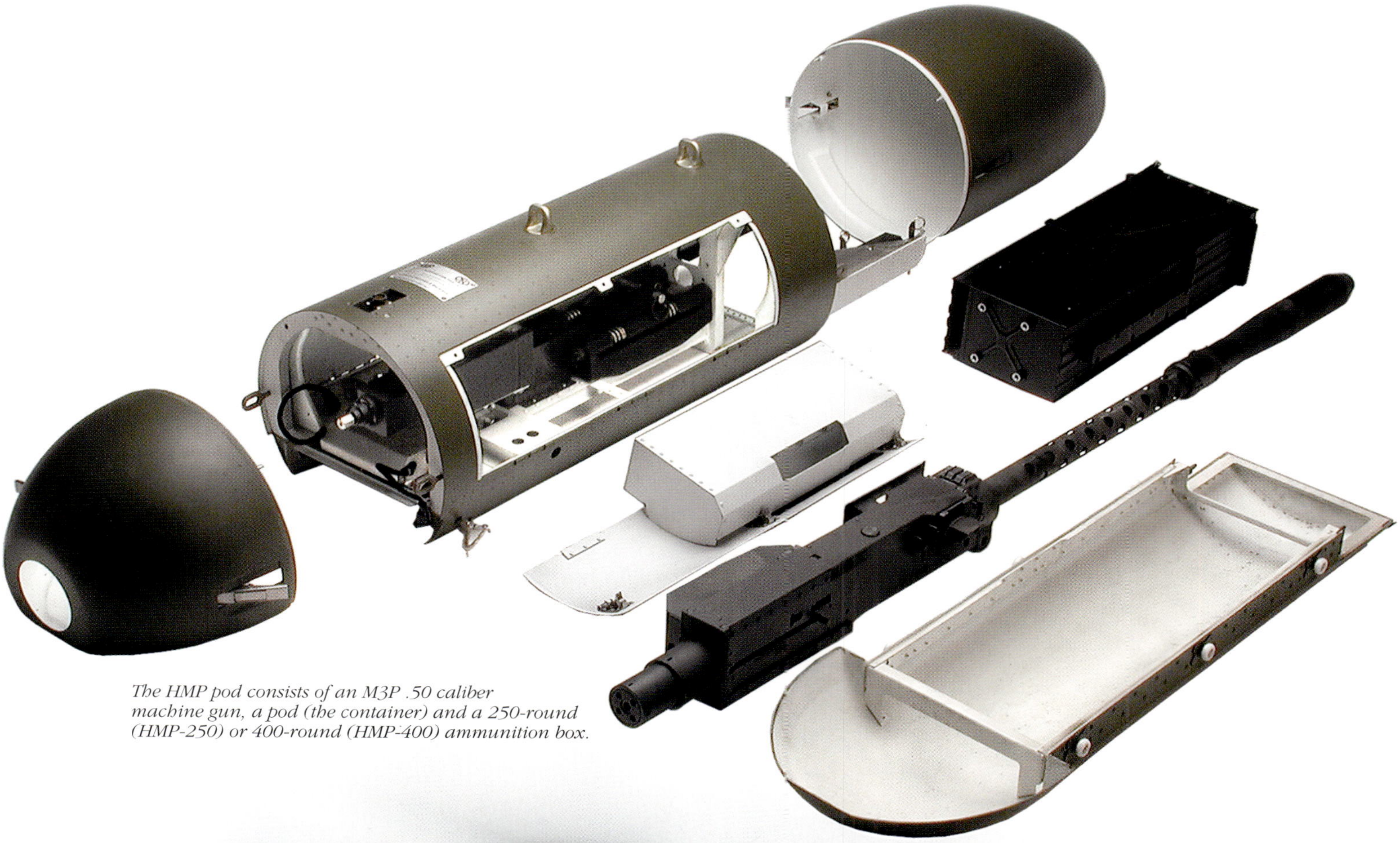

The HMP pod consists of an M3P .50 caliber machine gun, a pod (the container) and a 250-round (HMP-250) or 400-round (HMP-400) ammunition box.

Above, an RMP pod is mounted on either side of the Agusta A-109 helicopter. This is a combination including an M3P machine gun (HMP-250 pod) and three 2.75" (70mm) rocket launching tubes.

HMP-400 LC

LWP (Light Window Pintle) system mounted on a Cougar helicopter. This system includes a MAG 58M machine gun (7.62 x 51mm caliber) for applications in helicopter windows.

Two MAG 7.62 x 51mm machine guns mounted in the door of the A-109 helicopter, built by the Italian firm Agusta, acquired by the Belgian land army at the end of the 1980s. This system, called the 'turning system', allows the helicopter doors to be closed.

MAG 58M machine gun (7.62 x 51mm caliber) in a door-mounted application for helicopters.

ARROWS station mounted on a Belgian vehicle, the MPPV Dingo II, in a procession on 21 July 2006.

Applying the same logic as aircraft systems, FN Herstal decided at the beginning of the millennium to develop new terrestrial systems consisting of remotely operated gun turrets mounted on vehicles.
Once again, the objective was to combine the skills intrinsic to the company's core trade and to adapt them to the changing needs of the market. The growing concern to protect the operator, which emerged during the new international conflicts, actually opens up a new opportunity for development.

This new activity was launched in partnership and led to the development of the «ARROWS» turret. After winning a first contract to equip French armoured vehicles (VBCI – Véhicule Blindé de Combat d'Infanterie), FN Herstal was selected for two major armoured vehicle programmes for the Belgian army, the MPPV (Multi Purpose Protected Vehicle) and the AIV (Armoured Infantry Vehicle). These successes have furthered the company's expansion into this new and promising market for terrestrial systems.

Test firing the ARROWS weapon station, carried out at Elsenborn, Belgium, on a military land, and in Zutendaal.

Developed jointly by FN Herstal and Oerlikon Contraves (Canada), the ARROWS 300 turret is one of the most advanced in the world and its firing accuracy is unique. It is fitted with a CCD camera for use in daylight and an infrared camera for night use. It can be equipped with several types of machine gun (12.7mm or 7.62mm caliber).
ARROW stands for "Advanced Reconnaissance Remotely Operated Weapon System"

ARROWS gunner's station inside an armored vehicle, with control screen. The gun and the optics are controlled by means of a control handle which enables one-handed selection of day and night sight modes, acquisition, firing, zoom control and development of the image.

Some views of the weapons and systems assembly halls (Herstal - 2007).

FN Herstal benefits from an ammunition loading facility employing 60 workers, in Zutendaal (Belgian Limburg), where it can also test its weapons in all extreme conditions. Below, varnishing of the 5,7 mm cartridge cases.

Visual inspection and packing of the 5,7 mm ammunition.

A group of 4 turning/grinding centers producing the chamber for military barrels, linked to a 3D measuring machine (Herstal).

CNC lathes, each equipped with a barrel loading/unloading system (Herstal).

Aerial robot for supplying 3 numerically controlled lathes with barrels (Herstal).

Rifles

In the field of assault rifles, the challenge for FN Herstal at the end of the nineties has been to reassert itself in a segment where in the past it has been a global leader. This involved developing a new generation of rifle, more in phase with the evolution of the market: the F2000.

An individual weapon system of the "bullpup" type in the 5.56 NATO calibre, the F2000's performance, compact size and the modularity of its configurations to suit the mission set it apart from traditional weapons. Since its launch, it has been adopted by many customers, both special forces and armed forces. Furthermore, it is undergoing assessments in a number of countries.

New openings within police forces have also emerged, particularly in the American market.

F2000™ Tactical

F2000™ with optical sights and a triple rail handguard equipped with a grip.

The modular design of the F2000™ enables the standard weapon to be reconfigured in any number of versions adapted to specific missions. Here, the F2000™ is equipped with a grenade launcher and a prototype for fire direction.

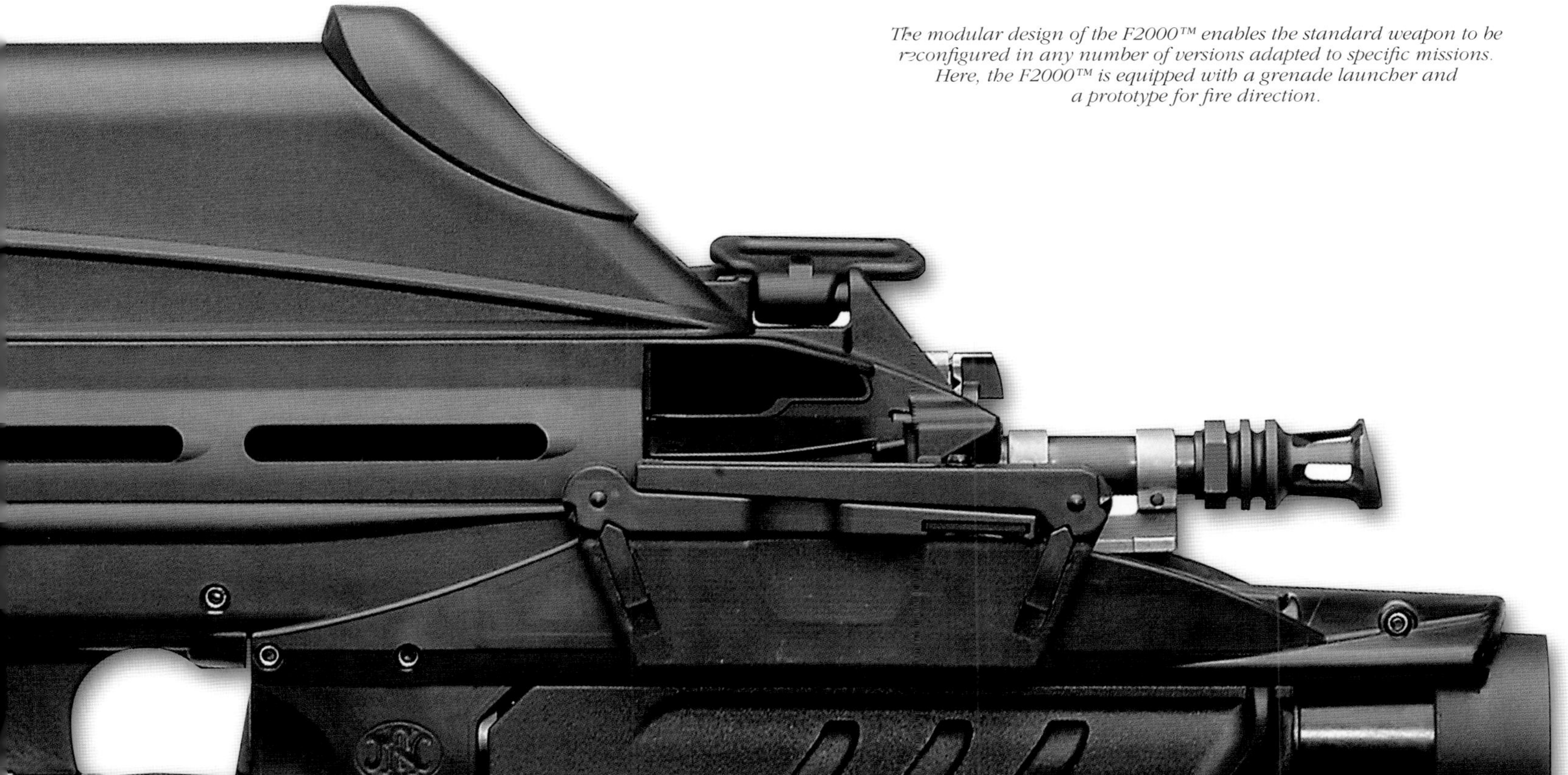

The F2000™ assault rile is based on the use of different modules and on the possibility of mounting multiple accessories adapted to different types of missions. The 'bullpup' design gives it very compact dimensions, even though it is fitted with a standard-length barrel. Above is illustrated the standard version of the F2000™ with a 40 mm LV grenade launcher.

Double page : F2000™ rifles assembly line (Herstal - 2007).

However, in early 2004, FN Herstal faced another challenge. The American army wanted to equip its elite troops, the US SOCOM, with a new family of assault rifles, the SCAR (Special Operations Forces Combat Assault Rifle).
FN Herstal won by a wide margin the invitation to tender issued by the US SOCOM, by offering the SCAR in two calibers : 5.56 and 7.62.

In January 2005, the US SOCOM signed an "ID-IQ" (Indefinite Delivery - Indefinite Quantity) contract with FN Herstal for the supply of the SCAR. For the company, this programme constitutes a major potential for development.

Thanks to its flexibility, its modularity and its reliability, this family of weapons – in parallel with the F2000 – represents a new opportunity to consolidate FN Herstal's position in the field of assault rifles, crucial for infantry weapons.

40mm grenade launcher module (or EGLM, Enhanced Grenade Launcher Module) which includes a rifle launcher, a fire direction system (here with a mechanical sight) and a stand-alone stock assembly.

SCAR H (7,62 x 51 mm caliber)

SCAR L (5.56 x 45mm caliber) equipped with a grenade launcher and a fire control system.

SCAR L (5.56 x 45mm caliber) equipped with a grenade launcher.

Test firing of the SCAR in its different versions.

Automatic phosphating line for weapons components (Herstal).

Computer-assisted metrology : three dimensional control of Gold shotgun receivers (Herstal).

Flexible line for 4 CNC horizontal 5-axis machining centers. A robotised system feeds 42 pallet machining tables (Herstal).

Each tool is preset by data stored on an integrated electronic chip. Right, control position for the FMS (Flexible Modular System) line (Herstal).

The 5.7 system

Just as it did in the 1950s and 1980s for the calibers 7.62 and 5.56 respectively, now standard for NATO, FN Herstal once again positioned itself as from the 1990s as pioneer of a new caliber, the 5.7. Although this opened up new opportunities because of its unique characteristics, which constitute a real revolution in ballistics, it would also entail, from the earliest launch phase, all the problems connected with the introduction of a new caliber, problems which the company would resolve one by one : international recognition of its advantages, logistics management, diversity of ammunition, etc.
Naturally enough, one of the milestones to cross will be the NATO standardization process for its caliber 5.7 x 28mm. This process will not yet be finalized in 2007, although positive signals have been received from NATO's competent authorities.

At the same time, FN Herstal gradually penetrated the international markets with its P90® machine pistol concept, enhanced a few years later with one of the most innovative pistols of its generation, the Five-seveN®, and at the same time a continual expansion of its range of ammunition, meeting the needs of both military and law enforcement markets.

Pistols

Internationally renowned for its HP pistol, FN Herstal owed it to itself to rejoin the race and once again position itself in the military and law enforcement markets with a new range of small arms combining technological advances with competitiveness. It was against this background that a line of "polymer" pistols, called FNP, was released in different calibers – 9mm, .40 and .45.

The Five-seveN ® pistol, the P90 ® submachine gun and the SS190 round have been designed jointly. Today, they form a unique weapon system that matches the requirements of users who have to operate in a rapidly evolving world.

The new FNP family of pistols represents the latest developments in handguns, continuing a long tradition of quality and innovation.
The FNP pistols were developed to adapt to later developments in different calibers, different operation modes – single action, double action, and dual action – and different sizes.
Using the very latest production techniques, the polymer framed FNP9 (30% lighter than steel) offers many ergonomic features for optimum adaptation to the various needs of Law Enforcement professionals.

Five-seveN® Tactical pistol equipped with a laser module and flashlight.

The P90® machine pistol and the Five-seveN® pistol are now used in over 40 countries throughout the world, by police forces as well as armies and special forces. The P90® machine pistol is illustrated here in its Tactical version with a superior Picatinny rail on which a range of accessories can be mounted, as here an in-line red point visor.

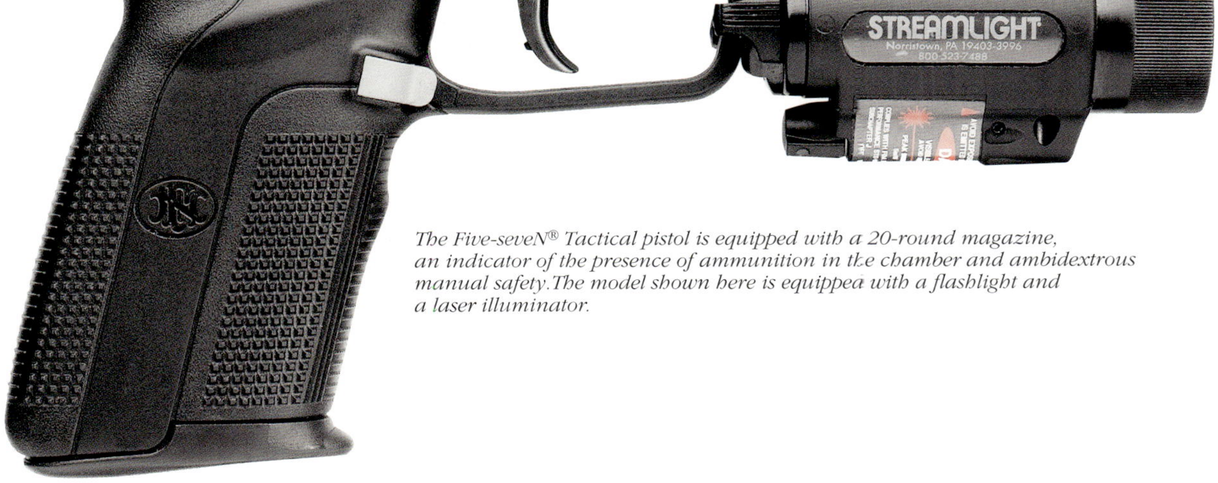

The Five-seveN® Tactical pistol is equipped with a 20-round magazine, an indicator of the presence of ammunition in the chamber and ambidextrous manual safety. The model shown here is equipped with a flashlight and a laser illuminator.

The US production tool

In 1976, FN Herstal signed its first contract to supply the American Army with the MAG machine gun (the M240 in the United States). The American administration had long since acquired the production rights (Technical Data Package) and had announced in the early 70s that, in accordance with the "Buy American Act", it intended to restrict the list of potential suppliers for its next order to those arms manufacturers who were established on North American soil.

Browning intended at the time to acquire an American production unit for its own needs. With the intervention of this new factor, FN Herstal decided first, in order to satisfy the conditions of the American Government's call to tender, to build a factory in Columbia, South Carolina.
The initial reasons for the choice were based essentially on the financial assistance offered by the local Authorities and the local availability of a qualified workforce.

It soon became apparent that FN Herstal had made the right decision as it gained the competition to its main competitors' home territory, thanks to the intrinsic quality of MAG58 machine gun. In 1980, FN Manufacturing started production of the M240 (US coding). A machine gun which has since been standardized for all American applications.
A succession of Herstal engineers worked with the emerging entity to help in the production launch phase, placing emphasis on the necessary industrial quality of the product.
The operation was profitable while annual volumes were high, but orders began to drop after a few financial years.
To ensure the company's future beyond this single contract, FN Manufacturing negotiated several contracts with the American Defence Department, in particular for the supply of bomb attachments for the Navy, Gatling machine gun parts for the Air Force and even - purely in the interests of profitability - parts for Caterpillar tractors.

Although marginal, most of these contracts were subject to the strictest quality controls (Statistical Process Control - SPC), which worked to the advantage of the factory's total quality management necessary for its strictly military contracts.

At the end of the 1980s, FN Manufacturing also began to produce conversion kits for the M16A2 American assault rifle and large volumes of spare barrels for the M16A1 and M16A2.

In 1988 this breadth of experience, and the levels of qualification achieved in the course of fulfilling the contracts, resulted in the company winning a tender for an initial delivery of 300,000 complete M16A2 rifles. Production started in 1991. Since then, the American subsidiary has produced over 600,000 units of this rifle, in its various configurations.

Also in 1988, after an initial production phase in Belgium, FN Manufacturing tendered for the supply of the second FN Herstal product to be adopted by the US Army, the MINIMI 5.56 (M249) light support machine gun, and the production of the MK19 grenade launcher.

Again, the company was the successful bidder for these two new challenges, but in the end it was decided not to pursue production of the grenade launcher, in the belief that the factory's production capacity would be saturated by the M16 and M249 alone.
Production of the M249 at Columbia began in 1991. To date, the company has supplied some 80,000 units to the American Army.

Since its creation, FN Manufacturing's activity has seen continual expansion, thus fulfilling its role as a major supplier to the biggest army in the world, in total synergy with FN Herstal.

The spade-grip, pintled mounted M240 D has been developped for use in military helicopters.

M16A2 standard configuration.

M16A4 with electronic sight and an M204 grenade launcher.

M16A4 with Picatinny rails and a forward mounted handgrip.

Photo taken on the occasion of the initial fielding and transfer of the M240H machine gun to the US 101st aviation unit (2005).

Both pages : members of the personnel and main building at FN Manufacturing (Columbia).

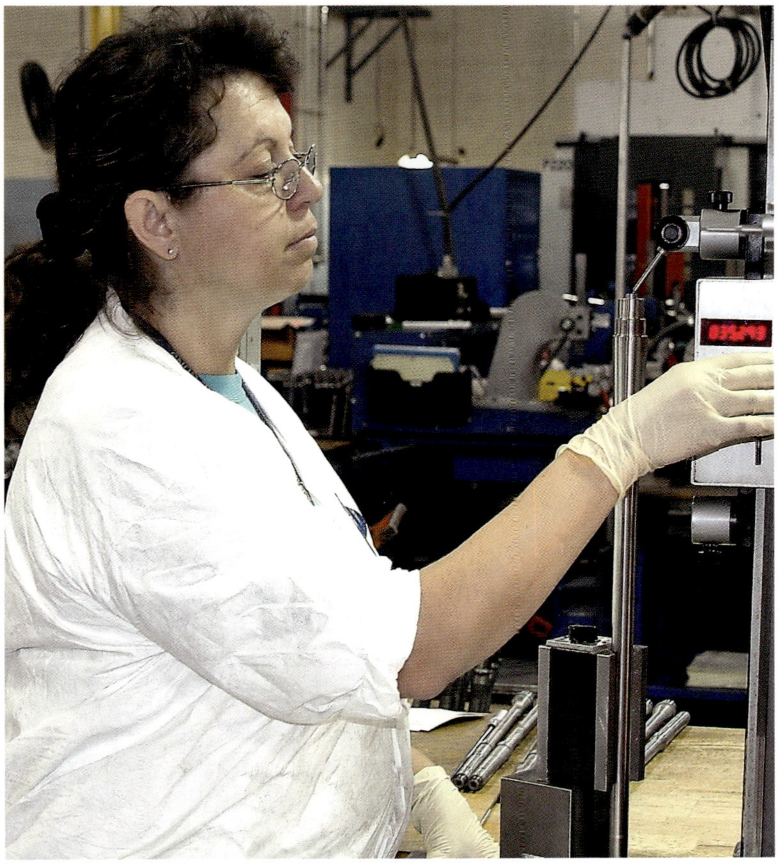

NATO troops train with the FN M240/MAG58 machine gun prior to their deployment to Iraq.

Watching for threats to his aircraft from the ground below with an FN M240D.

American soldier armed with an FN M249 Para assuming a covering position during operations in Iraq. (Photo courtesy of US Department of Defense).

Providing over-watch and covering fire with an FN M249 Para during a joint US-Iraqi operation. (Photo courtesy of US Department of Defense).

Providing covering fire from a rooftop position with a FN M240B machine gun.

Law enforcement agencies, the new customers

The late 1990s saw FN Herstal undertake a proactive approach to the law enforcement market, by capitalizing on synergies between the products of FN Herstal, Browning and Winchester and exploiting the distribution networks of the Group's two core areas of activity. Strenuous efforts were made to gain an understanding of the special characteristics of this particular customer and gradually to adapt the product ranges to its specific needs.
In the early stages, these efforts were centred especially on accessing the American market, where significant investments were allocated in terms of dedicated resources and adapted ranges and distribution networks.
Under the impetus of FNH USA, which had gradually put together a completely new team in Washington, during the 2000s this activity experienced significant growth and a recognized positioning for several of its flagship products. Because as a complement to the 5.7 range, FN Herstal developed, sometimes in collaboration with one partner or another, a full range of pistols, ammunition and rifles for law enforcement agencies, ranging from the pump action shotgun to high precision sniper rifles and less lethal weapons.
Finally, the situation in America led FN Herstal at the same time to adapt some of its military weapons for the hunting and sporting shooting markets. Known as "black" weapons and restricted to semi-automatics, these were highly popular among hunters and sporting shooters.

Moving forward from this initial phase, FN Herstal is continuing efforts to consolidate its penetration of this segment which is of such strategic importance to its future, in the USA and worldwide.

But it was as a result of a very different kind of thinking around the expectations of law enforcement agencies that the Herstal Group would, at the dawn of the twenty-first century, come up with two ideas which were likely to interest other clienteles : "less lethal" weapons and "secure" or "smart" weapons.

Police officer equipped with a 9mm FNP9-M DA/SA pistol .

The Five-seveN® pistol in its original version, equipped with a laser aiming module.

Illustrations made in 1999 by the American artist Dick Kramer, promoting FN Herstal's Defense and Law Enforcement products.

During training, a police officer armed with an FN TPS pump shotgun covers a suspect in a simulated felony traffic stop.

A SWAT officer armed with an FN P90 PDW rappels down the side of a building. The compact P90 can easily be placed out of the way for any activity, no matter how difficult..

The FN P90 PDW has become a favorite close-quarters entry weapon among numerous elite law enforcement units.

Training with the P90 Tactical.

The still pioneering Herstal Group

In the world of arms manufacture, the end of the 90s was characterized in particular by the appearance of the "less lethal" concept. Although the idea of stopping an action using systems which do not kill or do not cause serious injury is not a new one, we had to wait until the end of the twentieth century to see the emergence of technologies capable of achieving it. It was during this same period that the need emerged for a credible alternative to the use of firearms, both for law enforcement agencies increasingly faced with public rejection of unnecessary fatalities or serious injuries, and military forces involved in peace-keeping missions.

So it came as no surprise when the Herstal Group began to take an interest in this new niche in the mid-90s. Initially, its interest developed around a general study of the concept and the appropriate technologies. This was turned into reality in 2000 by the release of a product, the FN303, which was the fruit of collaboration with an American company.

This compressed air-powered launcher can accurately strike a target over 50 metres away, having a deterrent effect using a projectile which has been specially developed to reduce the risk of penetration and thus reduce the risk of serious injury.

Interest in this sector, which the Herstal Group knows to be a key one, was further strengthened in 2005 with the creation of a Centre of Excellence on Reduced Lethality in collaboration with the University of Liège and other Belgian and foreign partners, under the aegis of the Walloon Region. The clear objective set for this centre by the Herstal Group is to stimulate the concept of reduced lethality within the scientific community, the political world and among users. Through in-depth technological and sociological research, it will also promote the development of a range of products which are geared to the needs of users.

In this same spirit of less lethality, the late 90s also witnessed the appearance of the concept of secure weapons. Under the impetus of the National Institute of Justice (USA) and thanks to research sponsored by it, the Herstal Group became a true pioneer of this innovative approach. The concept of a secure weapon aims principally to limit accidents or inadvertent discharge of domestic firearms by means of on-board electronic systems which enable – or disable – the use of the weapon.

Although the technologies available and market regulations mean a saleable product is not expected before 2015 or 2020, this research is an avenue of exploration for the Herstal Group in the very futuristic field of electronic recognition of a firearm user.

Whether with less lethality or the concept of secure weapons, the Herstal Group is clearly embarking on a path of innovation in the face of twenty-first century challenges and the changes these are bringing to the world of arm manufacture and the use of weapons in our society.

These lines of evolution, demonstrated in products intended for the military and law enforcement sectors, simultaneously represent a potential for the development of secure weapons in the hunting and sporting shooting sectors, long considered to be more traditional.

The launcher's compressed air canister can be filled using a standard diving bottle.

The projectiles are specially designed to eliminate the wounding risk.

The FN303 is also available in a version which can be adapted to most assault rifles.

Peace-keeping missions require new strategies and more suitable tools. Conscious of these new requirements, FN Herstal developed the FN303. The first launcher entirely designed around the concept of 'less lethal', the FN303 enables the user to neutralize and apprehend suspects with maximum efficiency and minimum risk for them.

FN

a worldwide dynamic

Hunting and shooting in a changing world !

At the turn of the century, the dominant position of the couple Browning-Winchester on the world market of hunting and target shooting seems to be confirmed.
The two sister brands have «globalized» their supply sources, – a trend impossible to ignore – and offer products on the market with certain components originating from regions of the world which were not represented in this sector until then.
The demands for quality, finishing and quality control have often benefited these new suppliers, by progressively integrating them in the restricted circle of trusted manufacturers.
The market has adapted to contemporary lifestyles.
Rapid urbanization has transformed the traditional image of the rural hunter into the slicker city-dweller keen on active leisure pursuits, or even shooting competitions in some regions of the world.
In other regions, "anti hunting" ecology or the restrictive legislation on the possession of firearms have discouraged the enthusiasts.
In other places hunting has become – if it not already was – an increasingly expensive and exclusive leisure, often practiced by rich landowners on private grounds rather than by popular classes in natural spaces open to the public.
The climate has also changed and infectious diseases, like bird flu, have sometimes rarefied game, and thus limited the purchases of firearms on the markets.

The offer of leisure activities has multiplied, and a new type of leisure activities has appeared, which often moves children away from nature by attracting them to an increasingly virtual world.
Finally, families themselves have changed. More and more often they are "single parent" units and have limited the potential for the young population of emulating a lifestyle which integrates nature activities like that practiced by the hunter-parent.
Simultaneously, however, a renewed popularity of clay pigeon shooting has emerged, with practices rather different from the traditional "trap" and "skeet", or a nostalgia for the Far West, which created new shooting activities inspired by "Western" films and TV series of the Fifties and Sixties, and finally a growing concern for domestic safety which is a result of the tragic events of 11 September 2001.

Despite the global decrease of the number of hunters in the world – but there are still more than twelve million in the United States – established brands like Browning and Winchester, which offer, furthermore, sufficiently diversified ranges to meet the market requirements, did not suffer too much from this evolution in social behavior. But they had to adapt, by integrating new market niches which would soon become new development opportunities.

From 1998, the whole enterprise, both Browning International in Europe and Browning North America, started a re-launch plan around a single axis : "total quality". Through a more audacious product policy for example, but also a better management of distribution circuits and an improved customer service. This does not mean these concerns were missing until then but they need to be established as a true company policy at all stages of the trade, in all areas.

Browning-Winchester are, at global level, amongst the only ones to offer such a wide range of products to gunmen and hunters : over-and-under shotguns, semi-automatic shotguns, repeating rifles, semi-automatic rifles, pistols ... as well as technical products and "outdoor" equipment, accessories and ammunition.

The challenge of this proactive policy is to remain the general leader versus the competition which specialized more in a virtually unique products niche.

The 21st century will demonstrate whether these choices were right, and we will soon see the competition follow the consolidation of ranges policy.
In 1998, a new "products plan" was therefore established. The Management indicates the ambitions and all the teams, R&D, industrial and commercial work at its realization by a progressive renewal of the whole – or almost – portfolio of the products of the two brands.
New semi-automatic shotguns appear very quickly. The Gold of Browning first, existing in different versions, which was awarded the First Prize of the American Academy of the Shooting Industry the same year.
This shotgun, for technical reasons of fine-tuning, will not receive the success it deserved, but will be at the root of a family of performing semi-automatics, such as the Browning Fusion, Silver and Phoenix.
Then the Winchester SX2 and SX3, with the former also awarded by the Academy the next year.
New over-and-under will also be created : the Browning B525, fifth generation of superposed Browning, the Winchester Supreme and Select, each in different versions...
And others reappear, like the famous BAR rifle, in the versions "Short Trak" and "Long Trak".
The renewal is also felt in the release of completely innovating weapons like the "Acera" rifle of Browning, and last but not least, in 2003, the completely new Browning "Cynergy", a weapon with low lever and superposed barrels, considered revolutionary and with a truly innovating design.

In the global trade, the years 2000 are characterized by the opening of the markets, as we know. Although the North-American market remains virtually unchanged, a Europe with 27 states alters the size of the offer, in particular with the arrival of low cost manufacturers.
The consumer's requirements also evolve constantly – or are even "revolutionized". Customers want more service and more advice in particular.
Faced with this double challenge, Browning-Winchester acts and adapts its organization at the start of the decade. The competitive strength needs to be improved by restricting expenses that do not bring added value to the offer, and by investing massively in those that do. In other words, the distribution chain needs to be optimized.

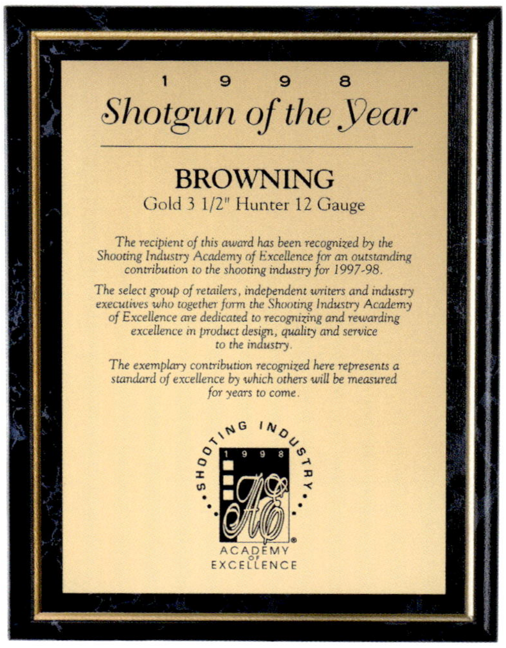

This new undertaking will, in time, become one of the most important projects of the civil branch of the Group Herstal, as it is a new concept of the profession that is being implemented and that distinguishes Browning and Winchester from their competitors.

The customer's expectations are at the core of this policy. They cover the whole distribution chain, from the customers to the production units, and indicate the performance and the customer satisfaction.
This chain – "sliced up" – is analyzed for added value. Progress groups are implemented and the whole staff is mobilized by the will of those who have understood that the company's success is at the end of the road.

In 2005, the results are there, with the recovery of Browning International in Europe, which sees its faltering financial situation for several years rectified and its market shares frequently increased. For Browning North America, the financial and commercial situation is stable and secure, despite adverse market conditions.

In 2007, the Browning Management updates its strategic plan and makes it even more ambitious. Its product planning commits it to fully innovative concepts and provides the customer with legendary but updated Winchester products. More than ever the objective of the Browning-Winchester couple is to offer even more pleasure to the hunter and gunman in the pursuit of their favorite leisure activity..

Photographic study for an advertisement for the Browning GOLD.

The year 1998 saw the introduction of the Browning Gold 31/2" shotgun, with a lengthened receiver and chamber capable of handling the heaviest magnum loads for turkey and high-flying waterfowl.

LongTrac

Developped in 1998 the reduced weight of the BAR Light was acheived by making the receiver of lightweight alloy.

Browning BAR EVOLVE
Launched in 2002, the BAR Evolve has a very elaborate look and a variety of innovations such as the Monte-Carlo stock, a more marked pistol grip and a slimmed-down tulip fore-end.

Browning BAR Light Battue De Luxe
A model with an antique silver nickelled frame with engravings of game scenes, embellished with animals in fine 24 carat gold.
The wood elements are of oil-finished European walnut (2002).

Browning BAR ShortTrac / LongTrac
In 2003, Browning launched the BAR ShortTrac which was the first semi-automatic rifle on the market developed specifically for the Winchester Short Magnum calibers.
These WSM calibers combined the power and accuracy of the classic magnum calibers with a cartridge of reduced size. The BAR LongTrac, for its part, offers conventional calibers while still adopting the same innovative design. Model presented: BAR LongTrac "New Elite" (2006).

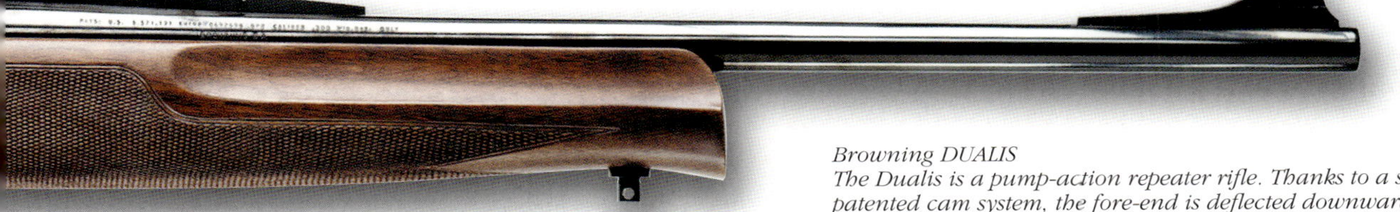

Browning DUALIS
The Dualis is a pump-action repeater rifle. Thanks to a special patented cam system, the fore-end is deflected downwards when it is drawn back. Its cold-hammer forged, free-floating barrel is manufactured at Herstal.

fusion evolve
phoenix
BACK-BORE TECHNOLOGY

Browning EUROBOLT
A bolt rifle with opening limited to 60° for quicker, smoother reloading, especially when fitted with a telescope sight.

Browning ACERA
A straight pull rifle, the ACERA is the only bolt rifle in the world with a bolt slide mechanism enclosed in the frame.

Browning GOLD Hunter 12 cal.
A weapon for those who want to fire heavy loads, especially when hunting waterfowl or migratory birds (2002).

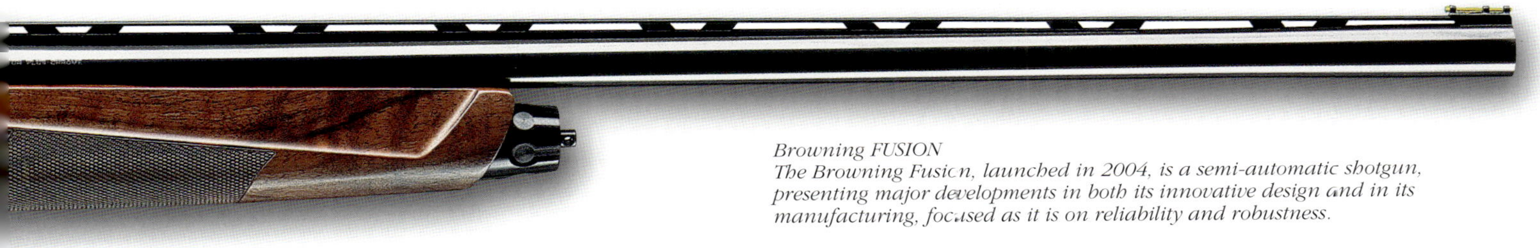

Browning FUSION
The Browning Fusion, launched in 2004, is a semi-automatic shotgun, presenting major developments in both its innovative design and in its manufacturing, focused as it is on reliability and robustness.

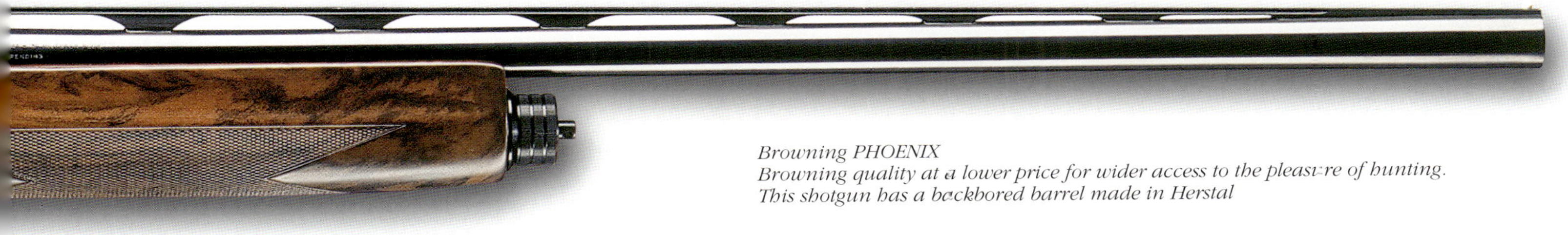

Browning PHOENIX
Browning quality at a lower price for wider access to the pleasure of hunting. This shotgun has a backbored barrel made in Herstal

Industrial partners

Browning and Winchester have always associated industrialists considered as real partners rather than simple suppliers to their production facilities. The efficient communication, joint reflection from the design stage, permanent striving for innovation and improvement of the products and modern industrial methods have enabled the two brands not only to be competitive on their markets, but to dominate the sector of outdoor leisure pursuits. Three amongst them, Viana, Miroku and ATI, have been the privileged historical partners of the two brands, and remain so to this day.

A new building for a new automated surface treatment line..

Browning Viana

Established in 1973 under the name of FN Viana, the "Fábrica de Armas e Artigos de Caça" is located in Viana do Castelo, in the North of Portugal on the Atlantic Coast, between Porto and the Spanish border.
In late 1973, FN Viana employed 84 persons, compared to 600 today.
The first order, using components originating from Herstal, consisted of the assembly of 1200 Trombone rifles, delivered in Belgium a year later.
From then on until today, the enterprise started living at the rhythm of the numerous developments of Browning products and their other versions : B2000, BAR, B80 (in cooperation with Beretta), semi automatic A500, GOLD, handguns BDA, HP 9mm and .40 SW, followed by the Fusion, Evolve and Silver, BAR Short Track and Long Track.

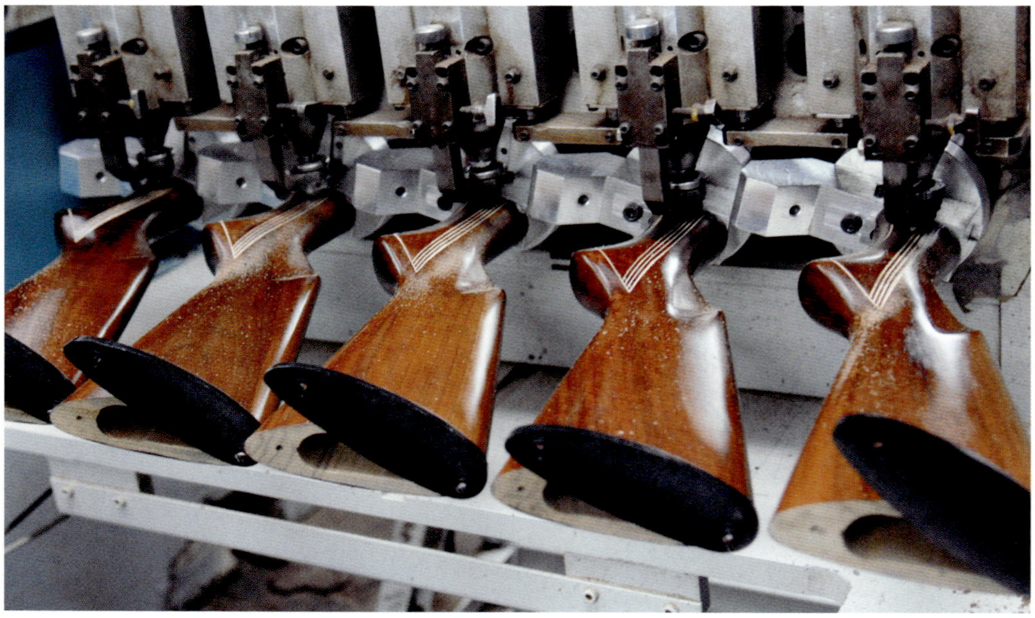

In 1998 Viana started the production of Winchester products, in addition to the Browning range: Semi Auto, SX2, SX3, Super X, stacked Select and Suprême as well as the SXR semi-automatic rifle.
Wood was soon added to iron, with a workshop in charge of the mechanisation, polishing and finishing of all the wood required for the local production.
Between 1978 and 1996, Viana also worked on the production of the Browning golf clubs, and issued more than one million items.

Over the last few years, major investments were carried out to adapt the enterprise to new products and new materials : automated surface treatment lines, painting lines, camouflage, engraving and laser marking procedures,...
Initially created for obvious competitive reasons, because Portuguese manpower was and still is less expensive than in the mother company, Browning Viana today is part of the enterprise's history, perfectly responding to the strategy of multiplied supply sources of the Group and one of the main factories of the Group Herstal for its civil activities.

Having concentrated for a long time on metal, Viana soon developed skills working with "wood", and took on the mission of mechanizing, polishing and finishing all woods involved in arms production.

New laser engraving and marking line.

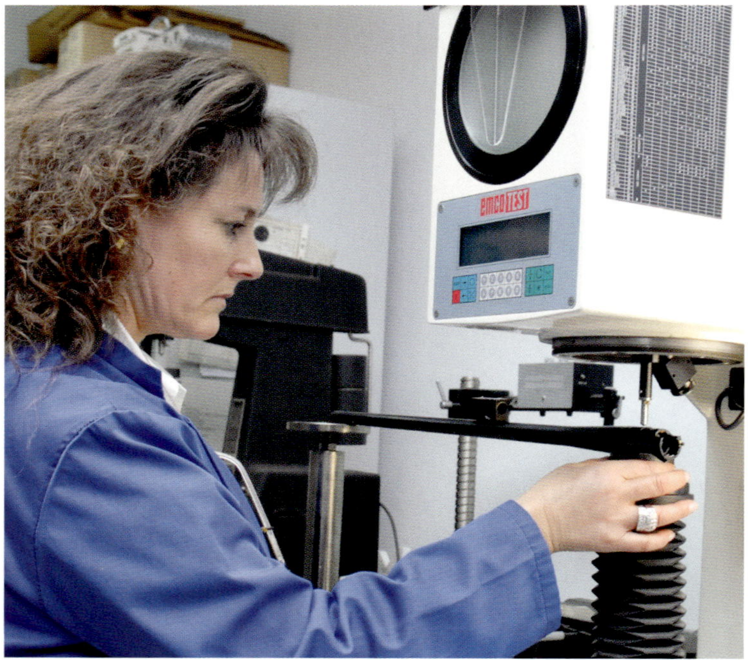

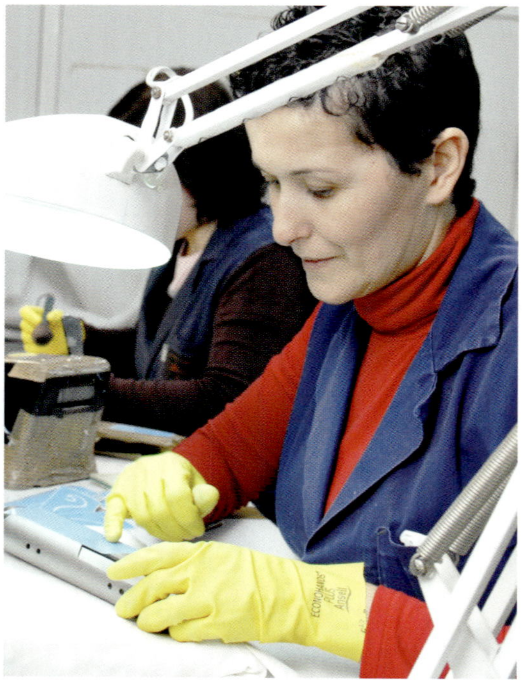

Double page : members of the personnel at Viana.

Browning
Viana

WINCHESTER
WINCHESTER

RVS

Miroku

As 19th century Japan was inspired by Western techniques to modernize its industry, a certain Karaji Miroku started the artisan production of weapons destined for hunting in the surrounding mountains.
In 1934, his son Bukichi extended the family production to pistols and harpoons for whale hunting. A giant harpoon still decorates the entrance of the company Miroku Firearms Manufacturing, located in Kochi, in the south of the island Shikoku, the smallest of the four main islands of the Japanese archipelago.

In the mid-Sixties, Miroku produced firearms under its own brand, but also for the famous American manufacturer Charles Daly, the grandson of John Moses, John Val Browning, who was looking to diversify suppliers and went to meet Miroku.

Japan benefited from a quality workforce, technical expertise and abundant raw materials, metals in particular.

John Val Browning appreciated the contact. It was the start of a new adventure...

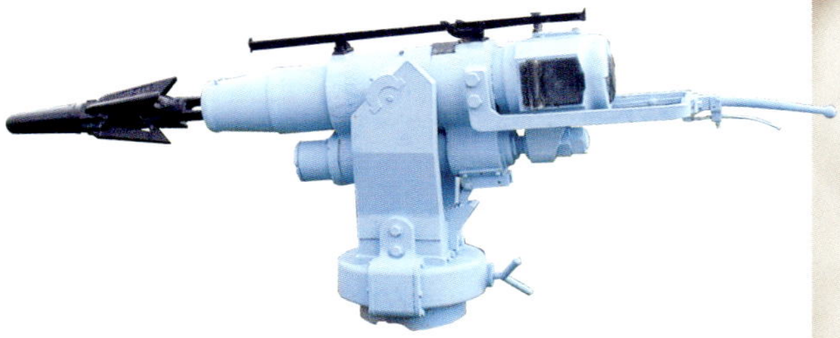

Above, from left to right : Kazuo Tsutsui, General Manager of Browning Japan, Katsuhisa Tanaka, Senior Managing Director, Yoshihiko Miroku, President, Shiroh Takemura (†), Chairman, Yoichi Shidei, Managing Director.

In 1967 followed the BT-99, the first shotgun manufactured by Miroku for Browning, the lever rifle BL-22 in 1969, the BSS (Browning Side by Side) in the same year, replacing the Anson model abandoned by the FN, the elegant Citori in 1970, the Browning Pump Shotgun (BPS) in 1973 and then in 1976 the legendary Auto-5, – all Belgo-American-Japanese joint ventures crowned with technical and commercial success, sometimes surprising the American consumers who at the time still doubted the reliability of Asian products.

The adventure continued, to such an extent that in the late Nineties Miroku dispatched more than 200,000 Browning firearms to the United States. Today Miroku is the main manufacturer of civil firearms in Japan, and the Group Herstal its main shareholder.

Bordered by the green mountains in the north and the Pacific Ocean in the south, Kochi abounds in natural beauty, and offers charm and tradition to the visitor, which undoubtedly were part of the attraction for John Moses Browning's son.

In a cultural environment that shows little attraction for weaponry products, the Miroku Group, also active in the sector of car equipment and power tools, has become a significant actor in the local economy, by providing quality employment to more than thousand employees, all also dedicated to their Belgian and American partners.

John M. Browning would undeniably be glad to hear that almost a century after his death the cooperation between the company he created and the Herstal Group would be so strong that both companies become one.

He would also be happy to know that the Herstal Group confidently and successfully continued the cooperation his offspring started with the Japanese manufacturer Miroku. The Herstal Group, Browning and Miroku are today a deeply united winning trio.

MIROKU

JAPANESE TECHNOLOGY

General view of Kochi.

Miroku MK60 Universal Grade I cal 12 M

Miroku MK70 Hunter Grade I cal 12 M

Miroku MK70 Sporter Grade III cal 12 M

Miroku MK38 Trap with an adjustable stock cal 12 M

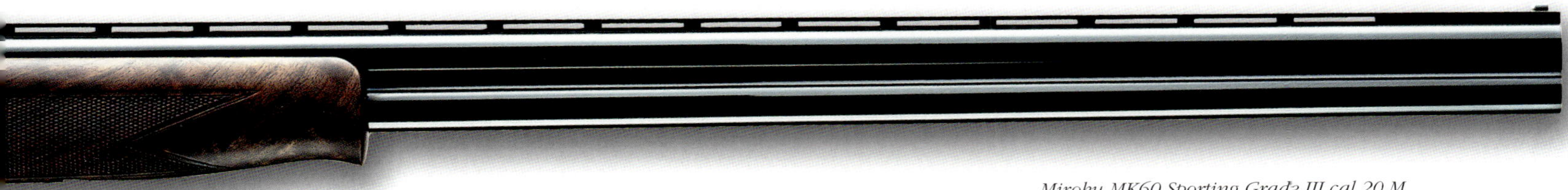

Miroku MK60 Sporting Grade III cal 20 M

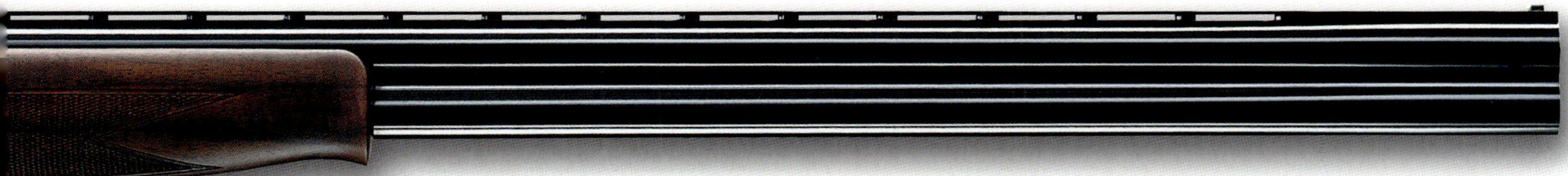

Miroku MK70 Sporter Grade V cal 12 M

BROWNING

Browning B525 Classic Ultimate Grade VI 12 gauge
A multi-use shotgun, integrating a 3" barrel capable of firing 70mm 2 3/4" cartridges and 76mm 3" cartridges. The model presented here has an antique silver action-frame and gold inlaid engraving.

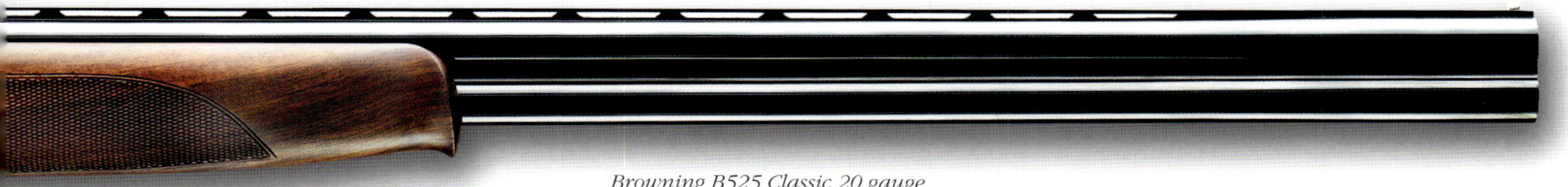

Browning B525 Classic 20 gauge
'Slimmer, lighter and easier handling' - these are the advantages of the 20 gauge. The lighter weight is a considerable advantage during a long day's hunting

Browning B525 LIGHT 12 gauge
Constantly striving to reduce weight while maintaining efficiency, Browning presents the B425 Light as the 'weapon of the 3rd millennium'. It features an 'aircraft' grade light alloy action-frame, reinforced with high resistance steel inserts.

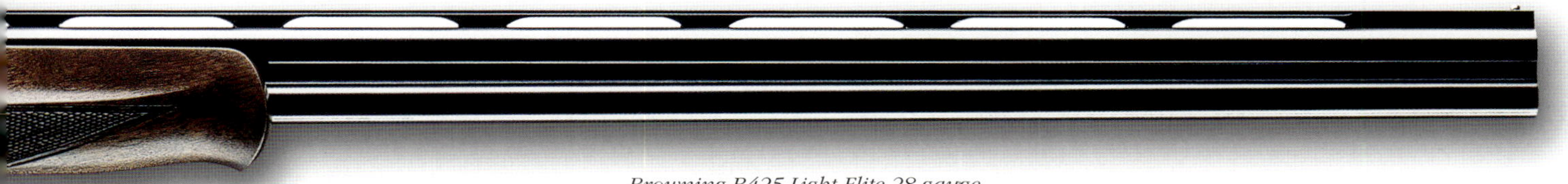

Browning B425 Light Elite 28 gauge
With its old-style caliber, the B425 Light Elite 28 gauge offers the finesse of the line and the appeal of traditional arms.

Browning B525 SPORTER Premium Gold 12 M gauge
Model adapted for Sporting and featuring a black action-frame decorated with gilt birds (2006).

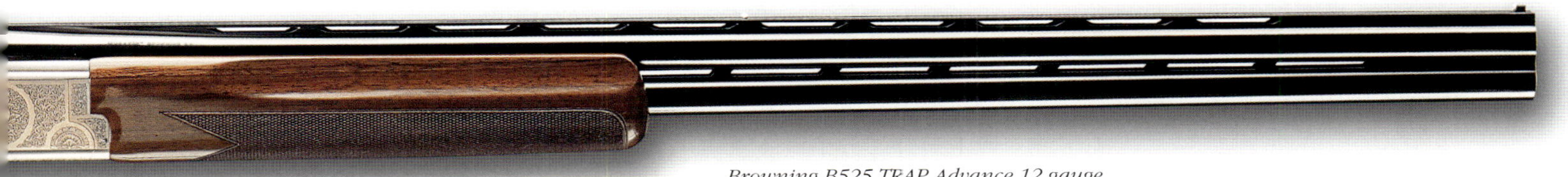

Browning B525 TRAP Advance 12 gauge
An over-and-under shotgun specially adapted for Trap shooting. The Advance version has English engraving and superior quality wood.

BROWNING
BROWNING

Browning Ultra XS Sporter
Over-and-under shotgun specially developed for sporting and equipped with an adjustable stock. Brand new engraving with gold inlay, improved wood grade and 100% titanium chokes made by Briley exclusively for Browning (2007).

Browning Ultra XT
Under the name ULTRA XT (T for Trap), this over-and-under shotgun is intended for all classes of trench shooting. Its adjustable stock enables the user to adjust the sight at all angles of the clays. New engraving with silver inlays and a brand new design. This model has its own special range of 100% titanium chokes manufactured by Briley exclusively for Browning (2007).

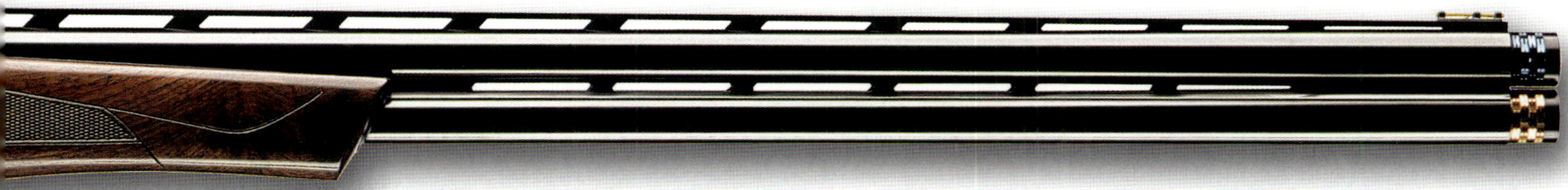

Browning CYNERGY SPORTER INFLEX
The Cynergy is an over-and-under shotgun with an ultra-low profile, giving it a revolutionary look. The 'Inflex' anti-recoil pad offers unmatched recoil absorption.

Browning CYNERGY COMPOSITE BLACK ICE 12M cal.
In order to obtain a significant weight advantage, the CYNERGY COMPOSITE's stock and fore-end are made of composite materials with rubberized inserts. Three adjustable combs delivered with the gun enable the pitch and advantage of the stock to be adapted to the characteristics of the user.

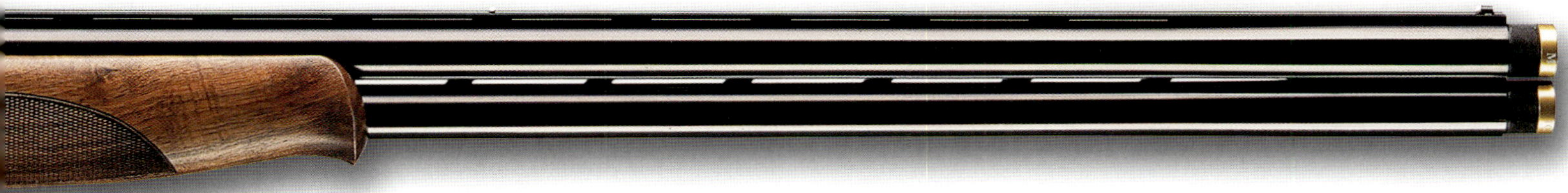

Browning GRAND PRIX SPORTER
The Grand Prix Sporter or GP Sporter is equipped with a cartridge ejection selector, which enables swapping between ejection mode or extraction mode. The action-frame is decorated with engraving by gold and silver deposition.

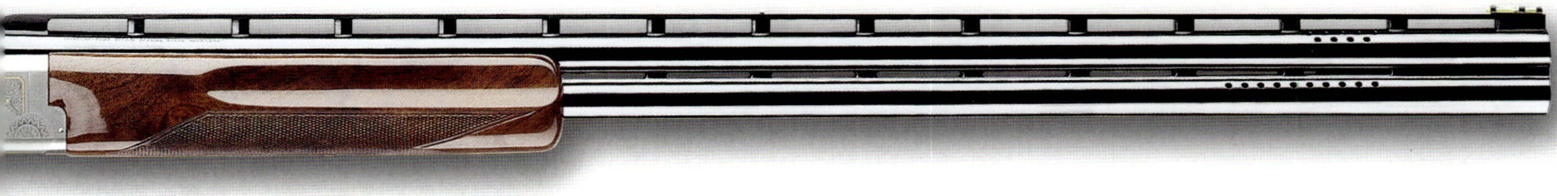

Browning CITORI XT Trap
Over and under trap shotgun specially designed to help the shooter its personal best. Gloss finish Monte Carlo stock. Receiver with gold accented, high relief engraving. Ported barrel with ventilated top and side ribs (2007).

Browning BAR MKII Safari
Autoloading rifle available in 8 different calibers. Steel receiver with blued finish and scroll engraving. Drilled and tapped for scope mount. Detachable box magazine. Select gloss finish walnut stock and rounded forearm. Recoil pad. BOSS® system.

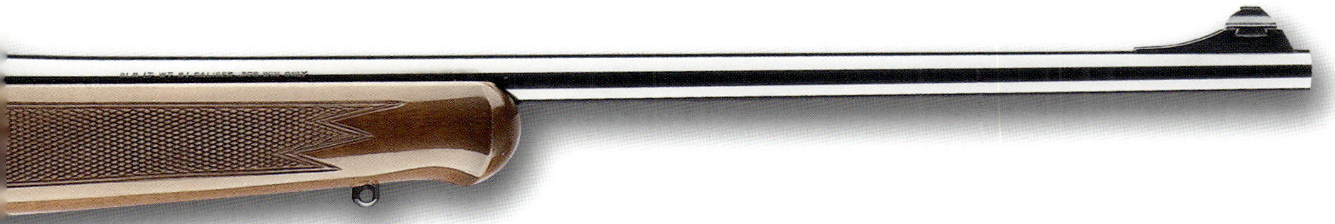

Browning BLR Lightweight
Lever-action rifle with gloss finish walnut stock and "Schnabel" forearm. Receiver in aircraft-grade alloy. Detachable box magazine. Adjustable sights. Drilled and tapped barrel for scope mount. Sling swivel studs installed. Recoil pad.

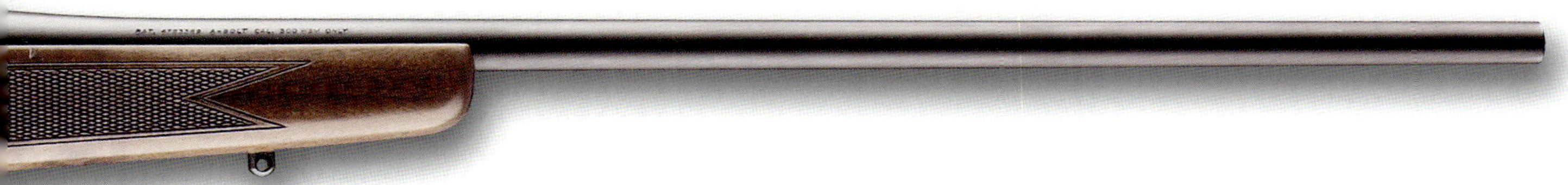

Browning A-Bolt Hunter
Bolt-action rifle with walnut Monte Carlo stock. Free floating barrel. 60° bolt lift. Adjustable trigger. Detachable box magazine. Sling swivel studs installed. Recoil pad.

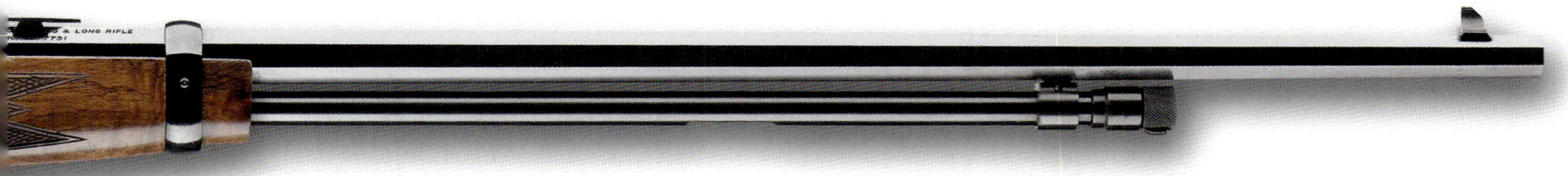

Browning BL-22 Grade II Octagon
Traditional lever-action rifle designed to handle 22 caliber rimfire ammunition (short, long and long rifle) in any combination in the tubular magazine. Grooved receiver for scope mount, scroll engraved. Satin finish American walnut stock. Adjustable sights with front gold bead.

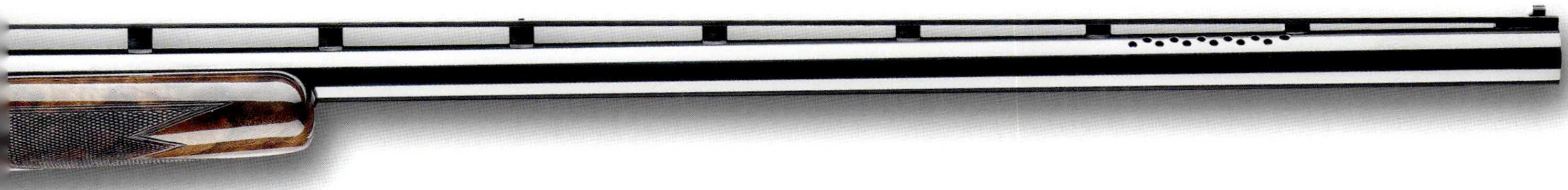

Browning BT 99 Golden Clays with adjustable comb
When it was introduced in 1969, the Browning BT-99 immediately gained a loyal following among both beginner and veteran shooters alike. Above, receiver with gold accented, high relief engraving. Gloss finish walnut stock and beavertail forearm. Ported barrel with high-post ventilated rib. Adjustable straight comb. Adjustable Gracoil recoil reduction system.

ATI - the amazing Buck Mark

Since 1985, the Browning "Buck Mark" range of guns and rifles has exclusively been manufactured by ATI, in Salt Lake City, at a few cable's lengths of the head offices of Browning Morgan.
The family company Arms Technology Inc., has started this cooperation with the production of the Challenger II, previously manufactured in Europe, which was a development of the design of the Colt Woodsman designed by John Moses Browning around 1915.

Today ATI produces more than 25 models of the semi-automatic pistol for target shooting. Renowned for their legendary reliability, these Buck Mark pistols with aluminium frame offer a stylish design and great shooting precision. The range includes models with polymer, walnut or rosewood butts, with barrels in carbon steel, stainless steel or alloy-coated. The Buck Mark rifles are based on the same design as the pistols and use the same chargers.

ATI is the ideal size with its 40 employees: large enough to use the most modern industrial technologies, and small enough to apply the flexibility rules of small and medium enterprises. It thus ideally fits in the Group Herstal producers unit.

Buck Mark Plus Stainless UDX
Receiver : alloy, matte blued finish.
Barrel : stainless slabside, matte finish (5½").
Black laminated Ultragrip DX™ ambidextrous grips.
Adjustable Pro-Target™ rear sight and
Truglo®/Marble's® fiber-optic front sight.

Buck Mark Plus Rosewood UDX
Receiver : alloy, matte blued finish.
Matte finish barrel (5½").
Rosewood Ultragrip DX™ ambidextrous grips.
Adjustable Pro-Target™ rear sight and
Truglo®/Marble's® fiber-optic front sight.

Buck Mark Target Gray Laminate Rifle
Receiver : alloy, matte blued finish.
Barrel : heavy bull (18"), matte blued finish.
Detachable box magazine. Stock : laminated gray
satin finish hardwood. Integral scope rail.

Buck Mark Bullseye Target Stainless
Receiver : alloy, matte blued finish.
Barrel : stainless heavy round fluted bull (7¼"), matte finish. Laminated rosewood grips.
Adjustable Pro-Target™ sights.

Buck Mark Bullseye URX
Receiver : alloy, matte blued finish.
Barrel : heavy round fluted bull (7¼"), matte blued finish. Ultragrip RX® ambidextrous grips.
Adjustable Pro-Target™ sights.

Buck Mark Camper Stainless
Receiver : alloy, matte blued finish.
Barrel : stainless tapered bull (5½"), matte finish.
Composite black grips.
Adjustable Pro-Target™ sights.

Buck Mark Camper
Receiver : alloy, matte blued finish.
Barrel : tapered bull (5½"), matte blue finish.
Grips : black composite. Adjustable Pro-Target™ sights.

Buck Mark Hunter Nickelé
Receiver : alloy, matte blued finish.
Barrel : heavy tapered round bull (7¼"), matte blue finish. Grips : cocobolo target. Adjustable Pro-Target™ rear sight and Truglo®/Marble's® fiber-optic front sight.

Buck Mark 5,5 Field
Receiver : alloy, matte blued finish.
Barrel : heavy round bull (5½"), matte blue finish.
Grips : cocobolo target. Adjustable Pro-Target™ sights.

Buck Mark Target Rifle
Receiver : alloy, matte blued finish.
Barrel : heavy bull (18"), matte blued finish.
Detachable box magazine. Stock : oil finish walnut.
Integral scope rail.

The failure of New Haven

Few events have had the same emotional impact in the history of the Herstal Group and on firearms enthusiasts as the closure of the Winchester factory in New Haven, Connecticut, as this factory was so intimately moulded in the industrial landscape of the Connecticut valley since 1860.

Indeed, by acquiring, in 1987, the production license of the Winchester brand held by the Olin Group, FN had inherited this factory where the legendary firearms used in the Conquest of the West had always been manufactured.
By taking over the Winchester brand, FN intended to complete the product strategy of Browning by adding another American reference brand, with the aim of dominating the market, by re-uniting two symbolic names with a common past.
In 1885, John Moses Browning sold one of his first licenses to Oliver Fisher Winchester for the manufacture of a one shot rifle, the Winchester 1885. Models 1886, 1892, 1894 and 1895, the rifles 1887 and 1897 as well as another dozen products would follow. We know the rest : Browning started working with FN in 1902, whereas Winchester continued its solitary road until the take-over by Olin in 1931, who transferred it 50 years later to the employees of the factory, and thus formed USRAC, US Repeating Arms Company.

In January 1991, the Herstal Group became the sole shareholder of USRAC. The factory was in an advanced state of disrepair, and the decision was quickly made to build a new one, opposite the old one, which was opened in October 1994 by the town of New Haven, the management of Browning and the Herstal Group, and of Giat Industries, then main shareholder of the Belgian Group.

One of the products launching this new era was a series commemorating the 100th anniversary of the lever-action rifle dear to John Wayne, Sitting Bull and Buffalo Bill : the 1894 model, the longest production in the armament history, followed – among others – by the 70 Model, nicknamed the "Rifleman's Rifle", and the pump rifle Model 1300.

As the investment which proved to be far too large for the absorption capacities of the market never became profitable, the decision was made in 2006 to stop the hemorrhage and the New Haven factory was closed. The updated Model 70 is today produced at FNM, in South Carolina, where it benefits from the military expertise of the Group, and enables FNM to slowly become the American pole for hunting and sport instead of USRAC, giving a new lease of life to the brand with the horseman.

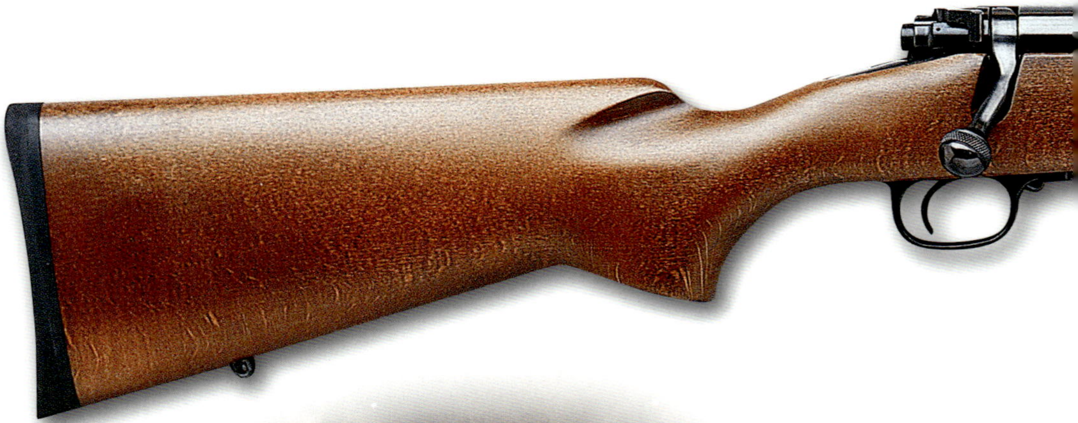

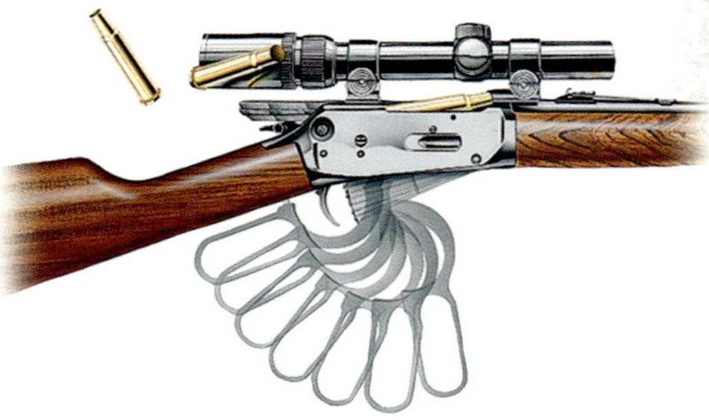

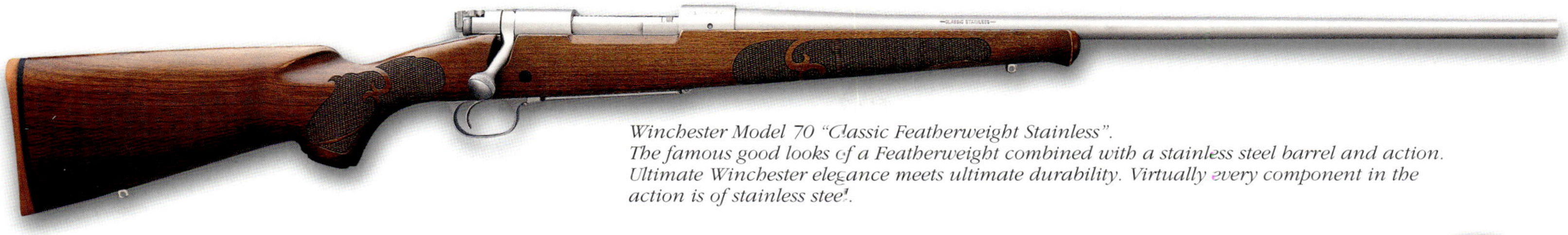

*Winchester Model 70 "Classic Featherweight Stainless".
The famous good looks of a Featherweight combined with a stainless steel barrel and action. Ultimate Winchester elegance meets ultimate durability. Virtually every component in the action is of stainless steel.*

Winchester Model 70 "Ranger™". The Ranger is a very efficient rifle for big game hunting. The satin-finished hardwood stock is made for hard work. It is fitted with adjustable sights and drilled and tapped for a scope.

Winchester 1895 Grade 1. A modern version of one of the greatest historical rifles of all time. The Model 1895 features a scalloped receiver, two-piece lever, Schnabel forearm and straight grip stock. The stock and forearm have an elegant satin finish. Chambered for .270 Winchester.

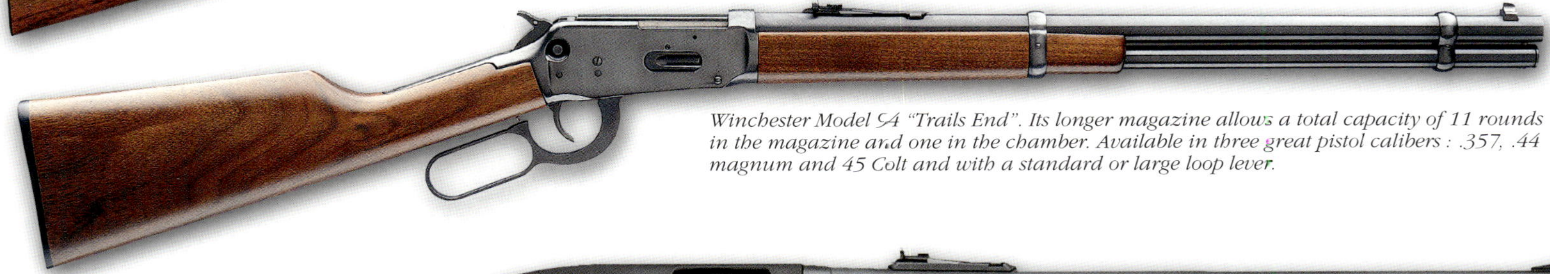

Winchester Model 94 "Trails End". Its longer magazine allows a total capacity of 11 rounds in the magazine and one in the chamber. Available in three great pistol calibers : .357, .44 magnum and 45 Colt and with a standard or large loop lever.

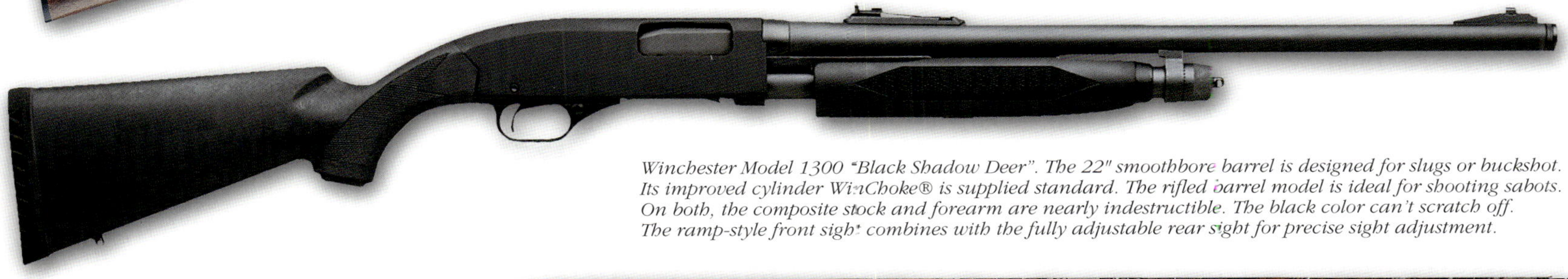

Winchester Model 1300 "Black Shadow Deer". The 22" smoothbore barrel is designed for slugs or buckshot. Its improved cylinder WinChoke® is supplied standard. The rifled barrel model is ideal for shooting sabots. On both, the composite stock and forearm are nearly indestructible. The black color can't scratch off. The ramp-style front sight combines with the fully adjustable rear sight for precise sight adjustment.

Winchester Model 1300 "Turkey Camo". The Advantage® camo completely conceals the stock, forearm, receiver and barrel. It's extremely durable on both the metal surfaces and on the synthetic stock. This model has extra versatility with its WinChoke™ choke system and 22" vent-ribbed barrel.

New Winchester show-room at FNM, Columbia, SC.

SX3
SXR VULCAN
Select Energy

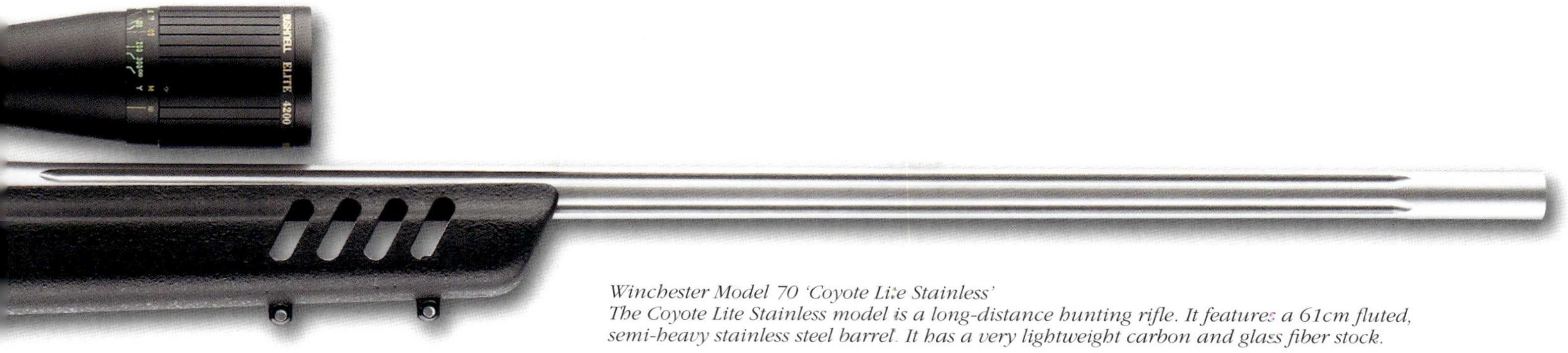

Winchester Model 70 'Coyote Lite Stainless'
The Coyote Lite Stainless model is a long-distance hunting rifle. It features a 61cm fluted, semi-heavy stainless steel barrel. It has a very lightweight carbon and glass fiber stock.

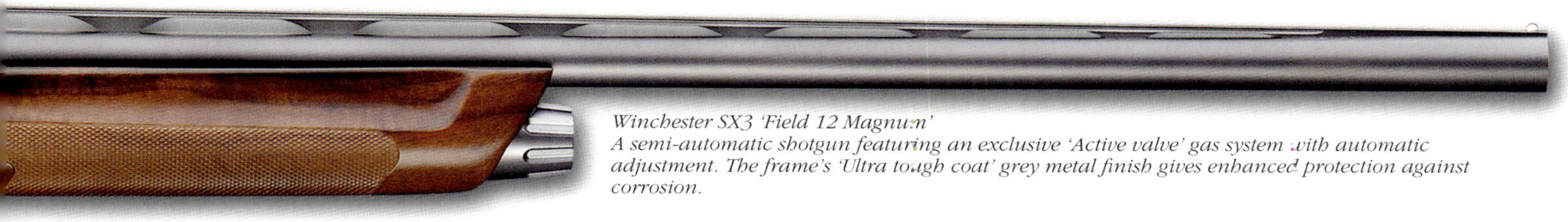

Winchester SX3 'Field 12 Magnum'
A semi-automatic shotgun featuring an exclusive 'Active valve' gas system with automatic adjustment. The frame's 'Ultra tough coat' grey metal finish gives enhanced protection against corrosion.

Winchester SXR VULCAN
A gas-operated semi-automatic rifle.
Available in calibers .270 WSM, .30-06 Sprg. and .300 Win. Mag.

Winchester SELECT ENGLISH FIELD
An over-under shotgun decorated with fine English style laser engraving.

Winchester SELECT ENERGY SPORTING SIGNATURE
Over-and-under shotgun for sporting, with an adjustable stock.

The Non-Gun partners

Browning has now been marketing "non-gun" products with the same quality standards of the brand. There are accessories (auditive and ocular protections, covers for firearms, cleaning products, luggage, ...), hunting knives, torches, a full range of technical clothing for all types of hunting and all climates, safes for firearms (in cooperation with ProSteel, Utah).
Products always credited with the label "The Best There Is."

With the aim of expanding this sector, and convinced that you can only excel in your core profession, Browning started in 1998 a programme of commercial licensing of the brand in the sectors of hunting clothes, shoes, archery, angling and camping, even offering the brand licence for limited editions of cross-country vehicles.

Since many years and in collaboration with Prosteel, Browning has been distributing a wide variety of safes. Opposite, the graphic design of the FN-Browning series safe pays homage to more than 100 years of collaboration between FN and Browning.

For North American outdoor activity enthusiasts, Browning offers a wide rande of camping material.

Windproof "Highland Wool" vest.

4-in-1 Parka in "Hydro-fleece®" fabric, quiet, waterproof, windproof and breathable.

Range Kit
with shooting glasses , softfoam earplugs and adjustable-fit earmuffs.

High-Grade Canvas/Pull-Up Leather case.
Solid woodframe canvas covered with heavy high-grade leather on borders. Buckmark logo stamped on leather patches. Two solid brass combination locks.

One of the Browning's Master Knifemaker Series : the Joseph Kesslar (second model). Joe is known for his integral guard "Brute de Forge" knife and reproductions of the classic Bowie styles.

Competition knife (10" blade length), produced for Browning by James Crowell and Reggie Barker, the only two men to have won an "American Bladesmith Society" (ABS) World Cutting Championship. If there ever was a first-class, unique knife to collect or use, the Browning Crowell/Barker Competition is that knife.

Browning, in association with "American Forest", offers a most unique collection of six limited edition Living Hystory knives. Each knife celebrates the great achievements of historic Americans. Represented here : Robert E. Lee, with a D-guard Bowie knife (blade length 8 1/4"), Crazy Horse, Chief of the Sioux Nation (blade length 4 1/2") and Dwight D. Eisenhower, Commander in Chief of the Allied Forces in Normandy, a knife patterned after the M3 knife, issued to many GI's who fought on D-day.

"Black Ice" 12 LED lantern. Unique dimmer switch allows light adjustment.

0,5 Watt LED lanterns. Removable top allows hanging upside down from included carabiner.

Night seeker cap light.
Clip-on design
Two white and one red LED.

Headlamp.

Black Ice combination lights : 3 Watt bulb for distance and 4 colored LEDs for close-task lighting.

Black Ice Xenon light
Designed with Xenon bulbs for maximum long distance performance (up to 120 yards for this 9V model).

"Tactical Hunter" series.
Xenon or LED ultrapowerfull ligths.

Top : Browning USA, Morgan. Center : Browning International, Herstal.

After the sale, the service...

Since the era when John Moses Browning repaired the firearms of his hunters and trappers customers himself in his workshop in Ogden, customer service is part of the tradition, both for Browning and Winchester. Today the after sales service is as simple as a free phone call or electronic mail.

Located in the geographic centre of the United States, the distribution center of Browning and Winchester (Arnold Distribution and Service Center) has been established since 1968 in St. Louis, Missouri, where it employs 86 persons, for administration, but also a team of gunsmiths and engravers in charge of maintenance and repair of firearms for independent gunsmiths and those belonging to the 800 Browning-Winchester partners-distributors in North America. It is also the worldwide quality control centre for the Browning and Winchester factories.

Located at the heart of Europe, a few kilometers away from the headquarters of the Herstal Group, the distribution centre of Browning International applies the same logics. It employs 100 persons, and is directly linked to the 900 Browning dealer partners which cover Europe and the rest of the world.

BROWNING®

Testing the products (Browning US, Morgan).

Machine and prototype shop (Browning US)

Repair and maintenance shop (Browning International, Herstal).

Distribution center (Browning International, Herstal).

BROWNING

The "Browning Custom Shop"

Throughout history, illustrious figures have sought to bear prestigious arms which were superior to all others by dint of their functional qualities, but also the opulence of their ornamentation.
Dazzling helmets, shields and armor embellished with gold and silver, or swords inlaid with precious stones are the epic forerunners of the embossed and damascened shotguns of the old aristocratic arms cabinets or of the jealously guarded display cases of today's avid enthusiasts. However, maintaining this alliance of the beautiful and the functional and perpetuating this tradition down the centuries has not been an easy matter in today's utilitarian civilization.

Some manufacturers have nevertheless achieved this feat. From 1926, although resolutely dedicated to mechanization in arms manufacturing, FN perceived the threat of a decline in one of the Liège region's most authentic artisanal traditions : arms engraving. This is why, under the direction of Maître Félix Funken, FN created a specialist workshop, capable of executing all the ornamental variations it intended to apply to the finishing of its civilian shotguns and pistols.
In the 1970s, this temple to engraving employed as many as one hundred and eighty people, all working away daily at their workbenches with hammer, chisel and graver. By taking this step, the Herstal company breathed new life into this ancestral art, which had been exemplified since the sixteenth century by the likes of de Bry, Valdor, Varin, Duvivier, Demarteau and dozens of others who followed them. Simultaneously, FN conferred on its hunting and sporting weapons – in particular the recently-created and celebrated Browning "over and under" – a unique aura which remains with it to this day. Technical and market conditions have certainly evolved since that time.
The cost of labor – particularly that of highly qualified craftsmen – has risen considerably, obliging manufacturers to resort increasingly to mechanical processes. Materials and working techniques, too, have sometimes changed.
But the fundamental features of the functional and beautiful fine arm remain immutable across the centuries. How is it possible, from an assembly of the best steels and the best walnut woods, to produce an instrument which is reliable, solid and elegant, decorated to the tastes of a highly varied clientele ? That is the question.

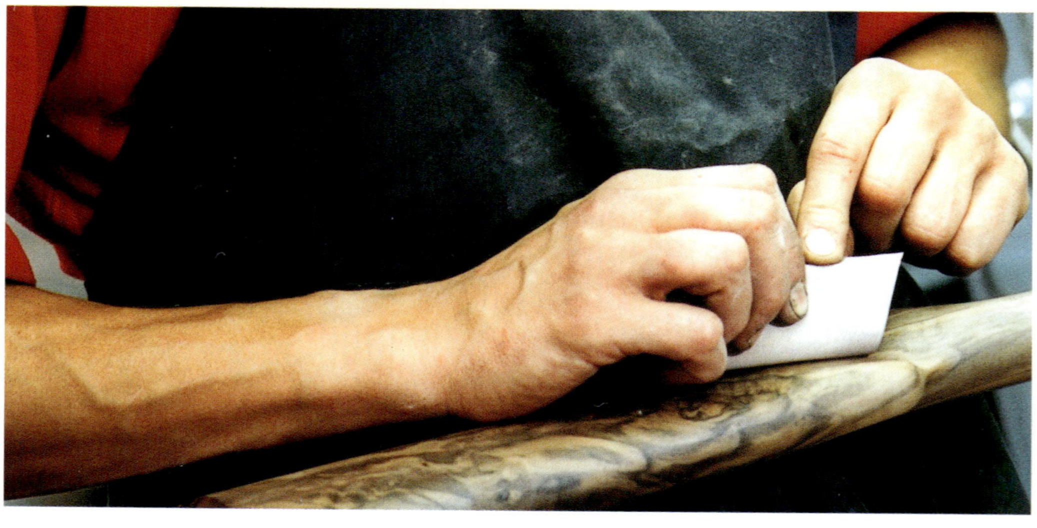

Furthermore, how is it possible to combine the necessary standardization – dictated by standardized mass production conforming to worldwide demand – with the subtleties of a crafted finish adapted to demands for customization?

This is the challenge faced every day in Herstal's fine arms workshop, known as the Custom Shop, where luxury arms are produced "à la carte" for customers, depending on their morphology and particular tastes. The workshop's activity is based above all on an organization of work which has prevailed in this trade for centuries : the various stages in a fine arm's production, assembly, finishing and ornamentation are entrusted to a succession of different specialist craftsmen. Each gunsmith brings to the task his expertise, experience, sensitivity and his own personal touch. This is why, in spite of observing the strictest community standards, each shotgun is actually a unique piece, rendered even more individual by the engraving, chasing and inlaying of its metal parts and by the way its stock is sculpted. Here, we find all the old craft names, suggesting a hint of gun oil, walnut chippings, lamp black and ... the satisfaction of work well done : trimmer, tipper, wood fitter, polisher, checkerer, walker, refiler, engraver, inlayer etc.... Through these craftsmen, customers would see the creation of the object of their pride and joy, from a catalog or on the basis of a special design. Some twenty-five craftsmen produce three hundred and fifty fine arms here every year, in the pure tradition of Liège, a tradition which has been revised and adapted to the latest methods and techniques by BROWNING.
It is the greatest workshop of its kind in the world.

Custom Shop
HERSTAL
-Since 1897-

Browning B25 - "Cheverny" side-plate engraving. Elegant stems of acanthuses leaves entwined on a tapestry of fine Royal English scroll engraving, wrapping the entire action. The style of this Royal English scroll is much finer than the traditional English engravings from which it is inspired.

On each face birds inlaid in fine gold sit slightly proud of the landscape, giving the impression that they are flying away from the gun. Ducks mallards and grey partridges have been selected for the scenes as they represent the most frequent choice of the hunters (Engraved by Dany Matagne. Created by Bernadette Romainville - 2001).

New materials new techniques

In a highly competitive environment, it is not enough to demonstrate creativity, nor merely to offer good products geared to the needs of the market. It is also necessary to control costs and lead times for development, launch, production and commercialization, and finally to generate profit margins which will finance the company's future.

It is here that a judicious combination of internal manufacturing, to retain control of major parts, the appropriate distribution of tasks between the Group's production units and turning to subcontractors or external specialists can prove fruitful.

In addition, the exploitation of new materials and new production methods has become inevitable. Indeed, in the field of small arms, traditional steel working is diminishing except for firing parts and barrels. Composite materials are gaining ground in military and law enforcement weapons and also among new generations of hunters and sporting shooters.

At the same time, global technologies for hot and cold moulding of parts are enabling the production, using injected powders, of high-precision, high-strength parts which are ready for immediate assembly. Their cost is much lower than that of their fabricated counterparts, even with modern machining centres which use traditional steel modelling methods. Many of the Herstal Group's products are therefore increasingly reliant on innovative industrial manufacturing processes.

This evolution in techniques and materials – taken in conjunction with the introduction of optical, electronic and digital systems and new thinking on ammunition (type, actuation, structure, content) to reduce weight and respect the environment – demonstrates the extent to which the title of this book must be taken today in its broadest sense, both in arms manufacturing as well as other traditional areas of precision mechanics.

Such is the extent to which this "Ars Mechanica", this "art of precision mechanics" is coming to incorporate the disciplines of physics and chemistry. In this respect, we are living through a revolutionary period and the company is putting considerable effort into mastering these new skills, without compromising its legendary gun making experience.

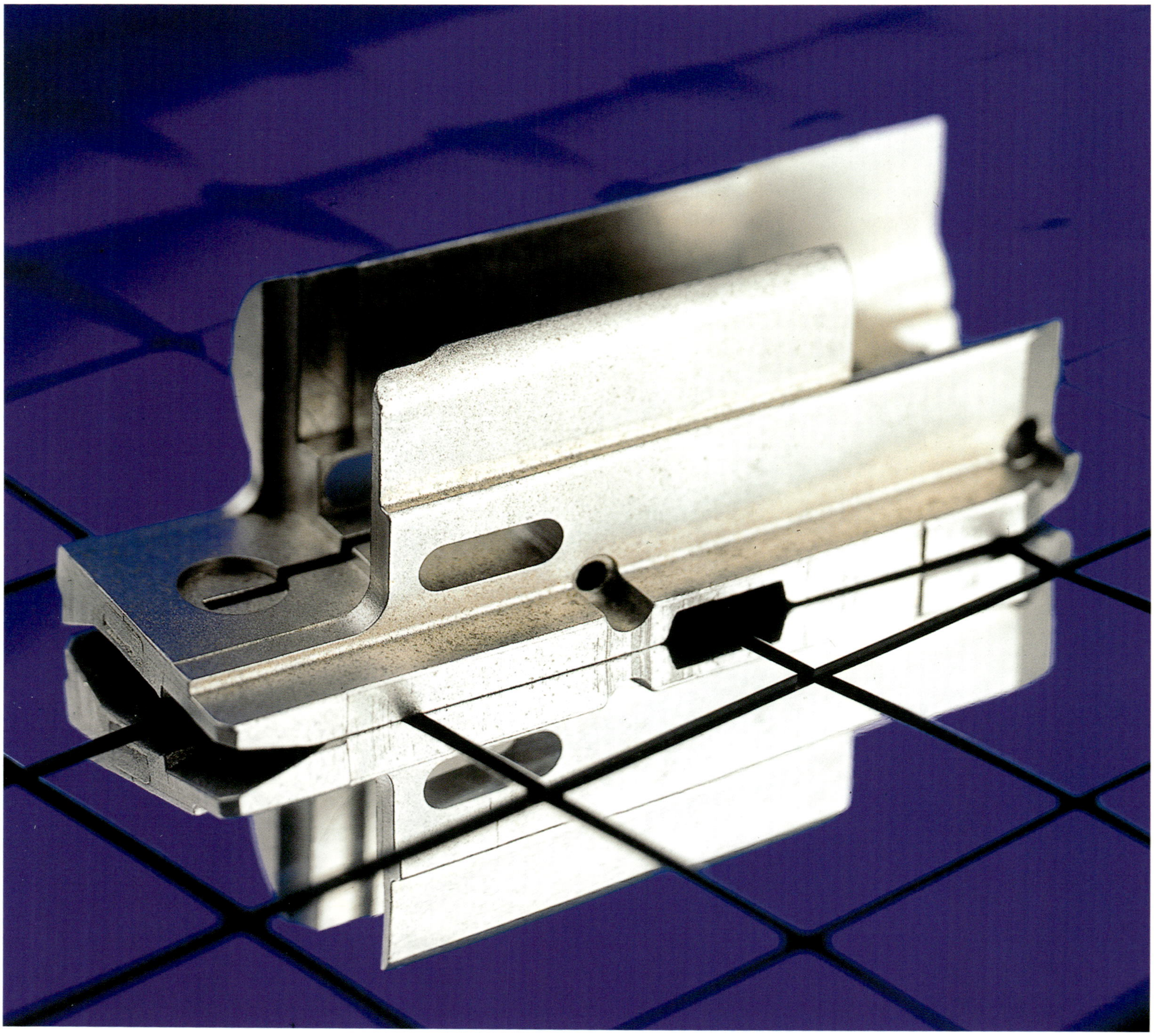

Breech of a shotgun, metallic powder injected using the QUICKSET and XFLOW processes : injection, binder removal, sintering using metallic powders.

Rapid prototyping uses 3D CAD software to solidify either metallic or plastic powder by laser, layer by layer.

Frame of the P90 injected in thermoplastic composite materials and its injection mould.

Using 3D CAD applications at the product development stage has consolidated rapid prototyping and digital simulation and involves the same motions for metal parts, and almost-complete sophisticated drafts for these can be planned in steel, aluminium and magnesium using global technologies such as metallic powder injection and thixomoulding.

Machine which converts the virtual 3D CAD model into the metal or plastic part by fusion, using a laser laying down successive powder layers.
A part being manufactured: the laser point fusing the powder layer clearly appears in red. A new powder layer will then be spread over the part, which the laser will fuse using a 2D motion. Little by little, the prototype part emerges.

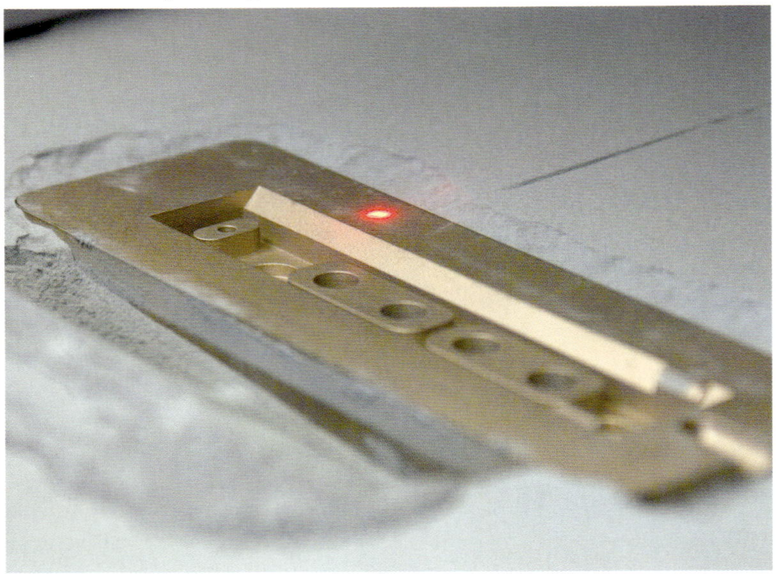

Prototype for the breech of the P90 injected in metallic powder by "Quickset".

Prototype for a hunting shotgun frame, made using the "THIXO" aluminium moulding technique.

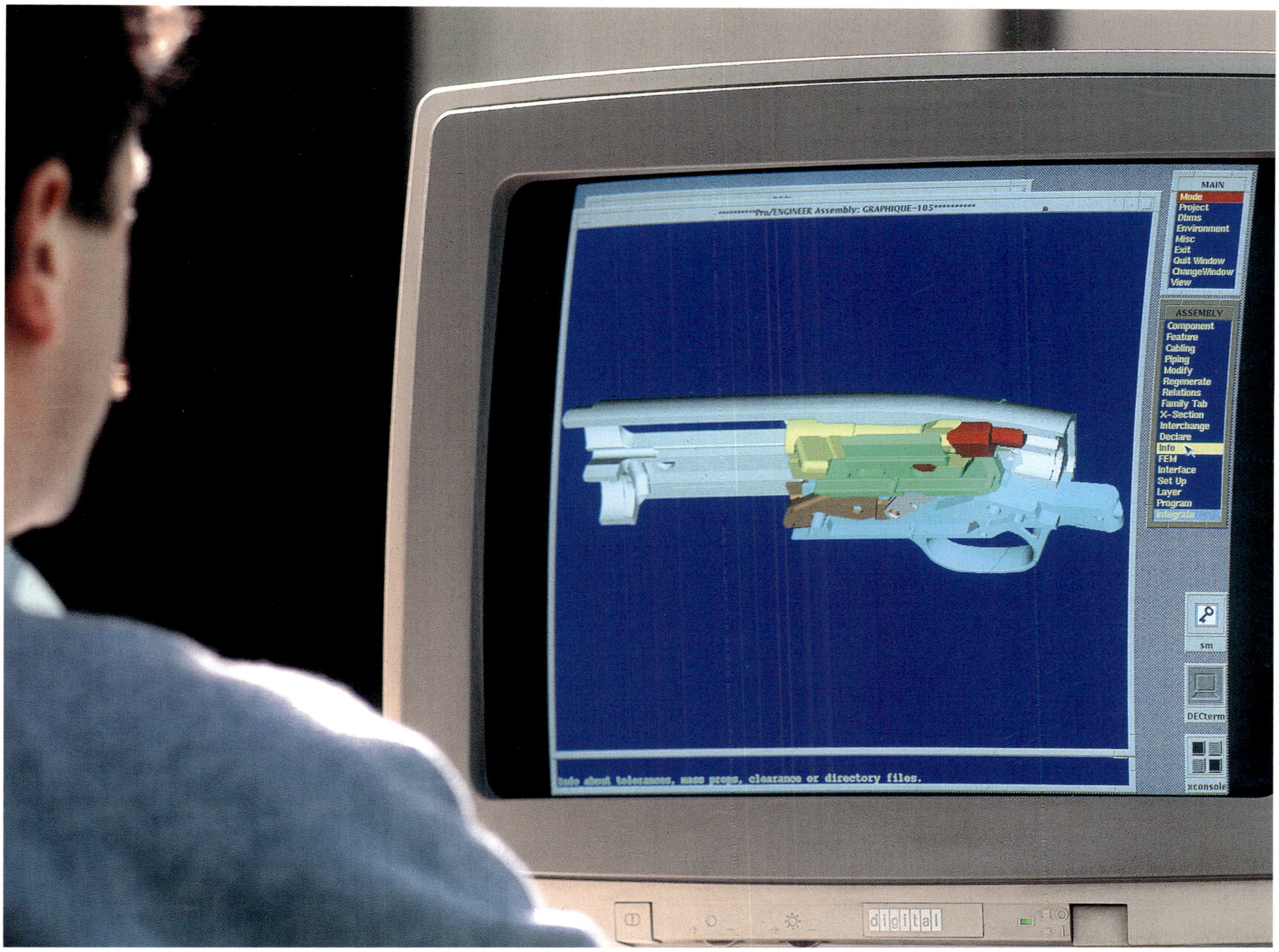

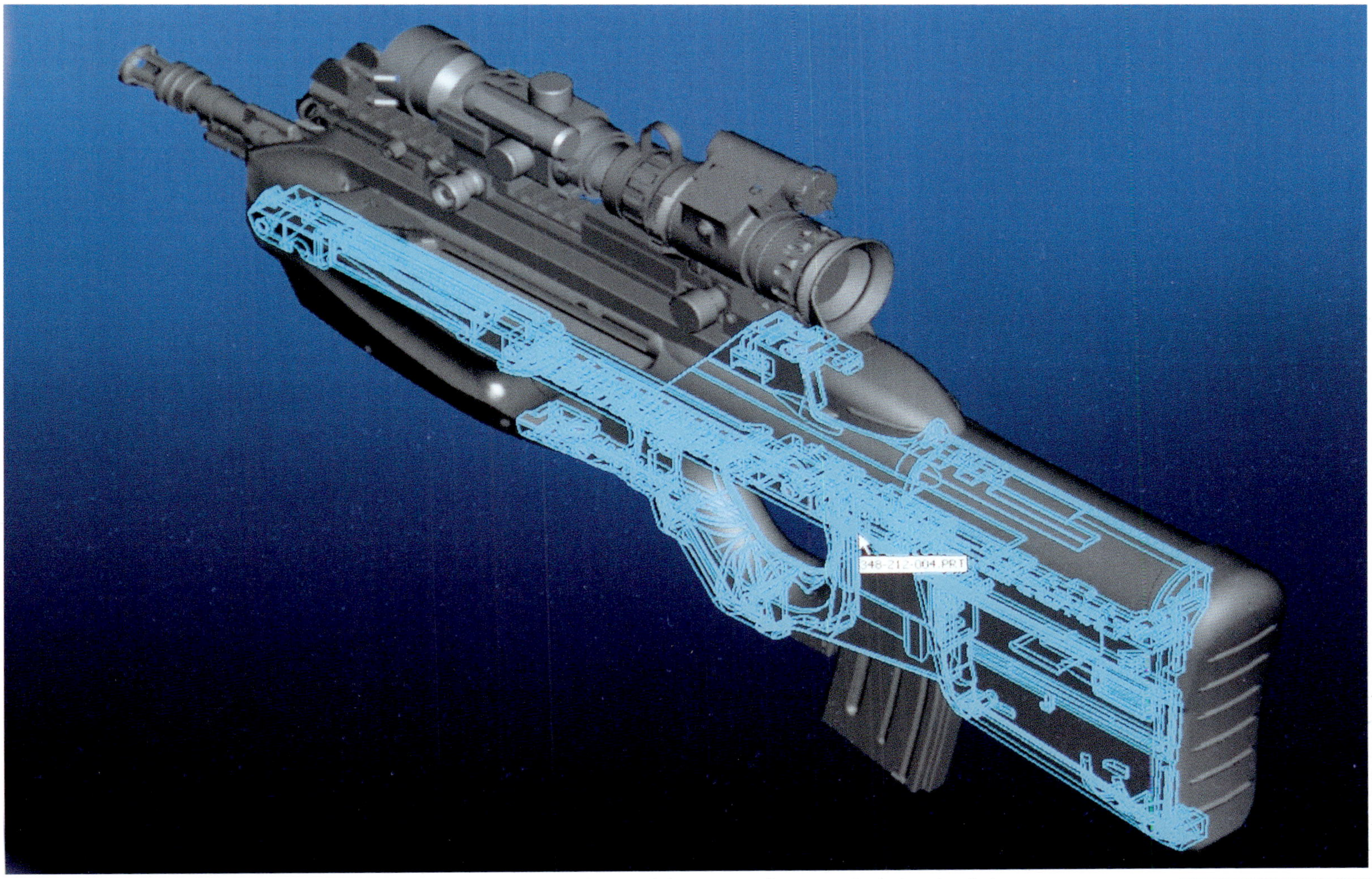

3D CAD software supports the full range of product development, from the initial design to performing calculation simulations, kinematics or the simulation of plastics injection.

Health and long life to Ars Mechanica !

Refocusing on its core business, embarking on a global strategy, renewing its product range, consolidating financial, commercial and corporate structures, modernizing its production assets across three continents, a clear vision of its raison d'être, vital for its future... all are transformations effected by the Herstal Group in the course of the 2000s, at the dawn of a new century and after a proud history of 120 years' standing.

120 years which have seen successes and failures, but always effort, perseverance, faith in the future, a steely determination to succeed and on occasion also creative genius on the part of the tens of thousands of colleagues who, over the years, have built the reputation of the Herstal Group throughout the world.

The upturn in the Group's fortunes seen in recent years would not have been possible without the unwavering support and trust of its shareholder, the Walloon Region of Belgium.

The Walloon Region has supported the company for ten years and is not standing in the way of building supportive partnerships in the future. Indeed, regardless of its potential, the financing of development work obliges the company to make continual investments which weigh heavily on its capital. This is the inevitable consequence of its gradually reclaimed role as world leader in civilian and military light arms.

The future of the company, and that of its 3,000 current colleagues, would also have been compromised if industrial relations within the company had not been marked, in spite of some difficulties, by responsibility and consensus around the company's overall strategy, thus making possible ten years of in-depth work to gradually and single-mindedly regain its fundamental equilibrium.

At this point of the book, it is now time to take stock and draw the chapter to a close. We trust that our successors will work towards the continuation of this "Ars Mechanica", which the Herstal Group embodies in Belgium and throughout the world.
It is for them to write the next chapters.

Until such time as they take up the pen that we now lay down, we have to say in temporary conclusion that the story we have just related seems to us to be an "extraordinary adventure"...

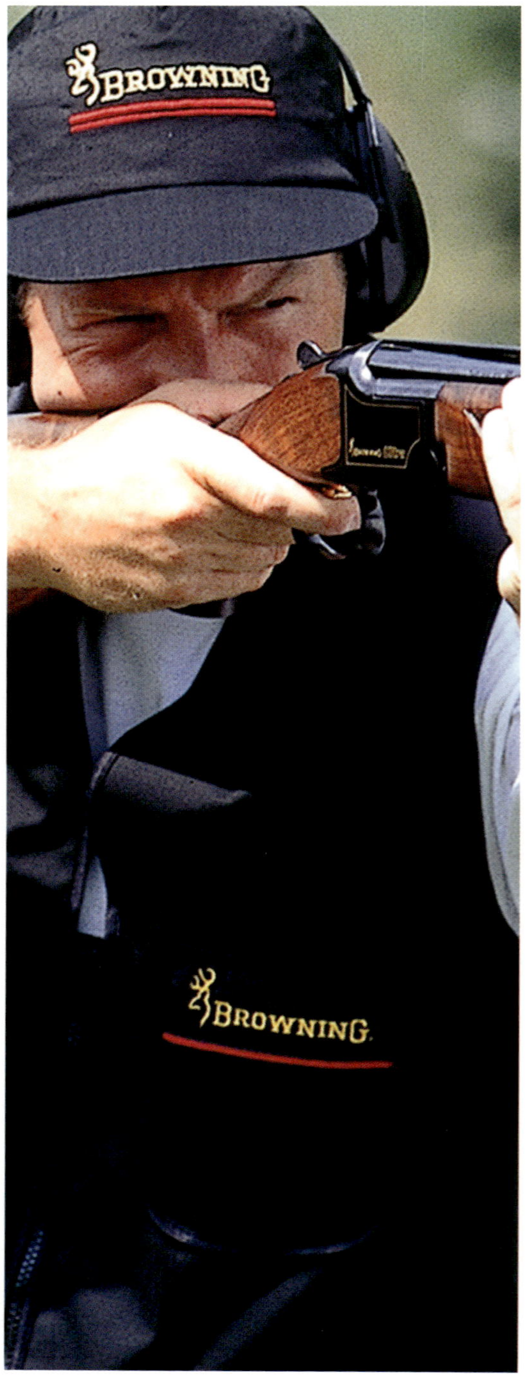

Aerial view of the Herstal Group at Herstal (2007).

An extraordinary adventure

How was it possible for a Liège company, founded to fill an order from its national army, to become over the years the world leader in small arms and one of the great names in hunting weapons ?

Unfavorable factors

At first sight, there seemed to be a number of unfavorable factors which could have prevented this destiny.
At the beginning of the 1880s, in spite of the expertise of the craftsmen of Liège, Belgium did not have the weapons it needed to arm its soldiers. And so it was necessary to acquire the manufacturing license for the Mauser rifle, leading to close cooperation with Germany, a cooperation which was to be annihilated with the onset of World War I.
In addition to this, the Belgian people were not a militarist people and could even be pacifist at times; there existed a very limited domestic market for both military and hunting weapons. Moreover, the Belgian Minister of Defense is only empowered, in terms of armaments, to equip the national forces, while industrial and research policy is devolved to the Regions which, although encouraging it, do not specifically prioritize the armaments industry.
Finally, the early Liège craftsmanship in gun making had gradually waned, because in contrast to other armament manufacturing centers abroad, Liège had been unable to maintain an environment in which the major manufacturers could call on a wealth of specialist sub-contractors. In Liège today, with the exception of Lebeau-Courally, Browning and FN Herstal are the only players in this sector.

Positive circumstances

On the other hand, a number of advantages were to favor the rise of this activity.
First of all, the centuries-old Liège tradition which forged the almost in-born art of gun making, and the institutions born of this activity such as the Liège Proof house and the Browning Custom Shop, that guardian of traditional art in Liège weapons.
The attachment of the people of Liège and of Wallonia to their "Fabrique Nationale", to the point in 1997 of "carrying" and supporting it rather than abandoning it to a foreign competitor.
The quality of those industrialists who succeeded in building up a factory around a local trade, the quality of their successors who transformed it into a globally reputed industrial Group, or that of its researchers who designed the principal military calibers of modern times following a veritable revolution in ballistics design and development. Besides, the names of John Moses Browning, Dieudonné Saive or Ernest Vervier have gone down not just in the company's history, but in that of armaments worldwide.
Finally, the loyalty of its personnel, from all nationalities and cultures, who subscribe to a highly specific "company culture", the originality, the importance and the exceptional – sometimes fragile – character of success demanding such a level of commitment and responsibility within the Group and the world at large, when the stakes impact on the security and defense of States, property and persons.

Two pillars of activity

On tracking the company's history, the fundamental principles of the company, requisites for its success, appear to be constants.
The military and civilian split in its activity has been a feature since the company's foundation, as part of a drive for expansion and stability. FN was not mistaken: the markets for both hunting and military weapons were entered thanks in particular to the close links forged with John Moses Browning, and they soon produced outstanding successes.

This joint approach has meant that the military factory in Liège (whose order book is by its very nature fluctuating, being linked to budgets for military spending) has been complemented by military and civilian components manufacture, thereby covering its fixed costs in the most efficient way. This was an essential advantage which explains the Group's attachment to unity. In this way, the two distinct activities have over time benefited from a complementary and interdependent approach.

A global approach

The Herstal Group could not have developed without a global approach. At first, this was naturally imposed by the markets.
Then, demands in terms of competitiveness prompted the industrial relocations of the 1970s : to Portugal, Japan and Brazil among others.
Worldwide distribution is today even wider, each of the Group's units playing its role in the success of the whole group, with all the advantages that the best skills and equipment can bring.
The Group's key skills base in terms of Research and Development and Industrialization is naturally located at Herstal, as is the Group's Senior Management. But special skills and the most adapted equipment, for performance and quality for example, are spread across the units according to their specific requirements.
In the 1890s, the Fabrique Nationale, depending as it did on an exclusive Mauser license, had found itself at one point in German hands, in the absence of an alternative. The Herstal Group learned from this experience and from that time on made it a principle to have ownership of its products, or at least to have a watertight agreement whenever it entered into a partnership.
The Group's profitability is at its optimum when it is a team player, and when the various cultures concerned respect and value one another and where they make up a coherent whole: this is today's reality, after several years of maturing.
In the hunting and sporting sector, this global approach is a real strength. The North Americans account for 70% of the market, the Europeans 20% and the rest of the world for the remainder. The 3,700-strong "Browning Partners" network is a significant advantage. Products are the subject of a global approach, while the requirements of individual markets are taken into account.
In the military market, equally distributed between Europe, the United States and the rest of the world, FN Herstal is able to prospect throughout the world thanks to its sales network and under the control of the Walloon Government. The United States is subject to a special system linked to the "Buy American Act", which led the Herstal Group to establish a manufacturing plant in South Carolina in order to supply the United States armies.

In the absence of a sizeable domestic market, given the size of Belgium, the whole world has gradually become the Herstal Group's stage !

Technological leadership

The Herstal Group deliberately seeks to remain at the operational and technological forefront in its markets. Indeed, because of the competitive environment in which it operates and its production costs, the Group is condemned to innovate... or disappear.
In the military sphere, technological superiority and reliability are essential factors for success.
In the Police and Law Enforcement sector, in order to reconcile order, security and respect for the public, weapons need to meet a variety of - sometimes contradictory – constraints and have to be perfectly adapted for the mission in question.
For hunting and sporting, weapons have to satisfy the increasingly demanding criteria of precision, reliability, design and cost.

FN Herstal, Browning and Winchester therefore have to anticipate developments on a continual basis: the number of projects in progress at the moment is an illustration of this.
This is not a risk-free strategy, given the size of investments made and the often lengthy market response periods.

In defense, it is thanks to these tenacious efforts that FN Herstal can today boast four top sellers: pistols, machine-pistols, rifles and machine guns, to which should be added light gun turrets, aeronautical and terrestrial remotely-operated systems, products for police forces and "reduced lethality" weapons. The slogan "tout notre métier, seulement notre métier !" (Our trade ! Only our trade !) is applied here in its full sense.

In hunting and sporting shooting, the outstanding successes of the auto-5 or the B25 over-and-under, still in use, are followed today by the successes of the semi-automatic BAR, the Cynergy over-and-under and a number of other products under development.
Thanks to the continuity of this policy, Browning has always been able to offer a catalog of regularly renewed, first-class products.
This is the basis of its success : "The best there is" is rooted in everyone's mind.
For its part, Winchester has invested heavily in its new over-and-unders or semi-automatics, perhaps at the expense of the traditional classics. It is currently putting considerable effort into reinvesting in these.
One last and key advantage is the fact that the Herstal Group intervenes directly in its markets. It never acts in the role of sub-contractor. This does however demand determination in its strategies and in the choice of its objectives.

To diversify or not to diversify ?

So throughout this long story, we can discern some constants in the strategies which were followed. These produced prodigious successes, and all with the same degree of tenacity!
The permanent industrial pressure stemmed from the uncertain character of the work arising from military orders and the temptation to compensate by diversification.
The initial diversification, a total success, was into hunting and sporting weapons. This is one of the company's foundations and it currently accounts for 40% of its activity. The partnership with John Moses Browning in 1897, then the acquisition of the brand and its commercial tool in 1977 have been a total success... and perfectly logical.
Not all the diversifications outside the Group's core trade have met with the same success: bicycles, motorcycles, cars, lorries … have all seen FN shine in these fields and the creation of some stunning products, but these activities, for lack of a sufficiently large domestic market, have been unable to survive world recessions and their consequences for market openings. In spite of the financial support of the defense activity, these products gradually disappeared one after the other.

In the same spirit, FN's back catalog is varied but surprising: typewriters, freezers, milking machines, looms, plastics recycling, escalators, aerial lifts, robotics etc. all economic failures of a diversification policy which was largely motivated by the concern to maintain employment… and sometimes the desire of the shareholder to change the company's direction.

This was followed by a great ambition to invest in the sports sector, exploiting the Browning brand and its distribution networks: fishing activities, now discontinued, are the best example of this, but also golf clubs, tennis rackets, surf boards etc.
By expanding its activities, the company hoped to reach a sufficiently high level to become one of the greats in the field of sport. The subsequent period demonstrated that it did not have the resources to achieve this, and the company then refocused on its core businesses, the Group's real fields of activity.

A major strategic orientation has been significant investment in the aerospace sector, through its engines.
FN is probably the only company in the world to have contributed to providing its country, Belgium, with high-level capacity in the field of aerospace propulsion. The authors of this strategy were visionaries, but resources were unable to match the size of the developments and FN Moteurs was eventually sold to the SNECMA Group, to become the magnificent Techspace Aero.

A company in radical transition

Throughout this book, we have touched on the fundamental pillars of our activity: Hunting-shooting and Defense, the Herstal Group's "temple columns".
This now profound transition is due to the fact that these pillars have been underpinned by two additional domains: onboard systems and law enforcement. The Defense activity is structured between the traditional activity of infantry weapons, on the one hand, and on the other, the new activity of onboard systems - light gun turrets for helicopters and armored vehicles, among the many weapons systems with their environments.

The important field of Law Enforcement has emerged, mid-way between Defense and Hunting, which is generating rapidly growing sales. The range has its roots in both civilian and military products, thereby providing a stimulus right across the Group's activity.

This evolution increases the complexity of the company, which has to rise to many challenges, but at the same time it provides the guarantee of risks better spread, the factories' order books being more secure and less fluctuating.

Add to this the transition in trade skills - first fundamentally arms manufacturing, materials specialists, then broadening into the specialisms of mechanical engineering, electronics and optoelectronics, systems architecture etc. – the transition currently in process is a profound one, sometimes stressful for the company and its personnel.
These ten years of hard work, product renewals and successful forays into new fields have resulted in the strong growth the Her-

'Vulcan's furnaces', Roman god of fire and volcanoes, forger of arms in ancient mythology. Canvas realized by Emile Berchmans (1867-1947),

stal Group's is enjoying today : a significant increase in factory production, recruitment of new staff. This is a positive sign which has to be accepted and built upon.

Also, this evolution is part of the growing strategic role of light arms in conflict situations. We have seen in the former Yugoslavia, Afghanistan and Iraq that conflicts now involve heavy weapons during the initial period of combat only. Then – rapidly and on an ongoing basis – comes the role of the infantry who must hold ground with light equipment and adapted weapons.

There are few companies in the world able meet this need : FN, global leader thanks to its Belgian and American establishments, does just that. This reinforces the company's strategic character and makes its decisions even more critical.

The Herstal Group's strategy has always been to be a global organization centred on Belgium, while retaining a presence in its principal markets in terms of marketing, research and development, production and sales.
The current strength of the dollar is a handicap for its European factories. However, this is in part compensated for by its international presence.

Finally, over the last ten years the Group has had the good fortune to enjoy the continued support of the Walloon Regional Government, which has approved a strategy, followed it through and placed its confidence in the management team.
In industry, constancy and tenacity are major advantages: it is therefore logical that results are assured.

So the Herstal Group is implementing a Medium-Term Plan, which is anticipating growth over the coming years and which should strengthen the company and the jobs which depend on it.

Responsibility, ethics and transparency

The Herstal Group is conscious of working within a strategic sector which imposes scrupulous professional ethics. The Group fully accepts this obligation. It strives to be totally transparent towards government institutions.
Moreover, the Group is at the forefront of major changes taking place in the relationship between civil society and arms. Previously designed for extremely violent uses, arms today have to be adapted to meet precise and measured objectives: in war, to combat the enemy effectively, but in a law enforcement situation to neutralize lawbreakers while minimizing collateral damage.

Belgium, Europe, the United States and its foreign customers have in the Herstal Group the advanced technical skills which should in the long term permit a real revolution in the deployment of arms.

Eventually, the Herstal Group would be nothing without its 3,000 employees.
One third of them are from Liège or its environs... a number of whom are from the Ardennes, no doubt a measure of tenacity !
One third are American, from South Carolina, Virginia and Utah, all strong, distinctive cultures. One sixth are Portuguese, from Viana do Castello, descendants of the great explorers of the world. Finally, there are its Japanese partners from Kochi and the mountains of the Iles Ryu Kyu, deep Japan.

All these men and women constitute the Herstal Group's strength, and remain loyal to it because of the policy which has been applied down the years: the Herstal Group wants to be a responsible employer in the Rhinish sense of the term. "Today's profits are tomorrow's investments and the jobs of the future !" This has led to a rich and diverse community, a community attached to its naturally decentralized company. Under a single, Belgian leadership, unity is based on a tradition of shared work, but also on a concerted strategy and mutual respect integrating specific and deeply-rooted cultures.

The indispensable step-by-step rhythm – imposed by the manufacture of high quality products – is the rhythm of our time. It has strengthened this company, a global leader in light arms. It now serves to broaden the Group's strategies, enabling it to become an important actor in the transitions occurring in its various areas of responsibility.

But its three priorities – technological progress, ethical rigor, customer service – remain fundamental values for the Herstal Group... and they are the key to its future.

Philippe Tenneson
Chairman and Managing Director

HERSTAL group
FN HERSTAL - BROWNING

to decorate the FN' pavilion at the Exhibition of Brussels in 1910. (Herstal Group Collection).

The Herstal Group Timeline

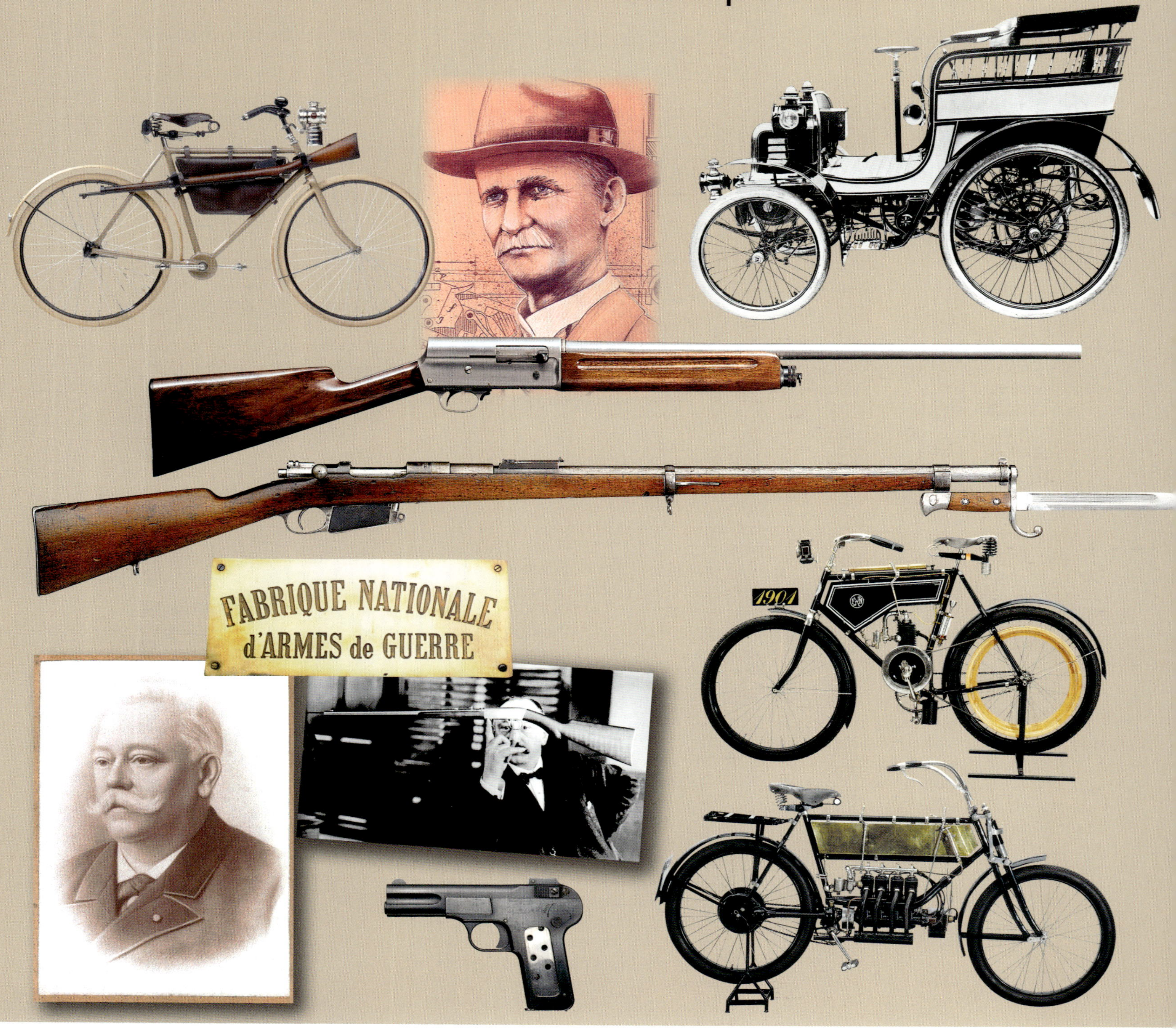

1889
Setting of "Fabrique Nationale d'armes de guerre".

1891
First Mauser rifle delivery to the Belgian Minister of War.

1896
FN is taken over by the German company Ludwig Loewe.

1897
FN meets John Moses and Matthew Browning in the United States.

1898
Production of the "FN chainless bicycle".

1899
Production of the first "1900 Model" Browning pistol (7.65 mm).

1901
FN starts production of cars and motorcycles.

1902
John Moses Browning comes to Herstal to order 10 000 automatic shotguns.

1905
FN is the first maker of in-line 4 cylinder motorcycles.

1907
Production of the "1910 Model" Browning pistol (9 mm).

1913
Production of the Browning automatic .22 carbine.

1914
FN is requisitioned by the German Army.

1918
"Société Générale de Belgique" takes over FN.

1921

Production of typing machines and Stenotypes FN-MAP in Paris.
Production of the BAR (Browning Automatic Rifle).

1925

Production of Browning Over & Under "B25" shotgun.
FN "1300" car wins the King's Cup at the Francorchamps 24 hours race.

1926

John Moses Browning dies of a heart attack in Herstal.
"FN 2000" car wins the "Tour de France Automobile".
Production of the first Browning machine-gun.
FN 350cc motorcycle beats 24 hours world speed record at an average speed of 65 m/h.

1932

"8 cylinder F" cars win first 3 places at the Francorchamps 24 hours race.
Production of FN lorries and special vehicles.
Dieudonné Saive sets up the "Baby" and "HP" (High Power) pistols.
Production of the FN Browning machine gun cal. 7.65 mm and 13.2 mm.

1933

"8 cylinder FN" car wins Liège-Rome-Liège race, and "Prince Baudouin" FN cars win first 3 places of the King's Cup at the Francorchamps 24 hours race.
Production of "twin machine guns" mounted on aircrafts.

1934

Production of tracer, armoured piercing and incendiary ammunition.
René Milhoux, on a 500 cc FN motorcycle, beats world speed record at a speed of 139.20 m/h.
M86 FN motorcycle wins Assen Grand Prix (The Netherlands).

1935

M86 FN motorcycle wins Monthléry Grand Prix (France).
FN stops producing cars.

1936

Production of anti-aircraft guns under Bofors license.

1937
Production of M12 FN side-car motorcycle, 1000cc twin-cylinder engine, forerunner of the Jeep.

1939
12 000 employees work for FN in Herstal.

1940
FN is requisitioned by the German Army and enters into "Resistance".

1944
FN headquarters are bombarded by V1 and V2 flying bombs.

1947
Dieudonné Saive sets up the LAR (Light Automatic Rifle) prototype.

1948
Production of the first semi-auto SAFN infantry rifle.
Belgian government orders FN Derwent jet engines to equip the Gloster Météor Fighter.

1951
Production of the "Ardennes" 4 wheel drive truck for the Belgian Army

1953
The Light Automatic Rifle (LAR) is adopted by the Canadian Army, followed by more than 100 countries.
Auguste Minguels wins the European Motocross Championship

1954
Production of Rolls-Royce Avon MK121 turbo-engine for Hunter Fighter.

1957
FN 7,62 mm ammunition becomes NATO's reference caliber.
Ernest Vervier sets up the MAG 7.62 mm gas operated machine gun

1958
René Baeten, on FN 500 cc motorcycle, wins the Motocross World Championship.

1960

Production of AS24, a parachutable motorized tricycle for airborne troops.
Production of general Electric J79 turbo jet engine for F104G Starfighter.

1964

Production of engine parts for Blue Streak Rocket.
FN employees are more than 13 000 in Herstal.

1965

Production of .22 T-Bolt carbine.
FN stops producing 2 wheels vehicles (1955 motorcycles, 1965 mopeds).

1966

FN "machine women" go on strike, claiming "Equal salary for equal work".

1967

Production of hunting ammunition with plastic shells.

1968

Production of FN 5,56 mm Light Automatic carbine.

1969

Browning US takes Japanese company Miroku as a second source.

1970

Fabrique Nationale d'Armes de Guerre becomes Fabrique Nationale Herstal.

1971

After war production of Auto-5 shotguns reaches 2 millions units.

1972

FN takes a 36% share in Italian Beretta company.
Cooperation with French SNECMA for the production of CFM56 commercial aircraft engine.

1973

Opening of Portuguese subsidiary FN Viana.

1975

Production of Viking engine parts for Ariane European launcher.

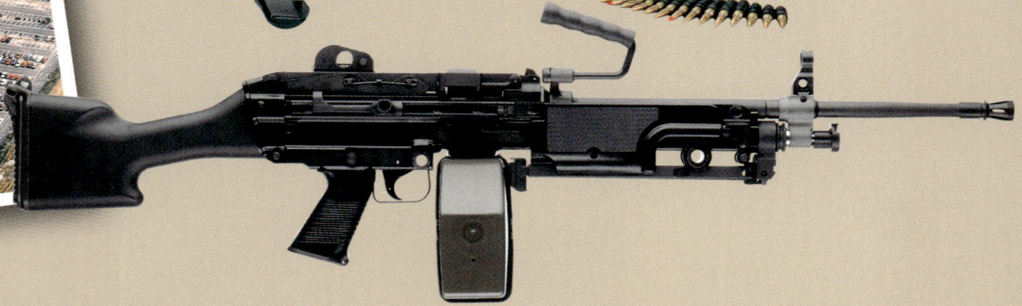
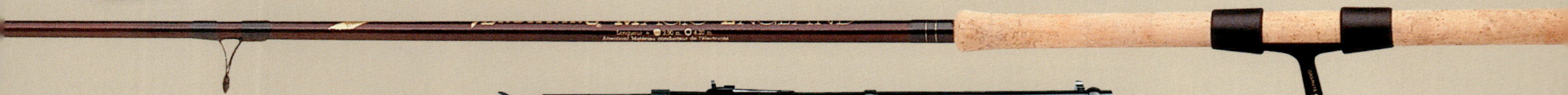

1976

7.62 mm MAG (M240) machine gun is adopted by US Army.
Production of Pratt & Whitney F100 engines for General Dynamics F16 Fighters to equip European Forces.
Browning enters the "non-gun" market with golf and racket sports equipment.

1977

FN takes control of Browning Inc., in the US.

1978

FN opens Brazilian subsidiary FN do Brasil. FN opens FN Moteurs facility for aerospace activities in Milmort (Belgium). Production of gun pods for helicopters and light aircrafts.
Browning enters fishing and archery markets.

1980

Browning enters sailboard market.

1981

Opening of FN Manufacturing Inc., US subsidiary in Columbia, South Carolina, and beginning of M240 mass production.

1982

FN 5,56 mm ammunition becomes NATO's reference caliber.
Beginning of MINIMI caliber 5.56 mm mass production in Belgium.

1987

FN takes shares in the USRAC-Winchester US company.

1988

5.56 mm Minimi machine gun (M249SAW) is adopted by US Army.

1989

FN celebrates its centenary. FN Moteurs is sold to the French SNECMA.
Société Générale de Belgique sells FN to GIAT Industries and the Walloon Region. FN becomes Herstal Group.
FN Manufacturing wins competition to produce Colt M16 rifle.

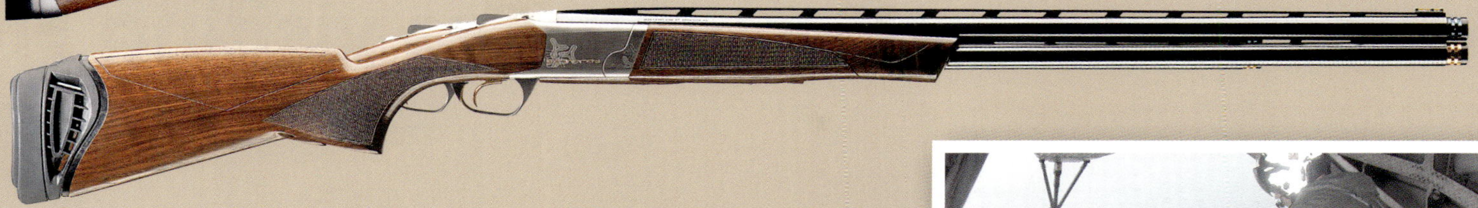

1990
Production of P90 submachine gun (5.7 x 28 mm).

1991
Beginning of MINIMI M249 production in Columbia.
Winchester wins Sport Shooting World Championship (also in 89 and 93).
Herstal Group takes over local barrel maker Delcour.

1994
Opening of new USRAC-Winchester facility in New Haven, Connecticut.

1997
Walloon Region becomes sole shareholder of Herstal Group.

1998
Browning Gold semi-auto shotgun elected Gun of the year by American Shooting Industry Academy.

1999
Opening of FNH USA subsidiary in Mc Lean, Virginia, near Washington.

1999
Winchester Super X2 shotgun elected Gun of the year by the American Shooting Industry Academy.

2000
Setting of a range of "gun systems" for air, land and sea vehicles.
Setting of modular assault rifle F2000 (5.56 mm).
Opening of the "law enforcement" Department.

Production of 7.62 mm FN MINIMI machine gun.

2003
Production of Browning "Cynergy" over-and-under shotgun.

2004
Setting of SCAR assault rifle in 5.56 and 7.62 mm calibers.

2006
Closing of USRAC-Winchester plant.
Creation of a "Pole of Excellence" to study less lethal concept.

Some well-known figures visiting FN

HRH Haile Selassie, Prince Heir of Ethiopia (1924).

HM King Fouad of Egypt (1927).

HRH Prince Leopold of Belgium (1928).

HRH Edward, Prince of Wales (1930).

The general Chief of Staff of the Lettic army (1930)

HM King Carol II of Rumania (1938).

HM King George II of Greece (1938) .

HRH the Prince Royal of Belgium, future King Baudouin I (1951).

HIH the Prince Heir Akihito of Japan (1953).

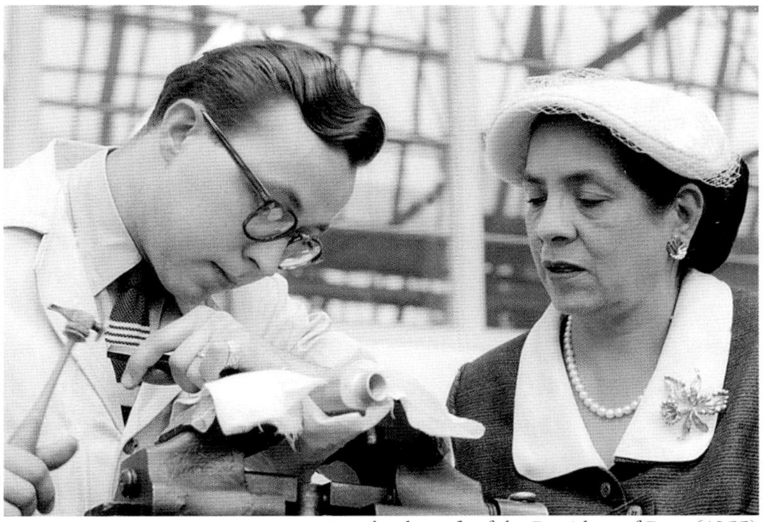
Visit by the wife of the President of Peru (1955)

Visit by the Emperor of Ethiopia and King Baudouin. (1959)

General de Gaulle, President of the French Republic, at the FN stand at the Le Bourget Aircraft Exhibition (1959).

Their Majesties Queen Juliana and King Baudouin (1960).

HRH Prince Bernhard of the Netherlands (1962).

General Mobutu Sese Seko, President of the Republic of Zaire (1964).

HRH Prince Albert of Belgium, at the stand of FN at the Munich Fair (1965).

HRH Princess Paola of Belgium (1968).

Georges Pompidou, President of the Republic of France (1971).

Bob Braithwaite, gold-medal clay-pigeon shooter (with Browning shotgun) at the 1968 Olympic Games in Mexico (1969).

Racing-car champion Jacky Ickx (1972).

The champion cyclist Eddy Merckx trying FN's old chainless bicycle (1974).

Mr. Horst Münzner, a member of the Volkswagen Board (1986).

His Royal Highness Prince Albert signing the visitors' book at the engine division (1979).

Didier Reynders (center), Deputy Prime Minister of Belgium and Minister of Finance with Philippe Tenneson (left), Chairman of the Herstal Group, and Philippe Claessens (right), Managing Director of FN Herstal. (2002).

Bob Browning Prince, great grand nephew of J.M. Browning, and his wife (2001).

Willy Demeyer, Mayor of Liège (2002).

The Browning family with "Moses", the Herstal Group decorated horse celebrating Belgium's 175th anniversary (2005).

Tom Korologos, Brussels US Ambassador (2005)

Jean-Claude Marcourt, Minister of Economic Affairs and Employment in the Walloon Government (left), with J. Stuto, (2002).

Sam Fox, Brussels US Ambassador (2007).

Chairmen

Allard BORMANS 1889-1892
Joseph JANSSEN 1892-1896
Charles del MARMOL 1896-1906
Victor FRIS 1906-1913
Ludwig HAGEN (Vice-Chairman) 1913-1918
Jean JADOT 1919-1931
Alexandre GALOPIN 1932-1944
Georges LALOUX 1944-1951
Paul GILLET 1951-1962
René LALOUX 1962-1970
André DUBUISSON 1970-1983
Marcel GOBLET 1983-1990
Albert DIEHL 1990-1995
Claude ELSEN 1995-1996
Philippe TENNESON since 1997

Chiefs Executive Officers or Managing Directors

Léon CASTERMANS 1889-1890
Joseph CHANTRAINE 1890-1895
Henri FRENAY 1896-1906
Alfred ANDRI 1906-1919
Alexandre GALOPIN 1919-1923
Gustave JOASSART 1923-1950
René LALOUX 1950-1963
Othon DRECHSEL 1963-1971
Michel VANDESTRICK 1971-1986
Bernard REGOUT 1982-1983
Albert DIEHL 1986-1988 (MD)
Jacques GERMAY 1989
Joseph LABYE 1990-1991
Albert DIEHL 1991-1995 (CEO)
Jean-Pol PHILOUZE 1991
Claude ELSEN 1995-1996
Jacques GENTGEN 1996-1997
Philippe TENNESON since 1997

Bibliography

A great many books and articles have dealt with the technicalities and the history of the gun making industry in the Liège area, notably about the Herstal Group, their branches and products. The same applies to Browning and Winchester. We refer here mainly to the following :

Concerning the Herstal Group and the arms production

A. FRANCOTTE, *Les relations de la FN avec la famille Browning de 1897 à 1914*, in Revue FN, n° 94, april 1962, pp. 5-7; n° 95, may 1962, pp. 10-12; n° 96, june-july 1962, pp. 6-7.

Entreprise, agent de progrès. Exposition scientifique et technique organisée à l'occasion du 150ème anniversaire de la Société Générale de Belgique, Verviers, 1972 (Marabout Université).

R. DAEHNHARDT - C. GAIER, *Espingardaria portuguesa-Armurerie liégeoise*, Porto, 1975.

C. GAIER, *Un grand nom de l'armurerie : René Laloux (1895-1981)*, in Le Musée d'Armes, n° 33, april 1982, pp. 17-20.

Ph. QUESTIENNE, *Une usine d'armement sous l'occupation ennemie. La Fabrique Nationale de 1940 à 1944*, in Le Musée d'Armes, n° 39-40, december 1983, pp. 11-44.

E. DE GROOTE, *De Belgische Mauser M. 1889*, in Militaria Belgica, 1984, pp. 43-49.

E. DE GROOTE, *De Belgische Mauser M. 1889. De versies voor niet-bereden troepen*, in Militaria Belgica, 1985, pp. 81-86.

C. GAIER, *La diffusion mondiale des cartouches de chasse belges il y a cinquante ans*, in Le Musée d'Armes, n° 47, december 1985, pp. 15-24.

E. DEMARET, *La FN et les canons Bofors*, in Le Musée d'Armes, n° 62, december 1989, pp. 1-19.

M.-Th. COENEN, *La grève des femmes de la FN en 1966*, Bruxelles, 1991.

R.B. STEVENS, *The FAL rifle. Classic edition*, Coburg (Canada), 1993.

Video report by Marie Anne THUNISSEN. Betacam numerical 16/9 - 52 mm - color, 1996 (The strike of FN female workers).

C. LIBERT, *Femmes-machines*, a film by Marie Anne Thunissen, s.l.n.d. [1996].

R. BRION - J.L. MOREAU, *La Société Générale de Belgique 1822-1997*, Antwerp, 1998.

L. GUILLOU, *Mauser. Fusils et carabines militaires*, 2 vol., La Tour du Pin, 1997-2004 (Mauser FN : vol. II, pp. 42-59) .

W. JOHNSON, *The FN-49. The last elegant old-world military rifle*, Greensboro NC, 2004.

Concerning the Herstal Group and the "non gun" production

Ph. DE BARSY - P. FRERE, *Livre d'Or du Salon de l'Automobile, du Motocycle et du Cycle*, Brussels, 1970.

Cycles et motocycles, des origines à 1935, Liège, 1971 (catalog of the exhibition of the Musée de la Vie Wallonne in Liège, 4 june - 17 october 1971).

G. GASPARD, *Les Demoiselles de Herstal. La motocyclette liégeoise des origines à 1940*, Liège, 1975.

G. GASPARD, *Les Dames de la Basse-Meuse. La motocyclette liégeoise de 1940 à 1965*, Liège, 1975.

G. GASPARD, *Au temps des phares à carbure. Les courses motocyclistes en Belgique avant 1915*, Liège, 1982.

O. DRECHSEL, *FN. Le service poids-lourds*: vol. I : 1935-1940 ; vol. Ia and b : 1945-1970 ; vol. IIIa : *les trolleybus 1932-1957* ; vol. III b and c : *activités diverses 1935-1958*, Herstal, n.d. (typewritten book).

O. DRECHSEL, *FN Turboréacteurs. 25 années de production*, Liège, 1976.

Y. & J. KUPELIAN - J. SIRTAINE, *Histoire de l'automobile belge*, Brussels-Paris, 1979.

Y. & J. KUPELIAN - J. SIRTAINE, *Soixante ans de compétition automobile en Belgique 1896-1956*, Overijse, 1981.

J. KUPELIAN - J. SIRTAINE, *Motos belges*, Overijse, 1983.

P.Y. THOMAS, *Design auto belge. Passé, présent, avenir. Des anciens carrossiers aux designers actuels* (catalog of the exhibition at "La Châtaigneraie", Flémalle, from 14 april to 5 may 1991), Liège, 1991.

G. DE BECKER, *Quand la FN avait deux roues*, Bastogne, 1993.

P. BARTHOLOMÉ and alii, 1949-1999. *Techspace Aéro. 50 ans d'histoire et d'aventure*, Herstal, 1999.

Concerning Browning

J.M. & M.S. BROWNING Co., *A history of Browning guns from 1831*, Saint Louis MO, 1942.

B. WEST, *Browning arms and history*, Santa Fe Springs CA, 1972.

EMJY - C. GAIER, *La vie passionnante de John M. Browning. Un récit complet en bande dessinée*, Liège, 1978.

EMJY - C. GAIER, *The American Gunmaker John M. Browning. A complete comic strip album*, Liège, 1978.

J. BROWNING, *Mountains rim my valley. A lean anthology selected by the wind*, Salt Lake City, 1988.

R.B. STEVENS, *The Browning High Power Automatic pistol*, new. ed., Toronto, 1990.

H.C. TYLER, *Browning .22 caliber rifles*, 1914-1991, new ed., Marceline MO, 1992.

J. BROWNING - C. GENTRY, *John M. Browning, American gun maker*, 2nd ed., Salt Lake City, 1994.

M. EASTMAN, *Browning. Sporting arms of distinction 1903-1992*, Fitzgerald GA, 1994.

Val Allen Browning, n.p.n.d. [1995] (Memorial booklet published by his children on the occasion of the one hundreth anniversary of his birth).

N. SCHWING, *The Browning superposed. John M. Browning's last legacy*, Iola WI, 1996.

G. GANGAROSA Jr., *FN Browning. Armorer to the world*, Wayne NJ, 1999.

J.M. BALLOU, *Rock in a hard place, The Browning automatic rifle*, Cobourg (Canada), 2000.

A.. VANDERLINDEN, *The Belgian Browning pistols 1889-1949*, Greensboro NC., 2001.

H.M. SHIRLEY Jr. - A. VANDERLINDEN, *Browning Auto-5 shotguns, The Belgian FN production*, Greensboro NC, 2003.

D. MILLER, *The history of Browning firearms*, Guilford CT, 2006.

Concerning Winchester

Y. CADIOU, *Monsieur Winchester*, Paris, 1972 ("Un homme, une arme").

D. BARNES et alii, *The history of Winchester firearms 1866-1980*, 5th ed., Tulsa OK, 1980.

T. TROLARD, *Winchester commemoratives*, Plano TX, 1985.

R. L. WILSON, *Winchester engraving*, 2nd ed., Palm Springs CA., 1989.

R.L. WILSON, *Winchester : an American legend. The official history of Winchester firearms and ammunition from 1849 to the present*, New York - Toronto, 1991.

R.L. WILSON, *Winchester: une légende américaine. L'histoire officielle des armes et munitions Winchester de 1849 à nos jours*, Editions EPA, 1991.

N. SCHWING, *The Winchester Model 42*, 2nd print., Iola WI, 1990.

N. SCHWING, *Winchester's finest: the Model 21*, Iola WI, 1990.

N. SCHWING, *Winchester slide-action rifles*, vol. I: Model 1890 & Model 1906, Iola Wi, 1992; vol. II: Model 61 & Model 62, Iola WI, 1993.

C. GAIER, *Winchester. Armes de légende. Fabulous firearms*, Liège, 1992 (Catalog of the exhibition at the Arms Museum, Liège).

H.G. HOUZE, *Winchester Repeating Arms Company. Its history and development from 1865 to 1981*, Iola WI, 1994.

D.K. BOORMAN, *The history of Winchester firearms*, Guilford CT, 2001.

La légende Winchester, un mythe patriotique et viril, Saint-Etienne, 2002.

Concerning the Liège gun making industry in general

J. PURAYE, *La gravure sur armes à feu au pays de Liège*, Liège, 1964.

J. PURAYE, *Engraving on firearms in the Liège region*, Liège, 1967.

F. BALACE, *L'armurerie liégeoise et la guerre de Sécession*, 1861-1865, Liège, 1978.

C. GAIER, *L'épreuve des armes à feu à Liège*, 1672-1972, Liège, 1972 (translated into japanese, Tokyo, 1986).

C. FEYS - R. SMEETS, *Les revolvers et les fusils Nagant*, Paris, 1982.

M. DE VLAMINCK, *De wapenindustrie in Belgïe tijdens de bezetting (1940-1944)*, in Revue Belge d'Histoire Militaire, vol. 26/3, september 1985, pp. 223-238; vol. 26/4, december 1985, pp. 289-324.

C. GAIER, *Cinq siècles d'armurerie liégeoise*, Alleur, 1996.

C. GAIER, *Five centuries of Liège gunmaking*, Alleur, 1996.

B. JOOS de TER BEERST, *Marques et poinçons d'armes à feu en Belgique. Handels en keurmerken op vuurwapens in Belgïe - Trade - & hallmarks on firearms in Belgium*, Antwerp, 1998.

K.H. WROBEL, *Drei Linien. Die Gewhehre Mosin-Nagant*, 2 vol., Schwäbisch Hall, 1999-2003.

Ph. & N. QUINET - DE SAEGHER, *Uldarique Marga, capitaine de l'Armée belge, inventeur prolifique, industriel méconnu (1854-1925)*, Brussels, 2003.

M. BEDEUR, *Mémoire d'une usine. Henri Pieper, un génie créateur 1867-1952*, 2 vol., Andrimont, 2003-2004.

M. DRUART, Bayard. *Les hommes, les armes et les machines du chevalier Pieper & Cie.* 1859-1957, n.p., 2004.

M. DRUART et G. GADISSEUR, *Le qui est qui de l'armurerie liégeoise (1800-1950)*, Biarritz-Paris, 2005.

G. GADISSEUR, *Les armuriers de Liège en catalogues (1892-1930)*, vol. I, Anglet-Paris, 2004 ; vol. II, Biarritz-Paris, 2005.

Generally speaking, gun-oriented periodicals have issued over the years, and continuously publish, quantities of articles dealing with FN Herstal, Browning and Winchester products. We have given up the idea of recording them here. We do recommend however to refer to the three following categories of periodicals owing to their particular relevance to the subject of this book :

• The house organs of the HERSTAL Group, published under various headings and configurations :
FN Revue (1953 to 1977)
Journal FN (1977 to 1989)
Infos GIAT Industries
(january 1993 to august 1994)
La Lettre de Herstal
(february to november 1995)
Herstal 2000 (june 1998 to may 1999)
Made in Belgium
(since june-july 1998).

• The Newsletter) of the "Browning Collectors Association", published in the USA since september-october 1979

• Le Musée d'Armes, published by the Arms Museum, in Liège since 1976.

Index

D

E

F

G

H

N

O

P

Q

R

S

T

U

V

W

X

Y

Z

Photographic credits

In the main, the illustrations and photographs contained in this book are taken from the photographic archives of the Herstal Group. Of these, those which bear on the period from 1981 to the present day are principally the work of Jean Davister and his team (for Browning/Winchester since 1981 and for FN Herstal since 1998).

Other sources have also been used, including :
F. Niffle: 10, 15 bottom
The Reims Museum : 10 bottom
The Skokloster Museum, Sweden : 15
Livrustkammaren, Stockholm : 15 top
Liège, fonds U. Capitaine : 17
Ito Josué, Saint-Étienne : 19
Liège State Archives : 21
Grand Curtius, Liège : 22, 23, 24, 26, 27, 28, 29, 32, 33, 34, 37, 41, 43, 136, 141, 214
Cockerill : 31
A. Vanderlinden : 39, 167
M. Druard : 41
Browning Firearms Museum, Ogden, Utah, USA. : 64, 141 top
J. Schaeger : 138
Royal Air Force Archives : 201
Media library of the Province of Liège. Fonds Desarcy-Robyns : 204, 205
Tracy Dungan : 211
Rodolfo Yui, Pérou : 241
The Churchill Museum, London : 243
Interavia : 252 bottom
H. Vanhaelewijn, Bruges : 349
United Technologies : 354 top
J. Huybens, with the kind permission of the 2nd Tactical Wing of Florennes : 355 top
Foto RORIZ, Viana do Castelo, 386 top, 387.
Foto JOCA, Viana do Castelo , 386 bottom, 508, 509, 510, 511
Planchard : 412, 413, 414
Techspace Aero : 420, 421
Giat Industries : 422, 424
Fernando Muscinelli : 426 bottom, 427 bottom, 446 top
US Department of Defense : 476, 477
Anthony Vanderlinden : 478, 479
David Bertinelli : 64, 512, 513, 514, 515, 516
Claude Tombeux (Photo Studio 9) : a number of photographs taken between 1978 and 1998

Every effort has been made to acknowledge all the copyright holders for works reproduced in this book. In the event that any copyright has been omitted, we ask the holder to contact the Herstal Group.

Contents

Printed completed by :

Snel
MORE THAN A PRINTER

SNEL SA
Vottem (Belgium)
on January 2008

NEW LOOK
COMMUNICATION SA

Phototypesetting, photoengraving, design and lay-out :
NEW LOOK COMMUNICATION SA